Wessex: A Landscape History

WESSEX

A Landscape History

Hadrian Cook

ARCHAEOPRESS ARCHAEOLOGY

ARCHAEOPRESS PUBLISHING LTD
Summertown Pavilion
18-24 Middle Way
Summertown
Oxford OX2 7LG
www.archaeopress.com

ISBN 978-1-80327-535-2
ISBN 978-1-80327-536-9 (e-Pdf)

This book is available direct from Archaeopress or from our website www.archaeopress.com

Contents

List of Figures and Tables

Preface

Mesolithic, Malthus and mangelwurzels

For over 30 years I have enjoyed teaching people about the British landscape, both within universities and delivering courses to adult learners. It is never easy, for the challenge is that nobody can write about landscape from a single-subject perspective. While there are many accounts of 'landscape', there remains much confusion around the concept because it is largely understood from a user definition of the term.

At one extreme, 'landscape' is determined solely by geology and landform and at the other it merits literary or artistic description. Wessex's own writer Thomas Hardy (1840-1928) ascribes human characteristics when describing landscape elements and uses the landscape to paint word-pictures of mood[1] as well as expressing the historic and archaeological aspects of the landscape[2].

What is important is that landscape is the outcome of natural and human factors whereby human beings worked both with and against 'natural forces' in a dynamic and assertive way. The drivers of this process encompassed anything from food and fuel production, construction, or other economic imperatives to the ritual or aesthetic considerations behind landscapes generally produced by elites. People's experience of landscape is therefore a kick-back from something created by us and (in human terms) typically over a long time.

The approach to landscape description becomes interdisciplinary, focusing on the interplay between historic factors and conventional 'landscape ecology'. Yet this book is also a response to regional *genius loci*. For within this large area we call Wessex, diverse landscapes are on offer, each (as Hardy would have noted) capable of imparting its own mood. Here are the heaths, woods, forests, river valleys, water meadows, downland, limestone and chalk uplands, clay vales, reclaimed wetland and coastlands - and those who live and work among them. That is the romantic stuff; there are also perceived 'facts'.

Relatively early in my career a senior colleague asked if I 'might help out' by teaching undergraduates the essentials of land use statistics. I was flattered because previously this area was covered by two of Wye College's long-past academic celebrities, Professors Gerald Wibberley and Robin Best[3]. Agricultural land use data has been collected systematically for England and Wales on a parish basis from the mid-19th century[4] and proved invaluable for official purposes of monitoring food production, setting agricultural policy prescriptions and looking at land use change. Actually, earlier situations had already demonstrated a need for detailed systematic information, especially when plucky little England, back to the wall, was fighting a certain dastardly Corsican. Times of national stress forced the establishment to

---

[1]   Thomas Hardy Society (n.d) <http://www.hardysociety.org/about-hardy> Accessed November 2017
[2]   M.J.P. Davies (2011) Distant Prospect of Wessex: Archaeology and the Past in the Life of Thomas Hardy. Archaeopress
[3]   S. Richards (1994) *Wye College and its World* 206, 225-6. Wye College press, Ashford
[4]   Agricultural Statistics (2017) <http://www.nationalarchives.gov.uk/help-with-your-research/research-guides/agricultural-statistics-england-wales/> Accessed May 2020

focus on management of natural resources such as timber production, armaments, and food security. This concern for auditing by state interests pre-dated Napoleon Bonaparte, however. For one thing, Samuel Pepys was deeply concerned about the condition of the English Navy following a devastating raid by the Dutch who burned the fleet in the Medway, something that would impact on national timber production.[5]

Back in the nineteen eighties, I took up my new teaching challenge but soon discovered why I had been passed this numerical buck. This land we all enjoy has a fascinating history with interesting detail (which illuminates landscape history) hidden in statistical information. Like swimming through cold water, statistical information is an acquired taste. Sadly, I found imparting such wisdom to (mostly) 20-year-old students challenging; there is a limit to interest generated by acreages of fodder root crops.

Rescue was at hand through colleagues in the environmental and historical communities who inspired me to place bare nineteenth and twentieth crop area and livestock data into a wider context. After all, I had personally completed courses in geology, geography, social science, archaeology and ecology and researched and taught soil science at various stages, as well as benefiting from classes delivered through the Workers' Educational Association.

Add a bit of 'how did they do that' and undergraduate students were hopefully now able to place a historical framework around their studies in ecological science, rural study, and agricultural management. No, the British countryside had not arrived by chance, rather economic interests, political imperatives, and social change driving pure graft had created something special. The landscape is a window through which we view the history of the nation as well. I was thus able to link with colleagues in related disciplines - and everybody seemed happy.

Predictably drivers of landscape form and function have been largely, although by no means exclusively, economic. How do you make decent a living? How do you expand production during times of growing population? What of famine? Plague? Will the grim predictions of Malthus (where population growth might outstrip food supply) be supplanted by increased production, better technologies or imported raw materials? Who might be creaming off the surplus value of others' labour? Thomas Hardy was a compassionate man, most concerned about the condition of the poor, and parallels with Karl Marx have been observed.[6]

I learned from, and taught with reference to, scholars such as Oliver Rackham, John Sheail, Marion Shoard, Joan Thirsk and Tom Williamson. I started by embedding stark observations (quantitative or otherwise) into a narrative around the English landscape. As his books emerged, I was inspired by Peter Brandon's writing on the South Downs[7] and Kent and Sussex Weald[8] in an integrative and engaging manner. Brandon really understood the human-natural dynamic of landscape history and gives insight that landscape writing for any purpose remains highly personal. While grateful he stopped at the borders of 'Wessex' leaving the field open for

[5]   H. Cook (2018) *New Forest: The Forging of a Landscape* Windgather press ch 7
[6]   S. J. Flynn (2016) The Return of the Poor Man: Jude the Obscure and Late Victorian Socialism The Hardy Review (18) 1 56-65 <https://cupola.gettysburg.edu/cgi/viewcontent.cgi?article=1073&context=engfac> Accessed May 2020
[7]   P. Brandon (1998*). The South Downs*, Phillimore, Chichester
[8]   P. Brandon (2003). *The Kent and Sussex* Weald, Phillimore, Chichester

others (!) I came to landscape writing from a similar place to him. Within Wessex, however, the work of historian Joseph Bettey and archaeologist Barry Cunliffe further provided invaluable accounts in support of this book.

Over time at Wye College my original three lectures expanded to five, then to ten. Eventually friends in ecology re-combined my contribution into courses explaining not only *how conservation management might be undertaken,* but *how we got here in the first place,* while friends in archaeology alerted me to the fantastic record preserved within the landscape. A plethora of historic maps, thematic maps of geology and soils, remote sensed and geophysical survey information, hydrometric information, documentary records, old photographs and more are now available.

By way of illustration, I recently visited a field just outside Salisbury where a university geophysical practical class was underway. Those leading the class were contemplating where, during a plausible 2000-year time corridor, one linear feature evident through survey and visibly manifest at the surface may be located, although a modern origin has since been ruled out. For another real-world example, contemplating the possible origins and maintenance of heathland one must consider: turf stripping for energy, over-exploitation through agriculture of vulnerable sandy soils, burning, grazing and vegetation cutting over many historic periods.[9]

The Malthusian nightmare applied to England of resource availability being outstripped by an exponential grown in population was largely prevented due to improved means of food production and expansion of production abroad. With that expansion went surplus population. Wessex in the meantime experienced the benefits of colonial expansion driving investment in its industries and infrastructure. The working classes at home may have paid for it through enclosure and being herded into poor conditions in the expanding towns; colonial 'subjects' of the Crown experienced still more grievous exploitation, including slavery.

Then there are the people themselves: for one, in Wessex as elsewhere, some people may migrate while others stay. How much similarity to other surrounding areas is displayed by the population at a moment in time is a legitimate question. Using a study published in *Nature,*[10] the *Mail Online* (26th July 2017) reported the results in terms of how much indigenous Brits are (more or less) like the (modern) French and Germans:

> Many people are a quarter German and 45% of their DNA is French.

Quelle surprise! One wonders where the others came from? They surely cannot all be EU nations?..... We move swiftly on.

It is fairly certain that Cheddar Man (who died around 10,000 years ago) has living relatives,[11] and this suggests some continuity from the earliest populations who migrated into Wessex following the retreat of the ice and continental-scale warming, although the time of their arrival might have coincided with a rapid decline in the existing insular population, for reasons

---

[9]   H. Cook (2018) *New Forest, op. cit.* 38
[10] S. Leslie *et al.* (2015) The fine scale genetic structure of the British population. *Nature* 519, 309–314. doi:10.1038/nature1423
[11] L. Barham, P. Priestly and A. Targett (1997) *In search of Cheddar man* Tempus ch 7

that are unclear.[12] On the other hand, the 'Beaker Culture' demonstrably had continental links. One fascinating example is the 'Amesbury Archer' skeleton (the man had lived *c.* 2300 BC) that is located in the Salisbury Museum. An oxygen isotope signature from his teeth suggest he grew up in central Europe, while a presumed younger male relative buried nearby was likely raised locally.[13]

My own genes left Wiltshire, Somerset and Gloucestershire on steam trains and would mingle with others before returning. These new infusions include Kentish Men (although this county was once included in an enlarged 'Wessex' as were some South Saxon forebears). For the exotic, there are Roman Catholic immigrants from southwest Germany in the mix, and one helpful family-tree enthusiast found the surname 'Lloyd'; I can claim some ancient British lineage. All this means that my own family heritage is redolent of the geographical, archaeological, and historic evidence suggesting people moving in and out of our region from surrounding areas. My children, whose west of England lineage is strengthened through my wife and who also have some Scottish heritage from her, keep moving around as they study and work. Wessex people have always been on the move one way or another.

This book is aimed at students, professionals and the interested general reader alike. It is about how people have shaped, developed and run their landscape. There are intimations for sustainable development and a strong message that we ignore our heritage landscapes at our peril. It is a personal synthesis of a mass on multi-disciplinary information. As I sign off for a tea-break, I do wonder if I may be remotely related to the 'Amesbury Archer' or maybe to Cheddar Man. I may never know, although their descendants and my own ancestors worked - and shaped - the Wessex Landscape.

Hadrian Cook, Salisbury, Mayday 2023

Acknowledgements

The author would like to thank those who supplied information or reviewed aspects of the text, often not pulling punches. Specific appreciation is offered to: Katherine Barker, Dr David Benson, Dr Peter Buckley, Dr Ian Cummings, Prof Madge Dresser, Dr Alan Dykes, Dr Amy Frost, Steve Hannath, Prof Alex Inman, Dr Howard Lee, Anthony Robinson, Sean Rice, Mike Schurer, Ben Heaney, Dr Kathy Stearne, Adrian Targett and Dr Nick Thorpe. Any errors or misconceptions remain with the Author.

Brief Bibliography

J.H. Bettey (1986) *Wessex from AD 1000*. Longman

J. Chandler (1987) *Endless Street*. Hobnob Press

H. F. Cook, Cowan M. and T. Tatton-Brown. (2008). *Harnham Water Meadows: History and Description*. Hobnob Press.

---

[12] McNish (2018) The Beaker people: a new population for ancient Britain. https://www.nhm.ac.uk/discover/news/2018/february/the-beaker-people-a-new-population-for-ancient-britain.html Accessed Aug 2021

[13] Wessex Archaeology The Amesbury archer (2011) < http://www.wessexarch.co.uk/projects/amesbury/archer.html> acc. Nov 2017

H. Cook and T. Williamson (eds. 2007). *Water Meadows: History, ecology and Conservation.* Windgather Press. Especially chapters by J.H. Bettey, K. Stearne, C. Taylor and R.Cutting and I. Cummings.

H. Cook (2018) *New Forest: The Forging of a Landscape.* Windgather Press

Cook H. and T. Williamson eds. (1999) *Water management in the English Landscape Field Marsh and meadow.* Keele University Press

Cowan, M (2005) *Wiltshire Water Meadows.* Hobnob Press.

B. Cunliffe (1993) *Wessex to AD 1000* Longman

B. Cunliffe (2012) *Britain Begins* Oxford University Press

M.J.P. Davies (2011) *A Distant Prospect of Wessex: Archaeology and the Past in the Life of Thomas Hardy.* Archaeopress.

Davis, T. (1813) *A General View of The Agriculture Of Wiltshire, second ed*

Introduction

Where and what is Wessex?

A massive change caused by people occurred around 4000 BC across Britain.[1] Mesolithic people who had hunted and made artefacts with exquisite 'microliths', forming arrow heads, sawing devices and even fishhooks, were, in ways unclear, persuaded by incomers to clear the 'wildwood' and start agriculture. These people, agriculturalists from Anatolia, mixed to some degree with the indigenous hunter-gatherer population during the earlier Neolithic although ancient DNA suggests a large replacement of population at this time.[2] There were furthermore likely regional differences of this wildwood within which Mesolithic people lived, migrated and hunted.

Pollen sequenced in wetlands as well as plant remains help reconstruct vegetation communities of the past.[3] In around 5000 BC, Wessex would essentially have been dominated by mixed deciduous woodland:[4] alder, oak, elm, and with small-leaved lime (*Tilia cordata*) dominant, although in the extreme west this may have given way to oak and hazel, where less fertile soils developed over older rocks on higher ground.[5] In so doing, Neolithic people changed a landcover dominated by trees (but not necessarily displaying a dense cover) to one dominated by animal and arable agriculture.

Landscapes do not happen by chance. The area we call 'Wessex' is the outcome of natural ingredients that have been worked over and over again by humans for millennia, and which today, more than ever, are contested spaces between anything from agricultural intensification and urban development to road construction. Legitimate questions include:

1. Where is Wessex?
2. What is Wessex?
3. When was Wessex?
4. To what extent has Wessex imparted identity to humans?
5. What can a study of the region tell us about the wider World?

Where is Wessex? For convenience, the region is defined as the counties of Hampshire (including the Isle of Wight), Dorset, Wiltshire, historic Somerset, and Berkshire. It borders two seas – the English Channel to the south, and the Bristol Channel, an extension of the Atlantic Ocean, famed for a high tidal range.[6] Neighbouring southern Gloucestershire and parts of Oxfordshire and Devon may also be alluded to.

---

[1] V. Gaffney, S. Fitch and D. Smith (2009) *Europe's Lost world, the re-discovery of Doggerland.* CBA Research report 160, York, 26-27
[2] S. Brace, *et al.* (2019) Ancient genomes indicate population replacement in Early Neolithic Britain. *Nature Ecology and Evolution* 3(5),765-771.
[3] J. Grove and B. Croft, eds (2012) The Archaeology of Southwest England <https://www.somersetheritage.org.uk/downloads/swarf/swarf_strat.pdf> accessed Sept 2021
[4] D. Blakesley and P. Buckley (2016) *Grassland Restoration and Management.* Pelagic Publishing, Exeter. Chapter 1
[5] O. Rackham (2006) *Woodlands.* Collins, London. Chapter 4.
[6] J.H. Bettey (1986) *Wessex from AD 1000.* Longmans, New York, 3.

The writer Thomas Hardy re-invented the region when writing of the area in the 19th century, changing the names of major places, i.e. Salisbury became Melchester, Dorchester became Casterbridge, Exeter became Exonbury. The County of Devon was in there and Stonehenge remained Stonehenge! This present volume, however, is about the origins and management of real semi-natural ecosystems, farming systems, and the associated settlements which, through diversity, have come to define the region.

What is Wessex? Wessex is diverse in landscape with varied geology, soils, and climate. To take one indicator, the region varies in height from sea level to over 500m above Ordnance Datum (mAOD) on Exmoor; the chalklands seldom rise above 200m. AOD is taken as Mean Sea Level (MSL) at Newlyn in Cornwall, between May 1915 and April 1921, outside Wessex indeed, but near enough, although sea level rise in 100 years places it out of kilter with the modern actual MSL, which is more than 200mm higher!

Elevations in metres AOD are in part a reflection of the underlying geology. As an approximate rule of thumb, the older the rock sequence (see Chapter 1), the harder it is against erosion by the elements and the sea. Across Exmoor, annual average rainfall ranges of typically between 1000 and 1900mm[7] are recorded; for the lowland areas of the south coast and the Somerset Levels this can be as little as 650-850mm per annum, for example some 800mm around Bournemouth.[8]

The agricultural basis for the economy of Wessex reflects the diversity of the wider landscape, and this is surely its economic strength – if one believes strength lies in diversity? The Exmoor uplands are largely grazed, the chalklands are complicated, but they have supported vast areas of arable agriculture in modern times. Other limestone areas, particularly the Cotswolds, include significant past and present wool production. This, after all, was the staple product of the Realm of England. Those vales enjoying better soils have been characterised as the 'Cheese Country', to differentiate them from the sheep and corn husbandry of the Chalk. Another stereotype is 'Zomerset Zider Apples'. Cod folk songs apart, this reference is to established horticulture industries in lowland areas. Behind such knowledge lies a plethora of information, once gleaned from observation, from estate records, or sifting through other historic maps and records, although latterly (for around 150 years now), detail is rooted in systematic survey and statistical collation.[9]

Sizable towns and cities also abound within Wessex and were supported by a rural hinterland. Part of Roman trade involved the export of grain, metals, and slaves from Britain. Although the focus here is less urban and more rural, the development of urban centres was commonplace from the *oppida* of the late Iron Age through to the Roman period, and on to the rapid growth of medieval and modern times. Salisbury (or New Sarum) did not exist as such before AD 1220 and yet was ranked within the top ten urban centres by population in England for much of its history, losing its pre-eminence (perhaps) in the 17th century.[10] Since then, it is Bristol that has enjoyed a comparable league table position and together with Bath, Portsmouth and Reading,

---

[7]   Exmoor National Park (n.d.) <http://www.exmoor-nationalpark.gov.uk/learning?a=122273> Accessed April 2012
[8]   Holiday Weather.co <http://www.holiday-weather.com/bournemouth/averages/> Accessed April 2020
[9]   Agriculture in the South West of England 2009/2010 (2009) <http://farmbusinesssurvey.co.uk/regional/commentary/2009/southwest.pdf> accessed April 2020
[10]   J. Chandler (1987) *Endless Street*. Hobnob Press, Salisbury. Chapter 2.

it recorded a population above 10,000 in 1801.[11] Elsewhere, on the south coast, overseas trade greatly stimulated Southampton, while Portsmouth soon realised its importance to the Royal Navy.

Hence, Wessex continued to owe much to its own imports, exports, and services. The region has always been the centre of a diverse economy. Theoretically speaking, the nature of the feudal system had meant that large estates could be acquired in return for military service (Chapter 3) and this pattern of landholding dominated the middle ages. Things were to change with the rise of a more capitalistic economy and new wealth included that derived through agricultural development (Chapter 5).

Here we must pause to reflect which centres owed their success less to 'worthy' agriculture, commerce or manufacturing, and those whose economy, growth and influence was boosted by slavery and slave trading in the Americas, typified by the Bristol merchant Edward Colston (1636-1721).[12] More notorious was 'Alderman' William Beckford (1709-1770)[13] and his son, also called William (Chapter 5). Locally, the resulting wealth was realised largely in Bristol and Bath.[14] Romans aside, while historians have made much of the negative impacts of in-country economic injustices resulting from factors such as enclosure, industrialisation, or urbanisation affecting the British or Irish poor, the disaster affecting millions of Africans is all too often overlooked as a source of ill-gotten economic gain.

When was Wessex? The roots of the West Saxons are far from clear. Human activity is well attested in the Old and Middle Stone Ages (the Palaeolithic and Mesolithic periods, respectively). There was significant activity in the Neolithic period as humans began to settle and farm, and it is presumed that a proportion of the population, rather than being local people who adopted agriculture, were migrants who arrived by boat. The routes in may have been from what is modern Northeast France, Belgium and the southern Netherlands, where there were already Neolithic cultures.[15] The crossing route(s) is not known, the shortest being around Calais-Dover, and there is evidence for associated settlement activity in Kent.[16]

There is also much activity from this period in Wessex, but no model exists for the adoption of the new Neolithic technologies: textiles (spinning and weaving), domestication of animals, cultivation and improvement of crops, enclosures, ceramics, stone building, monuments associated with funerary rites, and mining. Indeed, so important was the area in the Neolithic that Stonehenge was constructed, a significant monument for the island of Britain, and likely a place of pilgrimage from further afield.[17]

---

[11] J.H. Bettey (1986) *Wessex from AD 1000.* Longmans, New York, 230-231.

[12] M. Dresser (2007) *Slavery Obscured.* Redcliffe, Bristol, Introduction.

[13] Historic England (2020) <https://historicengland.org.uk/research/inclusive-heritage/the-slave-trade-and-abolition/>'; A. Frost (2008) Big Spenders: The Beckford's and Slavery. <http://www.bbc.co.uk/wiltshire/content/articles/2007/03/06/abolition_fonthill_abbey_feature.shtml> Accessed April 2020

[14] The Bath Scrinium (2013) <https://thebathscrinium.wordpress.com/2013/02/04/slavery-in-bath/> Accessed April 2020

[15] B. Cunliffe (2013*) Britain Begins.* Oxford University Press, Oxford. 139.

[16] N. Crane (2016) *The Making of the British Landscape.* Weidenfeld and Nicholson, London. Chapters 4 and 5.

[17] S. Viner *et al.* (2010) Cattle mobility in prehistoric Britain: strontium isotope analysis of cattle teeth from Durrington Walls, *Journal of Archaeological Science* 37.

There is some human DNA evidence for widespread replacement of population with the arrival of the '(Bell) Beaker people' from around 2900 BC.[18] Yet the survival within the present population of earlier hunter-gatherer populations is also attested.[19] Rich burials dating from around 2000 BC were once considered to be an early Bronze Age 'Wessex Culture', belonging to the first half of the second millennium and implying a cultural identity centred in the region. It is now considered to be an indigenous people developing more elaborate burial furnishings, rather than some new tradition arising from invasion.[20] However, continental contacts did exist, and there has been a tendency to ignore indigenous products in favour of the presumed exotic.[21]

The late Bronze/early Iron Age saw the arrival of technologies and defensive practices in 'hillforts'. By the late Iron Age we know the names of the dominant tribes – the Atrebates, Durotriges (who give their name to Dorchester), and Belgae (making Wessex a continental gateway); the Roman name for Winchester was *Venta Belgarum* ('Market Town of the Belgae'). These people created *oppida* and built impressive hillforts throughout the region, and enjoyed a thriving economy. The presence of re-constructed tribal boundaries is not conducive to any single Iron Age identity for Wessex.[22] It appears that Late Iron Age Wessex splits into an eastern part, where hillforts (for example Danebury) had been abandoned for *oppida*, such as Silchester, and a western part where hillforts continued to be occupied, for example Maiden Castle and Hod Hill. [23] Furthermore, long distance boundaries in the landscape evident from studies of field systems in Wessex would seem to add further weight to extensive subdivision of the landscape into territories.[24]

The Romans under Julius Caesar sent expeditionary forces in 55 and 54 BC respectively. They knew much about Britain and its people, such that after AD 43 they made a decision to subdue the region to exploit the already developed Iron Age economy. It is, however, now clear that some hillfort sites were neglected or abandoned by the 1st century BC, and the interpretation of the battle between the Romans and Durotriges at Maiden Castle in Dorset has been challenged in terms of the formerly believed ferocity of a Roman attack.[25] The Roman military commander Vespasian (later emperor) whose father had been a tax collector, did however, launch a campaign across what would become Wessex.[26]

[18] E. Callaway (2017) *Nature News* 545, 276–277. (18 May 2017) doi:10.1038/545276a; McNish (2018) The Beaker people: a new population for ancient Britain, https://www.nhm.ac.uk/discover/news/2018/february/the-beaker-people-a-new-population-for-ancient-britain.html Accessed Aug 2021

[19] K. Lotzof (2018) Cheddar Man: Mesolithic Britain's blue-eyed boy. <http://www.nhm.ac.uk/discover/cheddar-man-mesolithic-britain-blue-eyed-boy.html> Accessed April 2020

[20] Cunliffe (2013) *Britain Begins.* Oxford University Press, Oxford, 220.

[21] S. Needham (2010) 'A Noble Group of Barrows': Bush Barrow and the Normanton Down Early Bronze Age Cemetery. *The Antiquaries Journal* 90, 1-39. DOI:10.1017/S0003581510000077

[22] S. Rimmer (2020) Maps of Britain and Ireland's ancient tribes, kingdoms and DNA. <http://www.abroadintheyard.com/maps-britain-ireland-ancient-tribes-kingdoms-dna/> Accessed April 2020

[23] N. Thorpe, Pers. Comm. 2021.

[24] Historic England (2018) *Prehistoric Linear Boundary Earthworks: Introductions to Heritage Assets.* Swindon, Historic England.

[25] J. Last (n.d) Roman invasion at Maiden castle. <https://www.english-heritage.org.uk/visit/places/maiden-castle/history/roman-invasion/> Accessed April 2020 ; M. Russell (2019) 'Mythmakers of Maiden Castle' *Oxford Journal of Archaeology* 38, 2019 <https://onlinelibrary.wiley.com/doi/full/10.1111/ojoa.12172> accessed Sept 2021

[26] Encyclopaedia Britannica (2020) <https://www.britannica.com/biography/Vespasian> Accessed April 2020

The Roman army conquered the region. One strategic motive may have been to stop the Britons supporting the Gauls against the might of Rome, although internal political reasons associated with the Roman Empire may have incentivised Claudius to conquer, so as to demonstrate his military and organisational abilities, and hence silence his critics. At least as far as the *bandes dessinées* comic-book character *Asterix* (1966) is concerned, the Britons were *cousin germain* to the ancient Gauls.[27] Indeed, the Romans, *inter alia*, were concerned about the security of Gaul, threatened by an unconquered Britain of the day.[28]

That human occupation and settlement are ancient, there can be no question. Placename evidence suggests a blend of Brythonic ('Celtic') and Germanic (English) elements in naming certain settlements and topographic features. Anglo-Saxon names predominate, but a search for likely Brythonic placenames soon proves fruitful. Opinion suggests that the roots of the Royal house of Saxon Wessex lies in both British and Anglo-Saxon elements dating from the post-Roman Period. For one thing, the supposed founder (Cerdic) bore a name that may derive from the Brythonic name Caratacos (Caradoc), who ruled in the upper Thames Valley around Dorchester; the original people at this time were known as *Gewisse*.[29]

Wessex, as a named geographical identity, has been retro-fitted to the early Bronze Age by modern scholars – and by Thomas Hardy, as a gloomy, thinly disguised and real 19th-century region. It has never been a county or a planning region, although the name survives within the region as a well-known water service company. Neither, frankly, is it tidy from a border point of view. For one thing, the block of land comprising the counties of Hampshire, Dorset, Wiltshire, historic Somerset, and Berkshire straddle two sea areas (the south coast of England and the Bristol Channel). Otherwise there are untidy land borders that show little reference to physical borders, such as ranges of hills or rivers. For another, the variety of geology and landform is hardly a unifying factor. Neither is the region coterminous with ideas of the 'West Country' or even the 'South West', for much of Gloucestershire, Oxfordshire, Devon, As well as all of Cornwall are generally considered as being outside it.

To what extent has Wessex imparted identity? This question creates interest. In the 9th century, King Alfred came to signify central and southwest England as defined not so much against other Anglo-Saxon kingdoms (there was by his time close co-operation with Mercia) but against invaders – the Danes.[30] Hardy effectively resurrected it as a fictionalised region, and members of the Windsor family have since used the name in an aristocratic title. These back-constructions are perhaps created out of a notion of English identity, closely referencing Alfred.

Wessex is therefore at once an archaeological province, the core kingdom of a nascent idea of England, and a thinly disguised fictional region. On the basis of what has been already written, one imagines no clear pre-existing racial or cultural identity emerging. As far as archaeological, DNA, or isotopic studies on human remains are concerned, nothing clear has

---

[27] *Asterix in Britain* (n.d.) <http://www.asterix.com/the-collection/albums/asterix-in-britain.html> Accessed April 2020
[28] G. Hovell (2020) <http://www.historyextra.com/article/romans/roman-invasion-whose-side-were-britons-0> History Extra Accessed April 2020
[29] D. Parsons (1997), 'British *Caratīcos, Old English Cerdic', *Cambridge Medieval Celtic Studies*, 33, pp. 1-8; H. Hamerow C. Ferguson and J. Naylor (2018). The Origins of Wessex <http://www.arch.ox.ac.uk/wessex.html> Accessed April 2020
[30] D. Horspool (2014) *Alfred the Great*. Amberley, Stroud. 7-12

come forward that tells us anything defining for Wessex. Spatial resolution of larger studies has shown differences between the east and west of Britain, as indeed for the north and south. A study published in *Nature* suggests statistically different genetic features for the populations of modern Devon and likewise for Cornwall, but the bulk of Wessex remains undifferentiated from much of the remainder of England. Interestingly, this study suggests strong in-migration between the Mesolithic and the coming of the Romans, a mere 4000 years, for:[31]

> Significant pre-Roman but post-Mesolithic movement into south-eastern England from continental Europe, [shows] that in non-Saxon parts of the United Kingdom, there exist genetically differentiated subgroups rather than a general 'Celtic' population.

There is no clear 'Englishness', and likewise nothing that can definitively be described as 'Celtic' (in any case a difficult concept). Incorporation into the Roman empire caused not only individuals but also new communities from both within and without Britain to settle. In Winchester particularly, isotopic information suggests several individuals possibly originating from the Hungarian Basin and the Southern Mediterranean.[32] Differences in genetic make-up will reflect not only in-migration, but also geographic isolation and language in choice of marriage partners. Yet there are other, earlier points of interest. We do know that 'Cheddar Man' apparently has living relatives, suggesting continuity from earlier populations.[33] On the other hand, the 'Beaker Culture' demonstrably had continental links, for scientific analysis of the 'Amesbury Archer' skeleton suggests he grew up in central Europe.[34] Modern humans have moved within and without the region since at least the end of the Ice Age. Extrapolating from archaeology, the inhabitants of Wessex should have genes from a wide area across Europe.

The degree and timing of migration of Germanic peoples, once presumed largely in the post-Roman period (after AD 410), remains an area of scholarly and scientific debate. Genomes recoverd from burials in York suggest a strong influence from post-Roman Germanic settlement,[35] and a substantial continental population migration has been proposed for burials near to Cambridge.[36] Some favour more, some less, migration *vs.* acculturation. However, in the year AD 552, near to modern Salisbury, and led by Cynric, son of Cerdic, the Germanic invaders were victorious over the Britons and were likely significant in terms of the establishment of the Kingdom of Wessex. Later, the laws of King Ine of Wessex (ruled AD 688 to 726) prove that there were Britons among the population, although their status may not have been especially high.[37] Ine's time sees the first mention of 'shires', i.e. territorial areas of legal, fiscal, and military importance (the king strengthened the concept of military service), and provided a firmer basis for the 'concept' of Wessex overall.[38]

---

[31] S. Leslie *et al.* (2015) The fine scale genetic structure of the British population. *Nature Journal* 519, 309–314. doi:10.1038/nature1423

[32] H. Eckardt, G. Müldner and M. Lewis (2012) *A Long Way from Home: Diaspora Communities in Roman Britain* [data-set]. York: Archaeology Data Service. <https://doi.org/10.5284/1000405> accessed Sept 2021

[33] L. Barham, P. Priestly and A. Targett (1997) *In search of Cheddar man.* Tempus, Brimscombe. Chapter 7.

[34] Wessex Archaeology (n.d.) <http://www.wessexarch.co.uk/book/export/html/5> Accessed April 2020

[35] R. Martiniano, A. Caffell, M. Holst, *et al.* Genomic signals of migration and continuity in Britain before the Anglo-Saxons. *Nat Commun* 7, 10326 (2016). https://doi.org/10.1038/ncomms10326

[36] S. Schiffels, S. Haak, W. Paajanen, *et al.* Iron Age and Anglo-Saxon genomes from East England reveal British migration history. *Nat Commun* 7, 10408 (2016). https://doi.org/10.1038/ncomms10408

[37] History of England (n.d.) <https://thehistoryofengland.co.uk/resource/selected-laws-of-ine-688-695/ > accessed April 2020

[38] K. Barker (2020) Pers. Comm.; The History Files (1999-2020) Anglo-Saxon Kingdoms. <https://www.historyfiles.

By the 870s, the Christian Saxon king Alfred's kingdom was strong enough to hold the Danes at bay and nurture the roots of a notion of England. An expansionist Wessex by the time of his grandson Athelstan (reigned AD 924 to 939) stretched (in the south) from Cornwall to Kent and became subsumed in a unified England.[39] Alfred also initiated the assessment (for taxation purposes) known as the *Burghal Hidage*, a document listing towns, *burhs*, almost all in Wessex, each listed with a tally of hides (areas of assessment used for tax purposes) – areas for which the 33 listed *burhs* were responsible for the defence of the Kingdom of Wessex against the Danes. The idea was to provide the haven of a fortress for all within a day's ride, around 20 miles (32 km). To achieve this, new burhs, or re-fortified existing settlements were built or maintained. Alfredian *burhs*, furthermore, played an important role in relation to the management of trade, including agricultural and food markets.[40]

Needless to say, the Normans made significant political and economic changes that impacted on the landscape, including removing English landlords, but they remained numerically a very small minority in Wessex, though the development of cathedrals, fortifications, and the designation of Royal hunting forests were all clear demonstrations of their power.[41] Winchester was to lose out to London as capital of post-conquest England. Perhaps one conclusion we may draw is simply to say that the region is, and has always been, open to both influence and opportunists seeking to benefit from a potentially wealthy region. While agricultural land in Wessex may not everywhere be top quality, the climate, diversity, and proximity to markets (be they within Britain or accessible by sea) stimulated development. The region would always be open to economic development, welcome or otherwise, and this is reflected in the landscape. To the Normans, like the Romans before them, the promise of taxation formed the real draw and incentive.

What can a study of the region tell us about the wider world? By the Middle Ages we can be certain of a strong agrarian economy, matched by the size and importance of towns for manufacturing, commerce, and active trading through the ports.[42] Wessex contributed to both the English and wider European economies. The region may also be said to have had considerable industry before the 'Industrial Revolution' affected employment and productivity in the midland and northern regions of England, as elsewhere. Considering investment in transportation, for example, this led in turn to Roman roads, turnpike roads, canals, railways, and motorways, which proliferated to create an integrated transport network.

By early modern times, Wessex people traded externally, through ports on two coasts, grew diverse crops, raised animals, mined, manufactured textiles, made clothes and hats, glass products, made ceramics, tiles and bricks, engaged in developed transportation and hospitality, and developed specialist engineering, including clocks and light engineering; they brewed beer, made cider, made rope and tobacco products – to name just a few initiatives. In the mid-17th century, the agricultural writer and improver Walter Blith[43] (actually a Midlander)

co.uk/KingListsBritain/EnglandWessex.htm> Accessed April 2020
[39] T. Hall (2016) *Athelstan*. Allen Lane, London. Pages 3-10
[40] K. Barker (2020) Pers. Comm.; The History Files (1999-2020) Anglo-Saxon Kingdoms. <https://www.historyfiles. co.uk/KingListsBritain/EnglandWessex.htm> Accessed April 2020; D. Crowther The Burgal Hidage < https://thehistoryofengland.co.uk/resource/the-burghal-hidage/ > Accessed April 2020
[41] H. Cook (2018) *New Forest: The Forging of a Landscape*. Windgather Press, Oxford. Chapter 4.
[42] J.H. Bettey (1986) *Wessex from AD*. Longmans, New York. Chapter 2.
[43] W. Blith (1649/1652) *The English Improver Improved, or the Survey of Husbandry. Surveyed* (2nd edn). Printed for John

could advocate irrigation, land drainage, improvements to tillage and soils, the restoration of a healthy balance between arable and tillage, and promote 'wood by new plantation'. Interestingly, one of the 'six peeces of improvement' advocated by this former captain in the New Model Army is: 'By such enclosures as prevents Depopulation, & advanceth all Interests.'

As early as 1653 this progressive writer could infer that enclosure certainly need not 'advance all interests.' Enclosure of open agricultural land was seen as progressive by some, generally wealthier, members of society and the government. In Wessex, as elsewhere, enclosure excluded the poorer and less politically powerful from access to the land. A study of the Wessex landscape tells us about agrarian change and the transition from more rural to urban societies, across a diversity of geological and climatic provision. In wider terms, Wessex was (and is) of considerable national and international significance, and, as such, the region reflects both environmental and social change, without being urbanised to the degree of other regions.

A legitimate question now arises: *How might we approach a study of the Wessex landscape?*

An approach to studying the landscape

The literature of landscape study and research is varied, complex, and arguably worthy of a book in itself. That is not the purpose here. The objective is to explain how landscapes function, and why humans value them in different ways and to different degrees. The notion of 'landscape' derives from the aesthetic, but is applied in an analytical-scientific sense as well. For example, appreciation of paintings such as Dutch 'Old Masters' is concerned with 'the view'. *Landschap* is a Dutch word linked to concepts of scenery and countryside that gives us the English word 'landscape'.[44] Other writers attempt to link it to analytical frameworks, including ascribing numerical values to elements of the landscape. There must be another way!

Landscapes are changing/evolving entities reflecting economic imperatives and subsequently those of culture, of ritual or private agency. The wider 'productive' landscapes of field, forest, heath, meadow and so on, relate to making a living, and, by definition, are integral to human activity and well-being. Control of the means of production came to be the preserve of elites, and generally operated with powerful interests in mind. Elite groups created their own landscapes, such as their 'stately homes', set in grounds created by great landscape architects, such as Lancelot 'Capability' Brown, and used to display power and prestige. Many prehistoric landscapes (apart from field systems) are often of unknown purpose, although a religious significance may be inferred: the areas around Avebury or Stonehenge are examples. It is furthermore likely (but not proven) that these, too, are products of societies controlled by elites. However, the real work of creating and maintaining landscapes was undertaken by humbler people – ploughmen, graziers, cowmen, shepherds, drowners, marshmen, foresters, woodsmen, miners, quarrymen – folk who also made the Wessex landscape work.

So, the region, history and landscape of Wessex are complicated. Denis Cosgrove found the term 'landscape' in geographical usage 'an imprecise and ambiguous concept', yet it extends terms like 'area' or 'region' while being in common currency with sub-disciplines such as

---

Wright, London.
[44] The Free Dictionary (2016) Landscape < https://www.thefreedictionary.com/landscape> Accessed April 2020

environmental planning and design, as well as links to art and literature. Landscape denotes the 'external world mediated through subjective human experience' – more than the world we observe. Landscape is a social construct and hence prone to subjectivity. Cosgrove became cautious of the economic determinism of Marxian thought.[45]

Here, the present author is in no way pretending to present a narrative that is other than basically explanatory of how this amazing historical region works, incorporating natural resources, landscape ecology and social history. Subjectivity is permitted and themes are chosen that hopefully the reader will find helpful in navigating a complex subject area. We are able to improve on the (often) nebulous concept of 'landscape', such that it possesses certain characteristics.[46]

In this account five themes are selected:

- **Structure** – landscape has definable and describable component parts that exist in clear relation one to another.
- **Function** – landscape does something for somebody. This is typically food, or other natural resource production; the concept is anthropocentric.
- **Value** – what is done conveys economic or other identifiable value to somebody. Non-economic values tend to be ecological, heritage, or those involving social cohesion, including presumed ritual and religious landscapes.
- **Scale** in description will include habitat, field-scale, domestic, local, settlement, regional, river valley, upland area, and more. Scale not only gives expression to the components of the whole, but may indicate how significant some components are, if the landscape really is to be set in a wider context.
- **Change** – applies to all considerations over time. Obvious this factor may be, but it explains how we move in geological and human time.

Structure is what is observed on the ground. It may be fields of given crops, trees comprising a woodland or forest, communities of plants comprising a chalk grassland, heath, or constructed features associated with water meadows or field boundaries, and so on. There is seldom any controversy as to what a surveyor or a remote sensor may reveal. *Function*, on the other hand, is not always so straightforward. While the function of a field or meadow may be relatively obvious, the cultural context of prehistoric monuments is lost, making their interpretation difficult.

Value is arguably the most controversial factor to evaluate, for it raises many questions. What is the nature of the value? It may be a market value, perhaps growing trees for timber or crops for food. Components of certain valley landscapes, such as mills and their infrastructure, concern manufacturing. Non-market value is more difficult. Ironically a landscape imbued with aesthetic value, something that dates from the original concept of *landschap*, was often captured on canvas. The only 'value' is perhaps the price of an Old Master! However, aesthetic,

---

[45] D. E. Cosgrove (1998) Social formation and symbolic landscape University of Winsconsin Press xiv. <https://books.google.co.uk/books?id=NrD2-nJ52aYC&printsec=frontcover&dq=cosgrove+landscape&hl=en&sa=X&ved=0ahUKEwiHjvqc4b7lAhUwQRUIHf2VD40Q6AEIKTAA#v=onepage&q=cosgrove%20landscape&f=false> Accessed April 2020

[46] E.G McPherson, D. Nowak, G. Heisler *et al.* (1997) Quantifying urban forest structure, function, and value: the Chicago Urban Forest Climate Project. Urban Ecosystems 1, 49–61 https://doi.org/10.1023/A:1014350822458

historic, archaeological heritage, 'ecosystem service', and, recently, 'spiritual' values, are all defensible and invoked in conservation policy.

For example, we may conserve a wetland area for non-market reasons, such as a habitat for marsh orchids, or perhaps water voles, only to find that it is also providing 'ecosystem services', including clean water and flood detention. These properties may have monetary value ascribed to their function, with environmental economists striving to place market value. The cost of clean water is an example of something readily ascribed as a 'value'. *Scale* in this context might apply to a small, localised wood or wetland, it may equally apply at a regional scale, as, for example, the Somerset Levels and moors.

The New Forest, largely in Hampshire, is an example of an area with a complex *structure*. And there are landscape elements at many scales. This reflects the geology, topography and soils beneath, and today presents a mosaic of valuable habitats, including woodland, commercial forestry, heaths, wetlands (often of considerable local complexity), and, of course, 'wood pasture' – considered to be the largest surviving area remaining in Europe.[47] *Function* is a tad more problematic and liable to *change*. The Forest as a legal entity was created by King William I for his pleasure and to show the English that Norman rule came with serious intent. Charles II was said to be the last monarch to hunt there (although famously he had other leisure pursuits). However, the state effectively had an asset. Until iron battleships emerged in the 19th century, the area was valuable as a plantation supplying the Royal Navy. These factors conflicted with the locals' interests, who grazed animals and gathered fuel and timber and thus found themselves in conflict with the powerful.

The value to William I was in recreation and, in effect, political. No Norman ruler really gained much economically from the New Forest, although others did from its products – including timber, underwood exploitation, hunting (including poaching of deer) and high-end-value products, such as honey. To the exchequer (from about the time of Samuel Pepys) the value would become directly economic, and to the government it was both political and strategic; production of timber would be measurable in economic terms. The Royal Navy famously used the 'Hearts of Oak' to deter dastardly foreigners, of whichever nation.

Yet iron warships came to be the norm, and as increased transport, rising standards of living, and as environmental awareness became commonplace in a democratising 19th century, more people could enjoy the New Forest for recreation. Rights of commoners were assured and the state retained, often controversially, a direct market economic interest through the 'Office of Woods', which would become the Forestry Commission. Given time, aesthetic, historic and conservation interests were to win out, such that the present century witnessed the establishment of the New Forest National Park. Functions changed dramatically over time.

The landscape of Wessex is the outcome of human activity, much of it direct economic activity, whereby people maximised land use and activity across the landscape just to make a living. Simply stated, there are lessons here for sustainable development; for while the micro-economic commentary of Thomas Hardy dwells very much on the efforts of poor people to survive, even to 'better themselves', many (including the present author) are concerned

---

[47] H. Cook (2018) *New Forest: The Forging of a Landscape.* Windgather Press, Oxford. Chapter 2.

variously as professionals, volunteers, campaigners, writers, and the like, in conserving the Wessex landscape for its own sake. Alternatively, it performs some function that is neither obviously economic, but which provides environmental services. Restoration of peatlands, historic water meadows, the maintenance of heathlands or coppice woods, are all good examples that will be considered in this book.

This volume is concerned with the place where landscape history meets landscape ecology – in a dynamic involving geology and climate, ecological processes and human intervention in creating an intriguing array of landforms and associated ecosystems within the landscape. The focus is rural, with the material organised mainly in accordance with semi-natural ecosystems that developed in response to human and economic forces.

Chapter 1

The region that is Wessex

1.1 Wessex in geography

While 'Wessex' has several meanings in time, place, presumed cultural identity and as a political entity, for present purposes it will comprise the counties of Hampshire, Dorset, Wiltshire, historic Somerset, and Berkshire,[1] with occasional forays into neighbouring southern Gloucestershire, Oxfordshire and Devon. Wessex is a heterogenous region based on a varied geology, climate, soil distribution and river systems that will immediately affect natural vegetation and agriculture (Figure 1.1).

Wessex is irregularly hexagonal in shape with four of the region's sides land boundaries, the remaining two being the English Channel coast between Portsmouth and Lyme Regis, and the south-eastern shore of the Bristol Channel, from the Devon border to approximately Thornbury. Physically speaking then, Wessex stretches from Exmoor in the west to the edge of the North and South Downs near Petersfield in the east, and from the south coast to the Vale of the White Horse and the southern Cotswolds in the north. The two stretches of coast include major ports, such as Bristol, Avonmouth, Southampton, and Portsmouth.

The physical basis of Wessex may be characterised by area by using (as a guide) the National Character Area Profiles (NCAP)[2] that are numbered and located within the region (Figure 1.2) *viz.*:

> Each is defined by a unique combination of landscape, biodiversity, geodiversity, history, and cultural and economic activity. Their boundaries follow natural lines in the landscape rather than administrative boundaries.

1.2 Wessex in geology

Table 1.1 shows the geological timescale from the Precambrian Era (starting *c.* 4.6 billion years ago) to the present. In general, the oldest rocks in Wessex are of the Palaeozoic Era from the Silurian period (although these do not crop out across large areas) through to the modern Quaternary period, which may be subdivided into the Pleistocene ('Ice Age') and Holocene ('Recent'). A new interloper into the Holocene is often termed the 'Anthropocene' (the age of humans). This topic will be discussed later.

In geology and soil science, a distinction is made between 'solid' and 'drift' (or 'superficial') geology. Essentially, solid geology is everything from the formation of the earth through the pre-Cambrian periods and ages listed in Table 1.1, as far as (and including) the Pliocene, which

---

[1] J.H. Bettey (1986) Wessex from AD 1000 Longman New York, 2; B. Cunliffe (1993) Wessex to AD 1000. Longman New York, xv

[2] Natural England (2014). National Character Area profiles: data for local decision making. < https://www.gov.uk/government/publications/national-character-area-profiles-data-for-local-decision-making/national-character-area-profiles >Acc. April 2020.

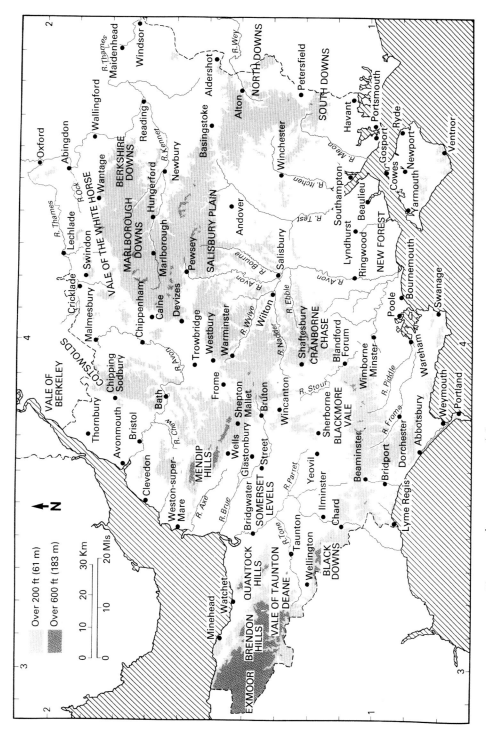

Figure 1.1. The Wessex Region (after J.H. Betty, 1986. *Wessex from AD 1000*, Longman, London)

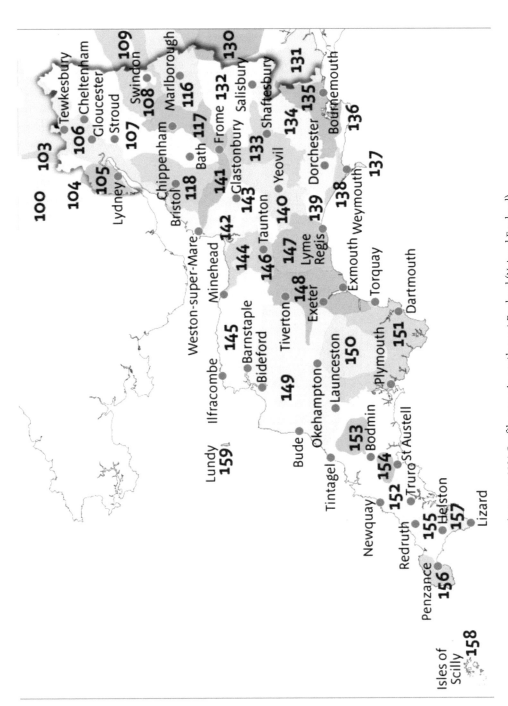

Figure 1.2. NCA Profile areas in south-west England (Natural England)

Table 1.1. Geological timescale

ERA	PERIOD	EPOCH / AGE	Million Years Ago	EVENTS
CENOZOIC *Age of Mammals* 65.5 mya – present day	Quaternary	Holocene	Today — 0.01 —	Ice Age ends Humans are dominant
		Pleistocene	— 1.6 —	Earliest Humans appear Ice Age begins
	Tertiary	Pliocene	— 5.3 —	Hominids (human ancestors) appear
		Miocene	— 23.7 —	Grass becomes widespread
		Oligocene	— 36.6 —	Mammals are dominant
		Eocene	— 57.8 —	Eocene – Oligocene extinction event
		Paleocene	— 65.5 —	First large mammals appear
MESOZOIC *Age of Reptiles* 245 mya – 65.5 mya	Cretaceous	Extinction of Dinosaurs	— 144 —	K-T extinction event Earth looks closer to present-day Flowering plants appear
	Jurassic		— 208 —	First Birds appear Pangaea splits into Laurasia, Gondwanna Dinosaurs are dominant
	Triassic	First Dinosaurs	— 245 —	Pangaea cracks First mammals appear Reptiles are dominant
PALEOZOIC 570 mya – 245 mya	Permian	Age of Amphibians	— 286 —	Permian – Triassic extinction event Pangaea forms
	Carboniferous		— 360 —	First reptiles appear First large cartilaginous fishes appear
	Devonian	Age of Fishes	— 408 —	Late Devonian extinction event First land animals appear First amphibians appear
	Silurian		— 438 —	First land plants appear First jawed fishes appear First insects appear
	Ordovician	Age of Invertebrates	— 505 —	Ordovician – Silurian extinction event First vertebrates appear
	Cambrian		— 570 —	End Botomian extinction event First fungi appear Trilobites are dominant
PRECAMBRIAN 4600 mya – 570 mya	Proterozoic Eon		— 2500 —	First soft-bodied animals appear First multicellular life appear
	Achean Eon		— 3800 —	Photosynthesizing cyanobacteria appear First unicellular life appear
	Hadean Eon	Priscoan Period	4600	Atmosphere and oceans form Oldest rocks form as Earth cools
Formation of Earth				

ended around 2.6 mya (million years ago).[3] In a physical sense, 'solid' refers to age, and not to how consolidated a sediment may be, although older, harder sandstones, limestones, igneous and metamorphic rocks are more easily imagined to be solid. Pliocene sands and gravels can be as loose and un-cemented as similar rocks from the later Pleistocene and Holocene. It follows that geological formations less than about 2.6 myr (million years) old are generally softer, and result from surface processes that are deposited from ice, water and air. They are generally 'in place' and said to rest upon 'bedrock' of the solid geology.[4]

Figure 1.3 shows the solid geology of the Wessex region. The western part of the Wessex region includes Exmoor and the Quantock hills that are located west of Bridgwater. These are composed of Upper Palaeozoic Era rocks (Table 1.1).[5] Virtually, the geological formations are sedimentary, although the Palaeozoic formations of the Mendips, Quantocks and Exmoor are strongly folded and faulted, and hence rather complicated to interpret when compared with younger strata.

In the eastern Mendip Hills, south of the Bristol Coalfield, in Figure 1.3, there is a small area of basaltic rocks, lava flows and tuffs of Silurian age – the 'Coalbrookdale Formation'. The core of Mendip comprises Devonian upper 'Old Red Sandstone' that also crops out along the Severn Estuary, including the coast between Clevedon and Portishead. South-west of Avonmouth. Carboniferous Limestone dominates the Mendips , providing many limestone 'karst' features, including gorges (such as Cheddar Gorge), sinkholes (or 'swallets') and caves formed by underground rivers for which the hills are famous.

The Avon Gorge at Bristol is also cut from Carboniferous Limestone and provides a clear section through the strata, influential in understanding the stratigraphy of the lower Carboniferous period.[6] Coal measures also crop out at the surface, but the majority of the Bristol and Somerset coalfields are 'concealed', that is lying beneath Triassic and other younger strata (Table 1.1). Due to extensive folding, faulting, thrusting and uplift during the Variscan 'orogeny' (mountain building phase), Permian deposits are absent in the western part of Wessex. However, these mountains eroded under arid conditions, which included wadi-type floods, so that lying unconformably over a range of older rocks is the 'New Red Sandstone' of the succeeding Triassic period, when the 'Dolomitic Conglomerate' was deposited across a landscape of folded Carboniferous and Devonian rocks.

The western part of the region east and north of Bristol (including southern Gloucestershire) is dominated by a range of rocks of the Jurassic period. In Figure 1.3, the key shows formations of type 'Inferior Oolite', Fuller's Earth, Great Oolite and Cornbarsh. Great Oolite yields the famous Bath Stone, and also 'Fuller's Earth' (see Chapter 2). These formations run from Stow-on-the-Wold southwards, beyond Bath, and largely form the Cotswold hills, There are also extensive areas underlain by 'Lias', predominantly a grey, well-bedded, marine

[3]   British Geological Survey (2020) Bedrock Theme <https://www.bgs.ac.uk/products/digitalmaps/digmapgb_solid.html > Accessed April 2020
[4]   British Geological Survey (2020) Superficial Theme <https://www.bgs.ac.uk/products/digitalmaps/digmapgb_drift.html> Accessed April 2020
[5]   G.W. Green, F.B.A Welch, R. Crookall, G.A Kellaway (1992) Bristol and Gloucester region 3rd ed. Geological Survey HMSO
[6]   Historic England (2017) *A Building Stone Atlas of Avon* <www.bgs.ac.uk/mineralsuk/mines/stones/EH_atlases.html> Accessed April 2020

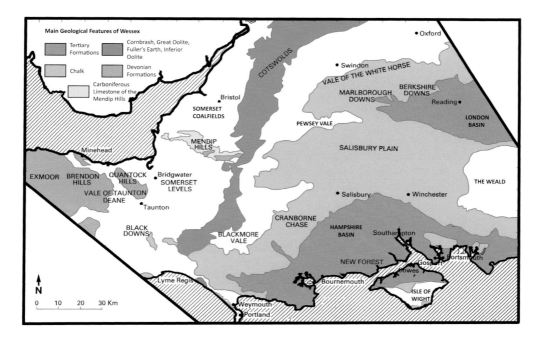

Figure 1.3. The regional geology of Wessex. (Ben Heaney)

calcareous mudstone and silty mudstone, some limestones and sands, such as the Bridport Sand Formation, largely of the Jurassic period, although the lowest part is from the Triassic. Large coastal alluvial areas mostly comprise the Somerset Levels and Moors (Figure 1.3). The area west of Bridgwater includes Lias rocks as well as the old, hard Palaeozoic rocks (such as gritstones and slates) of the Quantocks and Exmoor.

The area of Exmoor that is located within Somerset comprises rocks of the Devonian period, in round figures the oldest being 400 million years and comprising siltstones, mudstones, sandstones and conglomerates. For comparison, the Earth is around 4600 million years old. The limestones of Jurassic period and the chalk of the Cretaceous are, furthermore, important for water supply, constituting significant 'aquifers' that, by definition, are capable of supplying water in economically important quantities for human consumption, as well as acting like large sponges that keep the rivers flowing and sustain wetland areas.[7]

The southern and eastern portion of the region is dominated by chalk (Figure 1.3). The dominant solid geological formation of the area is a (generally) soft, porous, and fissured pure limestone, dating approximately from 100 to 66 million years ago. Forming an outcrop of approximately 4650 km^2, the Chalk aquifer is significant in maintaining flow in the main catchments: the Frome and Piddle, Stour and Allen, Avon, Test and Itchen. Smaller rivers drain the Chalk of East Hampshire and the Isle of Wight. Chalk water is therefore significant to

---

[7] An 'aquifer' here is taken to mean a rock formation capable of storing and transmitting water in economically or environmentally significant volumes.

agriculture including irrigation, public water supply, industrial development, domestic supply, gravel washing, fish farming and watercress production. Much of the water is abstracted from groundwater sources.[8] In the Hampshire Basin, the Chalk is over 490m in total thickness, reducing northwards across Wessex, being perhaps 330-400m thick in Wiltshire.[9] Otherwise there is a range of geology and soils of varying type and economic value. The poorest soils are in the New Forest, deriving from clays, sands, and gravels. The Chalk in the eastern part of Figure 1.3, forms the downland and Salisbury Plain, whereas the older rocks and coastal lowland areas of the Somerset Levels, west of the Cotswolds, provide enormous contrasts in the Wessex landscape.

The area inland of the south Coast in Figure 1.3 includes both the Hampshire Basin (Tertiary Period sediments) and the world-famous 'Jurassic coast', where the Lias forms the unstable cliffs around Lyme Regis. An overall easterly regional tilt on the geological sequence means the rocks become younger eastwards to the Isle of Portland, and onwards to Purbeck (located west of Bournemouth) and both of these areas are famous for supplying limestone for building (see Chapter 2). There is a steep, southern side-fold defining the southern limb of the Hampshire Basin syncline (downfold) displaying east-west trends of three anticlines (upfolds) along the cost running eastwards from Weymouth and into the Isle of Wight. These are steep-sided folds in cross-section, and resembling a step, are sometimes called 'monoclines'). The outcome is a strong east–west ridges in the upper Jurassic rocks and chalk across the landscape of southern Dorset, including the dramatic coastal features at Lulworth Cove and Durdle Door, as well as around the Isle of Wight.

The geological structure north of here is dominated by a 'syncline', or downfold in the Chalk that preserves younger Tertiary deposits of clays, sands and gravels, which dominate the Hampshire coast, the New Forest area, and the northern part of the Isle of Wight. To the north, the Chalk re-emerges to provide the upland areas of the Dorset Downs, Salisbury Plain, Marlborough Downs (north of Pewsey), and the area today referred to as the North Hampshire Downs (Figure 1.3).

Table 1.2. Dates comparing geological time, archaeological periods, and history.

Comparative dates for Wessex	
Age of the Earth	4600 million years (4.6 billion years)
Oldest surface rocks in Wessex (Silurian)	433 million years ago (mya)
Pleistocene ('Ice Age')	*c.* 2.6 mya to *c.* 10,000 years ago (ya)
Palaeolithic	*c.* 500,000 ya to *c.* 10,000 ya
Mesolithic to Iron Age	*c.* 10,000 ya to AD 43
Historic (Roman to Modern periods)	AD 43 to present day

---

[8]   D.J Allen (2019) *The Chalk aquifer of the Wessex region.* BGS, Keyworth <http://nora.nerc.ac.uk/id/eprint/522484/1/RR11002.pdf> accessed Sep 2021

[9]   I. Geddes (2003) Hidden Depths Ex-Libris Press Bradford-on-Avon; 50; P.M Hopson, A.R Farrant, A.J Newell. R.J Marks, K.A Booth, L.B Bateson, M.A Woods, I.P Wilkinson, J. Brayson and D.J Evans (2007) Geology of the Salisbury District. Sheet explanation of the BGS 1:50,000 Sheet 298 Salisbury, Keyworth, E. C Freshney and R.V Melville (1982) 4th ed The Hampshire Basin and Adjoining Areas HMSO London, 13.

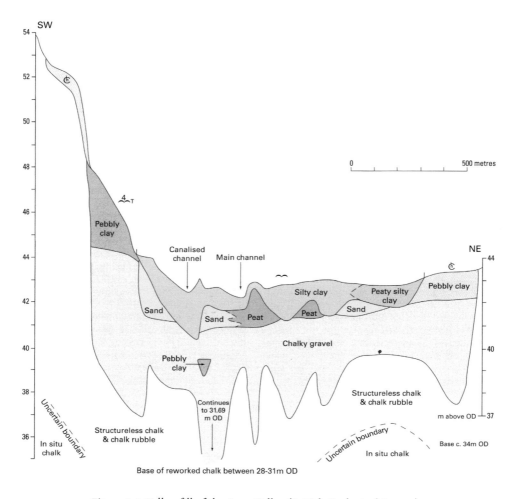

Figure 1.4. Valley-fill of the Avon Valley (British Geological Survey)

After the Tertiary came the Quaternary period, which includes the Pleistocene Ice Age (Table 1.1) and the Holocene, during which the Earth warmed once more. While the landscape of Wessex was affected by extremely cold conditions, it experienced neither ice sheets nor valley glaciers, being too far south and of insufficient altitude.

Figure 1.4 shows a generalised cross-section across the valley and valley-fill of the larger streams, including the River Avon in the vicinity of Salisbury. It also helps to explain the difference between solid and drift geology. Here the Cretaceous chalk (Table 1.1) is marked as being 'in situ' where relatively unaltered by surface processes. Processes such as freezing and thawing during the Pleistocene, linked with earlier rivers that transported material eroded from the downs and from younger Tertiary strata (Figure 1.4), created broken and rubbly chalk encountered beneath the 'chalky gravel'. The chalky gravel therefore represents a drift geological formation.

From an agricultural and construction perspective, drift deposits are very important because they determine the physical and chemical nature of soils present on the surface, and they also provide economically useful mineral resources, such as sand, gravel, brickearth and peat, where they overlie bedrock.

In general, there are three kinds of superficial gravel deposits: 'plateau' and 'river-terrace' gravels that form series of terraces above the modern floodplain; and 'valley-fill' gravels. Gravel deposits form the beds of many rivers in Wessex and are often regarded as part of river alluvium. Valley-fill and river terrace deposits are signified in Figure 1.4, using a '4' and a 'T' to signify the fourth river-terrace gravel, and are located on the southwest side of the diagram. River terraces relate to the present stream networks, their material locally derived and less abraded (worn away by friction or erosion) than the typical 'plateau gravels' found on hilltops. These deposits are common in Hampshire, including the New Forest area. They are located at elevations well above the present river system, frequently capping hills, and including well-travelled, well-worn material that has some origin in glaciofluvial material.[10]

River terraces were likely formed during the last 1.5 myr, meaning that the transport of coarser material by river systems occurred throughout much of the Pleistocene (Table 1.2).[11] Similar to the downs being covered by wind blown material, the upper horizons of river terrace deposits may contain appreciable wind-derived material deposited on the gravels of a pre-existing river in the particular valley that had sufficient power (probably associated with seasonal meltwaters) to transport gravel-grade sediment as part of the alluvial load. Some soils that developed on the river terraces around Salisbury are classified within the Hucklesbrook association, described as 'well-drained coarse loamy and sandy soils, commonly over gravel' and suitable for horticulture as well as arable farming and the extraction of underlying gravels.

The 'cover-loams' of silty loess composition over the gravels date from towards the end of the last glaciation (the 'Devensian') into the Holocene.[12] Aeolian (wind derived) sediments in southern England bury and preserve archaeological sites and their subsequent erosion has loaded rivers with sediment. Investigation of aeolian deposits provides insights into environmental and cultural change over a variety of timescales.[13]

Two large-scale expansions of the British–Irish ice sheet are recognised, which coincide with large glaciations that occurred across Eurasia and North America. The largest was the Anglian Glaciation of about 450,000 years ago. Over a period of some 50,000 years, most of Britain was

---

[10] D.C Findlay, G.J.N. Colborne, D.W. Cope, T.R. Harrod, D.V. Hogan, and S.J. Staines (1984). *Soils and their Use in South West England*. Harpenden: Soil Survey of England and Wales, Bulletin no. 14, 1:250,000 map, Legend and memoir, 11.

[11] C.M. Barton, P.M. Hopson, A.J. Newell and K.R. Rose (2003). Geology of the Ringwood district. British Geological Survey Keyworth ,22

[12] D.C. Findlay *et al.* (1984). *Soils and their Use in South West England*. Harpenden: Soil Survey of England and Wales, Bulletin no. 14, 13, 212

[13] M. Bell and A. Brown (2009) Southern regional review of geoarchaeology: windblown deposits Report 005- <https://research.historicengland.org.uk/Report.aspx?i=14721&ru=%2fResults.aspx%3fp%3d1%26n%3d10%26t%3dGeoarchaeology%26ns%3d1> accessed Sept 2021. English Heritage

covered by ice, extending to the northern outskirts of (what is now) London. A second, smaller glaciation that overall did not reach so far south, – the late Devensian – occurred between *c.* 35,000-15,000 years ago.[14]

The complexity of Pleistocene environmental change is attributable to dramatic swings in the Earth's climate. Not only does this impact on humans, animals, and plant distribution, but also on the hydrological system. Ice sheets during the Anglian and Devensian advances affected the landmasses that would become Britain and Ireland. This was as far south as a line approximated by the Thames-Severn estuary boundary for the earlier Anglian advance, and the area south of this (including Wessex) would have experienced a tundra environment. During these very cold periods, annual cycles of soil-freezing and thawing affected, broke up, and assisted downslope movement of surface rocks, such as the Chalk, although the subsoil often remained permanently frozen. It is possible that ice sheet advance during the Anglian affected the north Somerset coast, and although ice margins were generally some way further north in the Devensian, there was nonetheless impact from cold conditions.[15]

The absolute coldest period during the Devensian is referred to as the 'Last Glacial Maximum' (LGM, dated to *c.* 21,000 ya), and there was a Late Glacial cold period in England (*c.*13,000-10,000 ya).[16] The period approximately 14,700 to 12,900 ya (termed the 'Bølling–Allerød interstadial') was relatively warm and moist, before cold conditions returned for the last time.[17] These extreme variations in temperature allow abundant opportunity for freezing, thawing, and the transport of eroded material across the landscape. The Devensian ice sheet did not impact in Wessex, although the fluctuations in temperature over time certainly did. The 'typical' lowland British extra-glacial river during the cold phases of the Devensian was actively accumulating sediment along its course – the sedimentation being dominated by gravel deposition, with rivers adopting a braided or wandering mode.[18]

In Figure 1.4, such deposits are described as having the composition of 'pebbly clay' and this is distinguished from the head (designated 'Ch'), which comprises a heterogeneous group of superficial deposits that have accumulated by 'solifluction' (the mass movement of soil and surface rock fragments affected by alternate freezing and thawing during the Pleistocene), 'hillwash', and 'hillcreep' (surface-derived material from the downs). These deposits display a wide range of particle size, from gravel-grade, through sand, to clayey material.[19]

The Holocene (post-Pleistocene) *river systems* transport eroded, finer alluvial material, itself largely developed from wind-blown and silty loess deposits that once covered the land surface. This is termed 'brickearth', being once important in extraction for brickmaking, although

---

[14] British Geological Survey (2020). Quaternary Ice Margins <https://www.bgs.ac.uk/research/ukgeology/england/QuaternaryIceMargins.html> Accessed May 2020

[15] Exmoor National Park (n.d.) Geology on Exmoor <http://www.exmoor-nationalpark.gov.uk/__data/assets/pdf_file/0016/116440/Exmoors-Geology.pdf> Accessed April 2020

[16] P. I. Buckland *et al.* (2019) Mid-Devensian climate and landscape in England: new data from Finningley, South Yorkshire. *Royal Society Open Science* 6: 190577. http://dx.doi.org/10.1098/rsos.190577

[17] S.O. Rasmussen (2006). A new Greenland ice core chronology for the last glacial termination. *Journal of Geophysical Research.* 111 (D6): D06102. doi:10.1029/2005JD006079.

[18] Quaternary Palaeoenvironments Group (2008) Devensian (Weichselian) Late-glacial - Holocene (Flandrian) fluvial sequence as an analogue. <https://www.qpg.geog.cam.ac.uk/research/projects/interglacialrivers/analogue.html> Accessed April 2020

[19] P. M. Hopson *et al.* (2007) Geology of the Salisbury District. Salisbury, Keyworth, 25.

earlier studies show that water may also have played a role in the original deposition. In the Salisbury area, the environment of its deposition has been shown to have been cold, dating from the Pleistocene.[20]

Alluvial deposits in southern England are mainly deposited by rivers, with low channel gradients, low angle valley side slopes and display well-developed floodplains. That tend to carry fine sediment load, due to the low erodibility of their banks.[21] The resulting *alluvium* in the district is generally 4-5m thick, of which the top 2-3m is of the over-bank deposit type derived from the modern river system, and supplies significant soil-forming materials in the valley bottom. However, investigations in the Harnham Water Meadows show less thick modern alluvium, generally no more than 1m or so before gravelly material is encountered.

1.3 Wessex in soils

The soils of Wessex are complicated, and details of their formation and distribution are beyond the scope of this book. They are, however, well understood and described. Collectively, individual soil series are based on intrinsic properties and named after location type-profiles. The properties considered are essentially: parent material or *substrate* (e.g. organic material such as peat), where bedrock is within 80cm, thick drift geological deposits, gravelly substrates, and where the pre-Quaternary substrate is soft and thicker than 80cm (including this being due to weathering), and found at or within 80cm. The *texture* of the soil material (as percentages of sand, silt and clay, but also considering calcareous content and organic matter) and the presence, or absence, of material with specific and dominant *mineralogical properties*, including where there are extremely calcareous, saline, and sulphuric soils.[22] *Soil associations*, on the other hand, relate to landform and associated variations in the soil. Soil associations are named after the dominant soil series found during survey.[23] The scheme developed by the Soil Survey of England and Wales is thus a blend of practical factors encountered in management and their geomorphological occurrence.

Soils developed over alluvium are a good place to start considering regional soil variation, by reason of their complicated dynamics. The shallow alluvium near the surface forms the sandy, silty and peaty deposits found at, or near, the surface of the valley-bottom alluvium today. In any case, investigations of soil over water meadows is severely impacted by the construction of irrigated meadows over the pre-existing meads.[24] In practice, the finer alluvial material was likely removed, underlying gravels broken up, and the alluvium re-graded to create 'bedworks' (irrigated ridges) of the meadows and gravels used to make banks and causeways. The outcome for soil distribution on a mature water meadow system is the development of a variable alluvial 'topsoil' thickness (Figure 1.5).

---

[20] H. F. Cook (2008). Evolution of a floodplain landscape: A case study of the Harnham Water Meadows at Salisbury England. *Landscapes* 9 (1), 50-73

[21] G. Ayala (2009) Southern Regional Review of Alluvial Deposits. English Heritage Research Department Reports, 6/2009. <https://historicengland.org.uk/research/results/reports/6527/SouthernRegionalReviewofGeoarchaeology_AlluvialDeposits> accessed Sept 2021

[22] D. C. Findlay *et al.* (1984) *Soils and their Use in South West England.* Harpenden: Soil Survey of England and Wales, Bulletin no. 14, 54-59.

[23] D. C. Findlay *et al.* (1984) *Soils and their Use in South West England.* Harpenden: Soil Survey of England and Wales, Bulletin no. 14, 69-71.

[24] H. Cook (2018). River channel planforms and floodplains: a study in the Wessex landscape. *Landscape History* 39 (1) 5-24

The Harnham Water Meadows have characteristic floodplain soils, with textures that are shallow calcareous or non-calcareous alluvial loamy soils, over flint gravels affected by groundwater with areas of peat and alluvial soils. Mapping is based on the system of the Soil Survey of England and Wales,[25] with modifications that reflect specific locations displaying shallow topsoil. Most of the area comprises calcareous, fine loamy soils over gravel, representing fluvial re-worked aeolian deposits that originally formed soils on the interfluves upstream of the site (Figure 1.5).

For example, the Frome association commonplace in the river valleys around Salisbury includes a range of individual soil series including the Frome series (a common soil over calcareous alluvial gley,[26] and a fine, silty texture over calcareous gravelly river alluvium), the Wylye series, the Racton series, and the Gade series – all developed over alluvial floodplain streams draining the Chalk and its adjacent Tertiary age formations.[27] Not only does detailed soil mapping provide information of ecological significance, but it also sheds considerable light on the structure of the alluvial deposits, and hence the forms and genesis of the river system itself.

The calcareous Frome shallow phase, the non-calcareous Racton variant, and the 'terrestrial raw soils' of the Harnham series (particularly shallow, with 5cm or less of fine topsoil) all possess shallow alluvial topsoils. Harnham soils occur within certain watercourses, developed in fast-flowing watercourses leaving little or no fine topsoil. The Frome shallow phase soils, common in the south-west of the meadows, result from the construction of a bank along the river, or otherwise arise from their subsequent modification.

The disposition of the shallow, peaty 'Gade series' (calcareous *peat or humose soils* above 40cm) is of particular interest. The linear peaty features marked in black (Figure 1.5) relate to infill of drainage channels between the bedworks (see Chapter 4). Here, poor drainage would have permitted the accumulation of fresh peat since the construction of water meadows in the 17th century. However, blocks of Gade soils are mapped on the northwest side of the island, with larger blocks occurring along the eastern margin and to the south. These areas are interpreted as being re-workings of floodplain peat infills during the construction of the water meadows, which would thus predate their construction. Such peat accumulations are considered consequences of 'backswamp' conditions associated with anastomosing channels. The stability of such channels *between avulsion events* (that is when the river leaves its original channel, either partially or wholly during a flood and hence sometimes creating a new permanent channel) would be conducive to peat accumulation (see Chapter 4).

Soils formed away from alluvial valleys are next considered.[28] *Chalk downland* soils, where the chalk bedrock dominates soil formation. are frequently of '*rendzina*' type. These are shallow

[25] D. C. Findlay *et al.* (1984) *Soils and their Use in South West England*. Harpenden: Soil Survey of England and Wales, Bulletin no. 14.

[26] The term 'gley' in soil science refers to colours, typically grey, blue or green that arise from the reduction of iron in the soil. This is conveniently a marker of periodic, sometimes permanent waterlogging. Mottling implies a mix of gleyed and oxidised colouring (typically orange, brown or reddish) indicating seasonally waterlogging only so that the horizon in question may be alternatively reduced or oxidised.

[27] D. C. Findlay *et al.* (1984) *Soils and their Use in South West England*. Harpenden: Soil Survey of England and Wales, Bulletin no. 14, 182-185

[28] D. C. Findlay *et al.* (1984) *Soils and their Use in South West England*. Harpenden: Soil Survey of England and Wales, Bulletin no. 14, Legend.

SOILS AT THE HARNHAM WATER MEADOWS

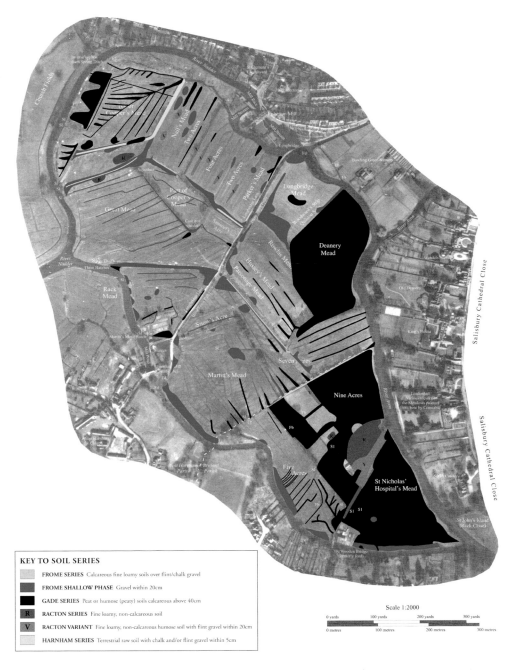

KEY TO SOIL SERIES

FROME SERIES Calcareous fine loamy soils over flint/chalk gravel

FROME SHALLOW PHASE Gravel within 20cm

GADE SERIES Peat or humose (peaty) soils calcareous above 40cm

R RACTON SERIES Fine loamy, non-calcareous soil

V RACTON VARIANT Fine loamy, non-calcareous humose soil with flint gravel within 20cm

HARNHAM SERIES Terrestrial raw soil with chalk and/or flint gravel within 5cm

Scale 1:2000

Figure 1.5. Example soil map: soils at the Harnham Water Meadows (the author, cartography by Sue Newman)

and 'lithomorphic' in kind, with an organic or organic-enriched topsoil that may tend towards becoming organic, even peaty, with soft unconsolidated material within 30cm of the surface. *Colluvium* is sediment that is eroded, transported and deposited on or near the base of slopes by gravity. Resulting from falls, slides, slumps and flows of water-saturated sediment, the resulting sediment my be a poorly sorted jumble of different particle sizes and rock types.[29] Examples of soil survey associations in England and Wales for the Wessex region include the Icknield soil association (a humic rendzina). Soils included in the Andover 1 association are brown rendzinas and, on account of a silty surface horizon, may have resulted from the erosion of a loess covering through agriculture.[30]

Deeper soils may be found in the downland valleys (generally dry valleys) and they include those in the aptly named Coombe 1 association, developed over chalk or chalky drift. These are not rendzinas but are termed *'typical brown calcareous earths'* and are well drained. These contain clay-enriched (argillic) subsoils. The Clay-with-Flints is a weathering residual deposit frequently encountered over the chalk, Upper Greensand and other strata. Clay-with-Flints typically comprises orange-brown and red-brown sandy clay with abundant nodules and rounded pebbles of flint. Angular flints are derived from the chalk, and rounded flints, sand and clay from Tertiary formations.[31] Soils developed on Clay-with-Flints resting on chalk constitute 'palaeo-argillic brown earths' and these pre-date the Pleistocene. These are often mapped as belonging to the Carstens association, described as 'well-drained fine silty over clays [horizons], and fine silty soils, often very flinty'. Alternatively, 'stagnogleyic palaeo-argillic brown earths' (as the name suggests) are developed on slowly draining subsoils, and where they are developed on Plateau Drift and Clay-with-Flints, soils of the Batcombe association, are mapped in the west of the region.

Soils developed on clays and shales that form the vales are very prone to poor drainage. These are typical of the 'Cheese Country' (see Chapter 5), being surface-water, gley soils that are seasonally waterlogged, slowly permeable and prominently mottled above 40cm. In Wessex, typical examples are the Denchworth association soils developed on Jurassic and Jurassic clay, and the Wickham 2 association soils developed from drift over Jurassic and Cretaceous clay or mudstones.

Soils that have developed over sand and gravel today typically form areas such as the New Forest and the Dorset Heaths (see Chapter 7). Although areas with specifically low water tables in permeable material ought to be dry on the surface, they are often not, to the point of allowing peat accumulation in hollows. The New Forest proper is economically marginal land, with many areas of clay from the Tertiary period, meaning that surface water gleys are commonplace and mapped generally as the Wickham 3 Soil association ('typical stagnogley soils'). Slowly permeable, seasonally waterlogged, fine loamy over clayey, and coarse loamy over clayey soils would be the cause. Landslips may locally be present, complicating the picture,

---

[29] K. Wilkinson (2009) Regional review of geoarchaeology colluvium. Report no 003/2009 < https://research. historicengland.org.uk/Report.aspx?i=14719&ru=%2fResults.aspx%3fp%3d1%26n%3d10%26t%3dGeoarchaeology%26ns%3d1> accessed Sept 2021 English Heritage

[30] J. B. Boardman (1992) Current erosion on the South Downs: implications for the past, in M. Bell and J. Boardman (eds), *Past and present soil erosion: Archaeological and Geographical Perspectives*. Oxbow, Oxford, 9–19.

[31] The British Geological Survey (2020). The BGS Lexicon of Named Rock Units <https://www.bgs.ac.uk/lexicon/lexicon.cfm?pub=CWF > Accessed April 2020

but in the New Forest coniferous and deciduous woodland may be present, or otherwise wet lowland heath is to be found.[32]

It is 'staglogley podzols' that develop on this sand and gravel geology. Soils with characteristic bleached and human-enriched subsoil horizons, directly underlain by a greyish or predominantly mottled horizon, result from periodic waterlogging. These include the Holiday's Hill association, developed on Tertiary (and Cretaceous) sand, loam and clay. They are naturally very acid, sandy over clayey locally, with peaty or humose surface horizons and generally supporting lowland heath habitats and coniferous woodland.

The Boulderwood association is developed on ridges formed by plateau gravel and river terrace drift. These soils are naturally very acidic coarse loamy over clayey horizons with bleached subsurface horizons, slowly permeable, some seasonal waterlogging and some shallow and very flinty soils supporting coniferous plantation, lowland heath and deciduous woodland and gravel extraction. The Sollom 2 association (a typical gley podzol) is develop over Tertiary sand and while similarly being podzolic, may be seasonally waterlogged but is well-drained where it occurs on slopes. Similar land uses include military training areas based on heathland to the west of Poole Harbour. The podzolic Shirrel Heath 1 association is associated with military use and conservation areas, largely islands in Poole Harbour.

Along the southern coast are soils developed on marine and river-terrace gravels, such as the Efford 1 soil association, well drained fine loamy soils, often over gravel, located around Christchurch and suited to a range of agricultural and horticultural crops. Set inland from Christchurch is the typical argillic gley Swanwick soil association, developed on sandy and loamy drift and supporting some agriculture, forestry, and wet lowland heath. Along the coast of Somerset there are localised soils developed such as 'typical sand pararendzinas' of the Sandwich soil association, comprising calcareous and non-calcareous, deep and well-drained windblown sand, and marine shingle-derived soils that tend to be unstable but include some wetland habitats in hollows.

Along the Somerset coast, localised occurrences of the Crwbin association (brown, ranker, lithomorphic soils similar to a rendzina) are mapped over Carboniferous limestone, but otherwise the Newchurch soil association is to be found, generally adjacent to the sea or not far inland, between the mouth of the Parrett at Bridgwater Bay, at the eastern limit of the Bristol Channel, and occurring well north of Bristol along the Severn Estuary. These soils are 'pelo-calcareous alluvial gley soils', deep, stoneless, mainly calcareous clays soils with groundwater and flooding controlled generally by ditches and pumps. These developed from marine alluvium, notably on the seaward side of the Somerset levels and Moors area. They differ from the soils of the Frome association by being largely reclaimed marine (rather than river) alluvium.

On Mendip, upland soils display the dominant Nordrach association, a palaeo-argillic brown earth, developed from aeolian silty drift (loess) over carboniferous limestone (see Chapter 6), which is well-drained and supports dairying. Over the Devonian sandstone there have developed the ferric podzols of the Larkbarrow association, otherwise the Mendip Plateau is

---

[32] D. C. Findlay *et al.* (1984) *Soils and their Use in South West England.* Harpenden: Soil Survey of England and Wales, Bulletin no. 14.

mapped as Milford association, a well-drained, brown earth developed on a range of Devonian sedimentary rocks and supportive of dairy farming. Around the periphery of the plateau, the Carboniferous limestone bedrock gives rise once more to the Crwbin association; also associated with dairy farming, it is capable of supporting herb-rich, limestone grassland.

The Jurassic limestone areas of Wessex (see Chapter 1), particularly the southern Cotswolds, are dominated by brown rendzinas of the Elmton 1 and Sherbourne associations. Developed on Jurassic limestone bedrock and clays, these are both capable of supporting cereals and grass production. The Elmton 1 association soils are overall better drained and potentially more versatile. Evesham 1 is the dominant soil association, common around Bath and to the south; it is a typical calcareous pelosol developed over Jurassic limestones and clays and is suited to cattle rearing, dairying, and winter cereal production.

The high areas of Exmoor (see Chapter 7) are dominated by dark brown or ochrous subsoil, with no overlying bleached layer.[33] Such is the Manod association developed on Palaeozoic-era slate, mudstone and siltstone, with bare rock outcrops found locally. Land use is woodland and cattle rearing. The Lydcott association is developed on Devonian reddish sandstones, supporting wet moorland and only low-intensity grazing, as well as woodland. These are ferric stagnopodzols with a peaty topsoil and iron enriched layer. Another commonplace soil association is the Denbigh 1, a brown earth developed on palaeozoic slaty mudstone and siltstones that supports stock rearing.

This account of typical soils in Wessex should prove instructive in understanding land use and economy.

1.4 Wessex and water

The region has a complicated geology. While virtually everywhere sedimentary rocks outcrop, their diversity is great. Aquifers include chalk, limestone, sandstones and gravels, while there are abundant impermeable strata that create largely surface runoff. Frequently several geological formations are observed in a single catchment.

Table 1.3 summarises the main rivers and associated geology referenced to the UK 'hydrometric areas' shown in Figure 1.6, which are bounded by natural river catchment boundaries. This diversity has provided for all of the water supply, transport, power and irrigation, as well as the aquaculture of watercress and fish within the region. For example, hydrometric area 39 is part of the enormous Thames catchment. The varied geology includes a range of aquifers as well as streams that rise in clay areas. Hydrometric areas 42, 43 and 44 are dominated by chalk-fed streams. This demonstrated the dominance of the Chalk to Wessex, as both groundwater resource and aquifer that maintains a relatively constant and even flow of water throughout the year. Hydrometric areas 45, 52, 53 and 101 are underlain by varied geologies meaning where they occur, aquifers supplying rivers and streams are diverse. In areas 51, groundwater plays a very small part in supplying surface water flows.

---

[33] According to the classification system of the Soil Survey of England and Wales <http://www.landis.org.uk/downloads/classification.cfm> Accessed April 2020. These are soils with a black, dark brown or ochreous subsurface horizon resulting from pedogenic accumulation of iron and aluminium or organic matter or some combination of these. They need not possess a typical 'bleached' horizon resulting from the downward migration of minerals and organic matter. This may seem confusing to those otherwise familiar with the concept of a 'podzol'!

Table 1.3. Hydrometric areas and key rivers in Wessex.[34]

Hydrometric area (Wessex only)	Main rivers	Dominant geology
39 Thames catchment, (west area)	Thames Kennet Loddon Lambourne Coln Churn Ray Windrush	Mainly chalk-fed rivers, however the Thames, Coln, Churn and Windrush rise in the Cotswolds from springs in the Jurassic limestones and the headwaters of the Ray, across Jurassic clays, as does the Thames for much of its distance.
42 Hampshire rivers east of the river Avon	Lymington Blackwater Test Itchen Hamble Meon	Predominantly chalk-fed streams that flow southwards to the Solent or English Channel across younger Tertiary strata. Exceptions being the Lymington and Blackwater rivers that rise from Tertiary sands, plateau gravels, or on clays
43 Hampshire or Salisbury Avon Catchment	Avon Ebble Nadder Wylye Bourne New Forest streams Dorset Stour Allen	Predominantly chalk-fed streams. Headwaters of Avon, largely from older Cretaceous rocks including Greensand. Nadder is similar, but has older Jurassic rocks, including surface runoff from clays in the upper river. The New Forest drains from a range of tertiary rocks and river-terrace gravels. The Dorset Stour rises on Jurassic limestones and clays in its upper portions. The 'Corallian limestone' is significant in terms of agriculture and water supply.
44 Rivers of south Dorset	Piddle Frome (Dorset) Asker Wey	Predominantly chalk-fed streams.
45 Rivers of east Devon	Axe (Dorset and Devon) Otter Tale Culm Exe Creedy Barle	A wide range of aquifers support these rivers, most notably Permian sandstones and breccias, and also the Triassic Otter sandstones and Budleigh Salterton pebble beds. The upper reaches of the Barle and Exe rise on Exmoor (see area 51, although there are some calcareous rock formations), while rocks lower in the catchments can introduce calcareous waters, making the hydrochemistry variable.
51 Rivers of west Somerset coast	Washford River Doniford Stream Horner Water Oare Water West Lyn East Lyn Other Exmoor streams draining northwards	Predominantly northward flowing and draining Devonian rocks, such as siltstones, mudstones, sandstones and conglomerates; limestone is largely absent.

---

[34] UK Hydrometric Register (2008). < http://nora.nerc.ac.uk/id/eprint/3093/1/HydrometricRegister_Final_WithCovers.pdf > Accessed April 2020

Hydrometric area (Wessex only)	Main rivers	Dominant geology
52 Rivers of Somerset Levels	Parrett Tone Isle Brue Yeo Axe (Somerset) Sheppey Congresbury Yeo Kenn	Rivers fed from diverse aquifers and draining the alluvial Somerset Levels and Moors. For example, the Parrett rises in Dorset from the upper Greensand and flows across a range of sedimentary formations. The Tone joins it, having risen in the Brendon hills (part of Exmoor), with other streams from the west draining the Quantocks, comprising Devonian non-calcareous sedimentary rocks. The Axe rises in the Mendip Hills to the north (Carboniferous limestone). Otherwise, much of the area is underlain by Triassic and lower Jurassic rocks that influence the rivers, e.g. the Yeo rising near to Sherborne in Dorset from largely calcareous Jurassic rocks.
53 Bristol Avon Catchment	Avon (Bristol) Frome (Bristol) Midford Brook Chew Boyd By Brook Marden Biss	The Bristol Avon and Bristol Frome rise in the Cotswold Hills; the Marden and Biss rise in the Chalk, flowing from the chalk and sandstone of the North Wessex Downs; the Chew catchment displays a mixed geology dominated by clay, while its headwaters are fed by limestone springs.
Part of Area 54 (to north of 53)	Little Avon	Limestone, sandstone and clay of Triassic and Jurassic age, including the Lias group of mudstone, siltstone, limestone and sandstone.
101 Isle of Wight	Medina, East Yar, Caul Bourne, West Yar	Predominantly chalk and Greensand-fed streams. In the north part of the Island, rivers and streams also arise from Tertiary sands and clays.

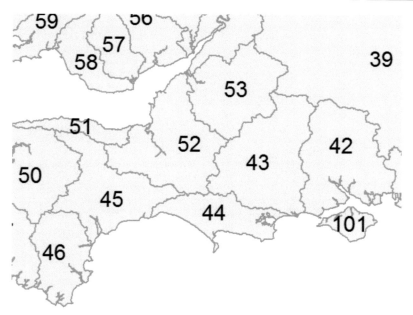

Figure 1.6. Hydrometric areas and key rivers in Wessex (National River Flow Archive https://nrfa.ceh.ac.uk/hydrometric-areas below; Catchment Data)

Table 1.3 demonstrates the complexity of the river systems. While practically all the geological formations are sedimentary in nature, their hydraulic and chemical natures vary greatly. Many aquifers are limestones, including chalk, imparting a calcareous nature to their waters. The geologically older formations of Exmoor contain little or no calcium carbonate and will impart a more acidic (low pH) nature to the water. Other geological formations, such as clays and shales, will not store or transmit water in useful quantities, so that most water falling on them will not infiltrate the ground but run-off to a river channel.

The topographically diverse landscape provides not only the visual amenity of hills and vales, of coastal lowlands, chalk downlands, limestone ridges, and the bleak, evocative uplands, but also a range of gradients and hydrological environments, enabling marshes, water meadows, water supply and river gradients that have long been exploited for waterpower. The coastline has left ample opportunity for economic development from fishing to ports, to holiday destinations, where both people and commodities could be moved.

1.5 Wessex in prehistory and history

Whereas Table 1.2 aims to set the three timescales in context: geological (that actually starts with the formation of Earth some 4.6 billion years ago, around one-third of the age of the universe!); archaeological (concerned with human prehistory. but moving into the historic period); and historic (which, strictly speaking, deals with time from the onset of written records). In Wessex this means not so much the start of writing – this started in the Iron Age with names struck on coinage – but with the narratives of invasion and empire by the Romans.

Table 1.4. Dates taken as 'Prehistoric' in this book.

Prehistoric chronology for England	
Pleistocene begins	*c.* 2.6 mya
Last Interglacial (Ipswichian) warm phase	*c.* 127,000 to *c.* 106,000 before present (BP)
Last Glaciation (Devensian)	*c.* 35,000 to *c.* 13,500 BP
Palaeolithic	ends *c.* 10,000 BC
Mesolithic	*c.* 10,000 – *c.* 4000 BC
Neolithic	*c.* 4000 to *c.* 2000 BC
Bronze Age	*c.* 2300 to *c.* 750 BC
Iron Age	*c.* 750 BC to AD 40

Hence, Table 1.4 takes 'time' from around 2.6 mya, when dramatic fluctuations in climate heralded the Pleistocene (or Ice Ages), into times defined by human activities, specifically technologies. One example of a warm period within the Pleistocene is the Ipswichian Interglacial. In Wessex, and in Britain in general, older archaeological finds are well-attested, including both modern people and Neanderthals.[35] Later, it is the transition from the Upper Palaeolithic to the Mesolithic (around the time of Cheddar Man) that is significant, for it is then that appreciable modern human *Homo sapiens* occupation began, as people followed

---

[35] Museum of Wales (2017) The Ice Age cave men of Wales <https://museum.wales/articles/2007-05-11/The-Cave-Men-of-Ice-Age-Wales/> Accessed April 2020

migrating animals into what is now Britain. There was significant use of the caves at and around Cheddar at this time.

By the Mesolithic, people who arrived by walking across the land bridge known as Doggerland, now drowned beneath the North Sea as the sea level rose,[36] possessed a sophisticated hunting technology useful for survival in the newly wooded landscape. Although permanent settlement was not so likely, it may have occurred in places such as Howick in Northumberland[37]. Widespread permanent settlement only really became possible with the arrival of agriculture in the Neolithic. The succeeding Bronze and Iron Ages reflect not only metallurgical advancements but are shorthand for a range of technological, landscape and deduced social changes that followed, that is until Rome broke the documentary silence.

Table 1.5 shows the familiar historic periods referred to in this book. Until the Norman invasion of AD 1066, when convention requires the names of ruling dynasties to be listed, the names of peoples are used, although other shorthand references enable us to talk of Medieval/post-Medieval/early modern, and so on. Despite a perfectly good calendrical system, terms like 'Georgian', or 'Victorian' remain in common usage. The present is best identified as the 21st century, as the term Elizabethan was always going to cause confusion! When considering environmental history, the term 'Industrial Age' becomes helpful. While highly imprecise, this notion links with recent conceptualisation around the 'Anthropocene', for it harks back to considerations of the technological periods in prehistory.

Table 1.5. Dates of the historic periods referred to in this book

Historic chronology for England	
Roman	AD 43-410
Anglo-Saxon	410-1066
(Danish Invasions)	840-1016
(Danish Kingdom)	1016-1042
Norman	1066-1154
Middle Ages	1066-1485
(Tudor)	1485-1603
'Post-Medieval' or 'Early-Modern'	1485-1800
Industrial Age	1760 to present

Defining Wessex is not easy, for it is not tidy or geographically distinct, even less so temporally. Somehow (as explored in the Introduction), cultural (material or otherwise) considerations seem to have driven any definitions we may have. While land boundaries fluctuated, the coasts were less effected, notwithstanding sedimentary deposition, reclamation, and coastal erosion. The integrity of our region would seem to lie in the interplay of two seas and historic

---

[36] E. Sturt, D. Garrow and S. Bradley (2013) New models of North West European Holocene palaeogeography and inundation *Journal of Archaeological Science* 40, 3963-3976 https://doi.org/10.1016/j.jas.2013.05.023
[37] C. Waddington (2003) A Mesolithic settlement site at Howick, Northumberland: a preliminary report. *Archaeologia Aeliana*, 1-12. < https://core.ac.uk/download/pdf/57527.pdf > accessed Sept 2021.

county borders, including a fascinating range of inland landscapes. It is people who ultimately made Wessex, from a mishmash of earth processes.

1.6 Coastal genesis

Because Wessex has coast on its north-west side and its southern margin and a resultant relationship with the seas that link with areas beyond England, it is helpful to consider some physical aspects of coasts to provide a context for both coastal management, the development of ports, and seaside towns. These coasts present a range of physical features and associated natural habitats, including cliffs, beaches, dunes, shingle bars, and saltmarshes, which have often been reclaimed for agriculture over the last two millennia (see Chapter 4).

One particularly dramatic event stands as a reminder of coastal complexity, the Bristol Channel floods that may have cost as many as two thousand lives.

On the Somerset Levels and Moors, records of flooding go back as far as the 1600s. In 1607 flooding was particularly bad, causing extensive damage on both sides of the Severn Estuary with great loss of human life, livestock, and damage to the agricultural economy.[38] With a

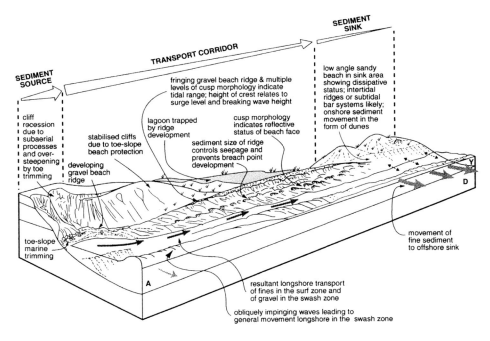

Figure 1.7. Schematic of coastal cell for sediment movement (After Orford 1986).
J. D. Orford, S.C. Jennings and D.L. Forbes (2001). Origin, development, reworking and breakdown of gravel-dominated Coastal Barriers in Atlantic Canada: Future scenarios for the British Coast. in *Ecology & Geomorphology of Coastal Shingle* (eds) J.R. Packham, R.E. Randall, R.S.K. Barnes and A. Neil Westbury, Otley

---

[38] K. Horsburgh and M. Horritt (2006) The Bristol Channel floods of 1607 –reconstruction and analysis. *Weather* 61 (10), 272-277. <https://rmets.onlinelibrary.wiley.com/doi/pdf/10.1256/wea.133.05> Accessed April 2020

typical elevation of 5-6mAOD, floodwaters exceeded 7m in places, making this a particularly catastrophic event. While the tidal range in the Severn Estuary is among the highest in the world, the nature of its funnel shape and the possibility of considerable storm surges from the Atlantic may all have had a part to play. (However, one opinion has even attributed this catastrophic event to a *tsunami*).[39]

Figure 1.7 shows an idealised along-shore coastal cell from the British coast. The transport mechanisms involved present linear features such as beaches and offshore barriers. Assumed is a cliff-based sediment source (which could supply a range of geological materials) and a transport corridor including a beach. This corridor terminates in a sink for fine sediment onshore (as dunes) or offshore (as a seabed shoal).

Drift alignment of beaches occurs where waves break at an angle to the coast. The swash (upward movement of water from a breaking wave and associated sediment) occurs at an angle, but downward return movement (the backwash) runs perpendicular to the beach. This pattern causes material to be transported along the beach as 'longshore drift'.[40] Where waves break parallel to the coast this is termed swash aligned, and the lack of longshore sediment movement is destructive to a beach or sediment bar.

Where the (longshore) sediment supply diminishes, the dynamics of the transport corridor alter sufficiently for the gravel beach or barrier to become a sink for sediment, where the gravel itself becomes locked in. Although both rising sea levels and changing wave behaviour (due to climate change) affect gravel bars, many British coastal barrier bars are in a terminal state, on account of becoming sediment starved, as they were once parts of larger systems. They become aligned with the swash, i.e. the upward movement of sediment caused be wave action, rather than being dominated by longshore drift. This makes them vulnerable to erosion and breaching.[41]

Where the longshore sediment supply diminishes, the nature of transportation changes creates new problems. There will be a new or renewed, or accelerated, erosion of the next vulnerable coastal rocks or sediments in the direction of the longshore drift.[42] For example, the bar at Porlock (see Chapter 4) has been abandoned to managed retreat, and new coastal habitats, such as saltmarsh, are in the process of being formed. On the south coast of Wessex, Chesil Beach needs to be repaired to maintain contact with the mainland.[43] Otherwise, there may already be sufficient supply of sediment for coastal habitat creation. For significant sand dune systems of considerable biodiversity interest, there are formations at Studland Bay in

---

[39] E. Bryant and S. Haslet (2002) Was the AD 1607 coastal flooding event in the Severn Estuary and Bristol Channel (UK) due to a tsunami?, *Archaeology in the Severn Estuary* 13, 163-167. <http://ro.uow.edu.au/cgi/viewcontent.cgi?article=1100&context=scipapers <accessed Feb 2018>

[40] For a basic explanation of Longshore drift see: Revision World (n.d.) https://revisionworld.com/gcse-revision/geography/coastal-landscapes/coastal-processes/longshore-drift Accessed April 2020

[41] J.D Orford, S.C Jennings and D.L Forbes (2001). Origin, development, reworking and breakdown of gravel-dominated Coastal Barriers in Atlantic Canada: Future scenarios for the British Coast in Ecology and Geomorphology of Coastal Shingle, in J.R. Packham, R.E. Randall, R.S.K. Barnes and A. Neil (eds) *British Shingles*. Westbury, Otley, 23-55.

[42] R. Silvester (1974) *Coastal Engineering, II. Developments in Geotechnical Engineering* (vol 4B, ch 3). Elsevier, Amsterdam, Volume 4B, Chapter 3. This is also a key reason why concrete sea walls do not work as erosion defences; they strongly tend to transfer the erosive potential of the sea to the land immediately beyond the end of a sea wall downstream of longshore drift.

[43] The Environment Agency (2014) <https://www.gov.uk/government/news/flood-defence-repair-work-completed-on-chesil-beach> Accessed April 2020.

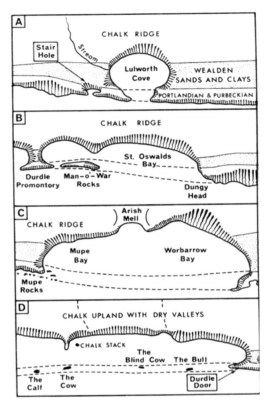

Figure 1.8. Concordant coastline: Bay development, Isle of Purbeck (After R.J. Small (1970) *The study of landforms.* Cambridge University Press)

Dorset,[44] and the Somerset coast at Berrow, near Burnham on Sea.[45]

The interplay of present erosional forces and the disposition of existing geological structures is referred to as 'structural control' on coastal form. This is important along both the Dorset and Somerset coasts. Over time, and with rising sea level, the Isle of Wight became separated from the mainland. It was originally joined to the latter, separated only by the 'Solent River', with a continuation of the Purbeck 'hogback' Chalk ridge that joined the Needles with the Chalk ridge on the north side of Swanage, ending at the coast with the sea stacks known as Old Harry Rocks.[46] This was breached by the sea, and today it runs east-west from Ballard Downs, forms the Purbeck Hills, provides the backing to Lulworth Cove, and runs to the sea, close to Swyre's Head.

Coastline form is a complicated interplay of geological factors, especially relatively hard and soft strata from the erosional point of view, of sea level, and the degree of erosive powers relating to exposure to the open sea, prevailing currents and winds, and geological structures. Depending on the relationship between orientation of the coastline and the structure(s) of the geological formations encountered (particularly folding and faulting of strata), geomorphologists are inclined to talk of concordant and discordant coasts. Classic examples are available from each Wessex coast and are illustrated in Figures 1.8 and 1.10 respectively.

Due to the strong east–west fold, which is best envisaged as a step in the Upper Jurassic Portland and Purbeck beds, and termed a 'monocline', the geological grain is approximately parallel to the coast. The hard Portlandian beds (the limestone used to build St Paul's Cathedral in London) form what might be thought of as a defensive wall against the sea. At a point where a stream would have breached the Jurassic limestone 'wall' and started to erode the soft Wealden (lower Cretaceous) sediments, the sea broke through and wave diffraction helped to

---

44   The National Trust (n.d.) <https://www.nationaltrust.org.uk/studland-bay/features/studlands-sand-dunes> Accessed April 2020.
45   Sedgemoor District (2020) <https://www.sedgemoor.gov.uk/article/991/Berrow-Dunes-Local-Nature-Reserve> Accessed April 2020.
46   SCOPAC (2004) Quaternary History of the Solent <http://www.scopac.org.uk/scopac_sedimentdb/quart/quart. htm> Accessed April 2020.

Figure 1.9. The chalk ridge above Mupe Bay east of Lulworth Cove, looking due east. In the middle distance is the chalk cliff at Arish Mell, in the distance (right) the headland is Worbarrow Tout on the east side of Worbarrow bay (the author).

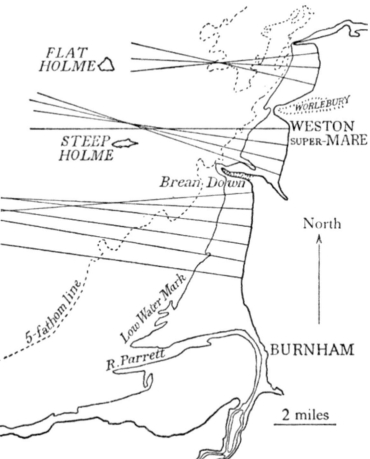

Figure 1.10. Discordant coastline between Bridgwater Bay and Weston super--mare, Somerset (2 miles is c. 3.2 km) (public domain)

form the near-circular Lulworth Cove.[47] An evolutionary sequence may be discerned (Figure 1.8), commencing at Stair Hole and the associated caves (Figure 1.18A). In Figure 1.8B there are remnants of the Jurassic limestone wall above sea level, and its associated caves, caused by marine erosion in Lulworth Cove proper to features such as St Oswald's, Worbarrow and Mupe Bays. Eventually, chalk crops out and forms the back of Lulworth Cove, which becomes the next formation to resist the sea (Figures 1.8C and 1.8D).[48]

Figure 1.9 shows the view looking east from above Mupe Bay towards Worbarrow Bay (Figure 1.8C). The first promontory is a chalk cliff, the more distant one shows the steeply northward dipping Portland, Purbeck and Wealden strata at the eastern end of Worbarrow Bay. Essentially marine erosion has breached the Jurassic 'wall' and is attacking the Chalk at this point. Figure 1.10D shows Durdle Door to be a remnant of this 'wall'.

Turning now to Somerset, the fold trend of Mendip is east–west, the same as the Dorset coastal example described; this fold trend affects not only Jurassic and Cretaceous strata, but also the older rocks of the Devonian and Carboniferous age. The folding belongs to the Variscan (late Carboniferous-Early Permian) orogeny (crustal deformation), when there was a serious collision of tectonic plates, something that would be repeated during the Tertiary, causing the formation of the Alps. This event also affected the younger strata along the Dorset Coast, when the likely re-activation of old faults in the Variscan basement rocks created the monoclinal structures in the Mesozoic strata of Purbeck and the Isle of Wight.

In the Bristol-Mendip region this resulted in no surviving deposition of Permian beds so that the Triassic Dolomitic Conglomerate deposited on and around Mendip (see Chapter 1) and in the Avon Gorge sits unconformably[49] on strongly folded and faulted Devonian and Carboniferous strata. This interesting deposit (in the modern geological literature termed the Mercian Mudstone Marginal Facies) is a weathering, residual 'breccia' (rock including irregular pebbles) resulting from erosion, under desert conditions, creating outwash fans and valley infills of the Mendips, like the wadi deposits of deserts today. North of Mendip, in the vicinity of Bristol, the fold trends switch to northeast– southwest, and almost north–south; these affect the geological structure of the Bristol and Somerset coalfield (see Chapter 1) by defining synclinal basins in which the coal measures occur.[50] Figure 1.10 shows an example of a discordant cliff line from Somerset. The 'rays' are lines drawn perpendicular to the coastline that serve to demonstrate the shape of the coastline, and where bays produce a focus for these due to softer rocks being eroded. The presumption is that these beaches are oriented at right angles to the approach of dominant waves.

The relationship to what is locally a predominantly east–west arrangement of folds and thrusts (low-angle faults) caused by the Variscan deformation phase, and cut by the modern coastline of the Bristol Channel/Severn Estuary presents harder strata. This is mainly the

---

[47]   T. Badman, D. Brunsden, R. Edmonds, S. King, C. Pamplin and M. Turnbull (2003) *The Official Guide to the Jurassic Coast.* Coastal Publishing, Wareham.
[48]   D.C. Jones (1981) The *Geomorphology of the British Isles: Southeast and Southern England.* Methuen, London. Chapter 10.
[49]   In geology an 'unconformity' occurs where there is a break in continuous deposition for millions of years. There may be different geological structures below and above the unconformity where there has been uplift, folding (or faulting) and erosion before the deposition of newer and younger geological deposits.
[50]   C. M Barton, P.J Strange, K.R Royse and A.R. Farrant (2002) *Geology of the Bristol District* and accompanying 1:50.000 sheet 264, British Geological Society, Bristol.

Carboniferous Limestone beneath Brean Down, Worlebury Hillfort, and the headland of Middle Hope to the north, forming prominent headlands, whereas the softer Triassic and Jurassic strata in between, form the bays. South of Brean Down, these softer formations back the coast, including the sand dunes, and there are considerable mudflats within the tidal ranges.

1.7 Landcover change

In principle, natural vegetation (the overall structure and composition of floral communities) is a response to factors such as climate (water and temperature), organisms including animals, topography, soil parent material and time for community development.[51] This has developed during the Holocene, although worldwide, so dramatic has been human-induced environmental change, that scientists have proposed the term 'Anthropocene'. This new geological epoch followed from, or incorporates, the Holocene (see Chapter 2). This is a time from which it is recognised that human activity impacted on planetary systems, including land use change, biodiversity, and influenced climate.

Re-colonisation by natural vegetation of the post-Pleistocene landscapes probably occurred sporadically as the climate was unstable, including a brief return to arctic conditions around 8500 BC, in a cold snap lasting some 200 years. Overall, this meant the open tundra landscape of (what is now) southern England, with its short grasses and herbs grazed by herds of large animals such as reindeer, was replaced with pioneer tree species of birch, pine and juniper; this covering of trees being termed the 'wildwood'.

There is much information about vegetation and vegetation changes to be gained from the study of pollen diagrams. The branch of palaeontology (the study of fossils) thus employed is 'palynology' – literally the study of pollen grains, or other spores found in archaeological and geological deposits. From the landscape point of view, pollen is generally identified through microscopic studies of peat bog, soil, or lake deposits. From this it is relatively simple to identify the presence of species of tree, grass, heather, or other plants, although it is far from straightforward to ascribe relative abundance or its contribution to land-cover. Once this is combined with other palaeo-environmental information, and with a means of identifying a date (typically ^{14}C-derived dates obtained from organic matter), and/or analysis of tree rings from wood samples and old buildings ('dendrochronology'), a picture of environmental change over time becomes possible.

Other valuable supporting information may be gleaned from remains of land snails, beetles, fossil remains of larger creatures, plant remains (such as roots, seeds, leaves and stems) and 'microfossils', such as diatoms (a kind of algae), and more.[52] To be of significant use in environmental history, fossil remains must be abundant, widespread, and easily collected during sampling – meaning the smaller they are the better. In peat deposits, a coring device is typically used, and the pollen abundance graphs are displayed in parallel. Relative, if not absolute, abundance of pollen can be calculated and displayed. Figure 1.11 gives an example of the use of such data. Although it is not in itself a pollen diagram, it displays the proportions

---

[51] Canadian National Vegetation classification (2013) <http://cnvc-cnvc.ca/view_article.cfm?id=180> Accessed April 2020.

[52] Natural England (2011) Environmental Archaeology <https://historicengland.org.uk/images-books/publications/environmental-archaeology-2nd/environmental_archaeology/> Accessed April 2020.

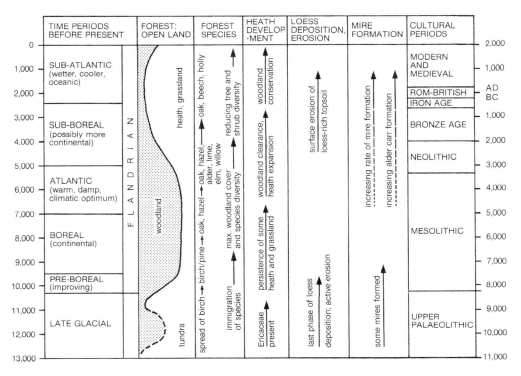

Figure 1.11. Timeline for the New Forest and its vegetation (from C.R. Tubbs, 2001. *New Forest, the history, ecology and conservation*, New Forest Ninth Centenary Trust, Lyndhurst)[53]

between woodland and ground-level species of plants, linking this to human technological periods, climate change and historical dates.[53]

Humans would have entered the Wessex landscape following the rise in temperature that heralded the end of the Pleistocene from about 9,700BC. Their technology changed from larger, cruder Palaeolithic-age tools for hunting in open landscapes, to the finer, expertly crafted Mesolithic implements of hunter-gathers, at some point around 8000 BC (Figure 1.11). Rising sea levels separated first Ireland and then Britain from the continent, perhaps with the assistance of a *tsunami* channelled southwards down the North Sea, and which might have destroyed 'Doggerland', the name given to submerged land on the site of that sea. All this occurred around 6000 BC, leading to the separation of Britain from the continent – from now on, new migrants to the islands would have to travel by sea.[54]

Cheddar Man lived around 8000 BC, placing him early in the Mesolithic.[55] He would have hunted and gathered in and around Cheddar Gorge, in a wooded environment that would have been changeable during the preceding centuries, as the transition from the Late Pleistocene

---

[53] C.R Tubbs (2001). *New Forest, the history, ecology and conservation*. New Forest Ninth Centenary Trust, Lyndhurst.
[54] N. Crane (2016) *The making of the British Landscape*. Weidenfeld and Nicholson, London. Chapter 3.
[55] K. Lotzof, (2018). Cheddar Man: Mesolithic Britain's blue-eyed boy. < http://www.nhm.ac.uk/discover/cheddar-man-mesolithic-britain-blue-eyed-boy.html> Accessed April 2020

cold was not smooth. Earlier skeletons are dated to the preceding Upper Palaeolithic, however their staple diet of reindeer was replaced by forest-dwelling elk, red deer and wild boar, which could be hunted individually using a spear or bow and arrow. Beneath the Somerset Levels lies a buried landscape drained by the rivers Axe and Yeo, across which Mesolithic peoples roamed and set up temporary camps. Since then, sea level rise caused a large estuary to silt up, and peat accumulation buried older sites and created the modern levels, although these new and productive environments provided fish, shellfish and birds, which would have readily been exploited by humans.[56] Other animals hunted in the region included pike, mallard, goldeneye duck, crane and large, herbivore roe deer and aurochs.[57] An adult aurochs stood up to 2m high, grazed largely on floodplains, and would have been relatively successful in conflicts with wolves. Aurochs were hunted by, and would have to compete with, humans for low-lying fertile areas once agriculture became established, thus they eventually became extinct.[58] Humans would domesticate these to produce modern-type cattle. Domesticated cattle were largely introduced to Britain from continental Europe, although there was probably some subsequent interbreeding with indigenous aurochs.[59] Separating humans from native woods and forests presents a false dichotomy. When Europeans explored and conquered large areas of the planet during the colonial period, they made presumptions around the nature of the interaction between humans and forests, and these presumptions continued into the 20th century.

Where there were humans, forests would have been altered by their activities. For example, in the Americas, what has been called 'the pristine myth' presumed a handful of Native Americans who hardly affected forests. In practice there was a dynamic involving the construction of roads, reservoirs, buildings, fish weirs, and often quite intensive agriculture systems. Fire was used to clear areas to extend grassland for grazing.[60] Possession of 'nature's abundance' is essential to the sustenance of humans.

Post-glacial southern England would have been (re-)populated by humans moving from the south, and who likely followed herds of animals upon which they depended for food and clothing. While the 'wildwood' developed from the arctic tundra, there would be a change in human technology towards the more sophisticated tool kits of the Mesolithic (10,000 BC - 4000 BC). The Mesolithic period included not only the ability to hunt and gather in a wooded landscape, but also transition towards a food-producing society.

Clearance of the land for the development of agriculture necessitated a more settled population (there would likely be a great reduction in personal mobility) and greater investment in labour. The ability to increase the birth rate created a positive milieu, whereby more land had to be cleared as the 'frontier of cultivation' expanded.[61] While the exact form of the transition to agriculture remains unclear, we infer a higher requirement of social organisation, eventually supporting more people occupied in *consciously managing the landscape*; the social changes

---

[56] P. Barham, P. Priestly and A. Targett (1997) *In search of Cheddar man.* Tempus, Stroud. Chapter 5.
[57] N. Crane (2016) *The making of the British Landscape.* Weidenfeld and Nicholson, London. Chapter 10, 25-26.
[58] N. Crane (2016) *The making of the British Landscape.* Weidenfeld and Nicholson, London. 146-7.
[59] S.D.E. Park., D.A. Magee, P.A. McGettigan, *et al.* (2015) Genome sequencing of the extinct Eurasian wild aurochs, *Bos primigenius*, illuminates the phylogeography and evolution of cattle. *Genome Biology* 16, 234. https://doi.org/10.1186/s13059-015-0790-2
[60] C.C Man (2005) *1491: The Americas before Columbus.* Granta Books, London. Chapter 1.
[61] P.J. Reynolds (1987) *Ancient farming.* Shire Archaeology, Aylesbury. Chapter 3.

would have been profound. In temperate regions, clearance of interfluve regions between river valleys typically (but not always) dates from at least the Neolithic period – in southern England from about 4000 BC. Then, land was cleared for agriculture and this would have triggered, among other problems, soil erosion that accreted on the floodplains as alluvium and would have affected the rivers, loading them with sediment.[62]

On the nature of the supposed postglacial 'wildwood', there remains much debate although there were important differences across Britain and Ireland.[63] While the dominant species of much of Wessex was likely to have been small-leaved lime, followed in abundance by hazel, oak and elm, occurring as areas of lime-wood and areas of hazel wood, where not shaded out by lime.[64] Further west, in more upland areas, including Exmoor, oak and hazel would have come to dominate the tree cover. The term 'high-forest' implies a mature climax vegetation that became established, by stages, after the climate warmed.[65] Familiar tree species migrated northwards, reaching a maximum development around 5000 BC. This was the 'wildwood' – but how open was it?

Early models for developed, climax 'wildwood' over Britain, could suggest that 'a proverbial squirrel could have leapt from Land's End to Ullapool [Northwest Scotland] without putting a paw on the ground'.[66] Although promoted by Oliver Rackham, this view derives from earlier researchers on the historic vegetation of Britain and has been termed the 'Tansley model', after the pioneering ecologist Sir Arthur Tansley.[67]

But how likely was a dense, continuous cover with most crowns of the trees touching? Increasingly the presence of, and activity by, both large, grazing herbivores and humans within forests worldwide has been recognised. In addition, it must be remembered that trees have finite lives and will die eventually, leaving gaps, and that natural events, including 'wind-throw' and lightning strikes, are frequent over time. While the ability of native British broadleaved woodland to burn has been challenged,[68] there remain mechanisms that will break up the forest cover.

The ecologist Frans Vera caused controversy when he developed 'the theory of the cyclical turnover of vegetations'.[69] With some subsequent modification,[70] his theory may be summarised:[71]

---

[62] H. Cook (2018) River channel planforms and floodplains: a study in the Wessex landscape. *Landscape History* 39 (1) 5-24.
[63] H. Cook (2018) *New Forest: the Forging of a Landscape.* Windgather Press, Oxford. Chapter 2.
[64] O. Rackham (2006) *Woodlands.* William Collins, London, 67. Elm was to decline dramatically across Europe from 4000 BC, probably from disease.
[65] In ecology a climax community is where populations of plants or animals remain stable and exist in balance with each other and their environment. A climax community is the final stage of succession in balance with the local climate, soils and topography.
[66] Based on the saying (anon): 'from Blakeney Point to Hilberee: a [red] squirrel could jump from tree to tree'. Blakeney is on the Wash and Hilbere Island is at the mouth of the Welsh Dee. Supplied by G.P. Buckley
[67] Rackham (2006) *Woodlands.* William Collins, London, 61. Elm was to decline dramatically across Europe from 4000BC, probably from disease.
[68] O. Rackham (1987) *The History of the countryside.* J.M. Dent, London, 72.
[69] F.W.M Vera (2000) *Grazing Ecology and Forest History.* CABI, Oxford.
[70] K.J. Kirby (2003) *What might a British forest landscape driven by large herbivores look like?* English Nature Research Report 530, Peterborough.
[71] A.C Newton, E. Cantarello, G. Myers, S. Douglas and N. Tejedor (2010) The condition and dynamics of new forest woodlands (2010) in A.C. Newton (ed.) *Biodiversity in the New Forest.* Pisces Publications, Chapter 13.

1. Break-up phase representing the transition from woodland back to open habitat. The canopy opens out as trees die and vegetation shifts from woodland to grassland and ericaceous species.
2. Park Phase, largely open landscape with a thin scatter of trees left from a previous grove, vegetation is mostly heathland and grass.
3. Scrub Phase, where thorny shrubs effectively exclude grazing herbivores, permitting overtopping and development of new trees.
4. Grove phase, which is tree-dominated and where a closed tree canopy shades out the shrubs and large herbivores can return. Regeneration is prevented. Eventually the canopy will break up and return to (1).

Specialised large herbivores, such as deer, bison, aurochs and wild horses comprise the grazing mammals (assisted by some bird species such as jays) and they are active in all phases except (3). Thorny shrubs, such as blackthorn (*Prunus spinosa*) and hawthorn (*Crataegus monogyna*) excluded grazing herbivores, resulting in areas where young trees (both shade tolerant and intolerant species) in turn regenerate. In this 'regeneration window' events, such as hard winters or drought can cause the grazing animal populations to drop, aiding regeneration.

Opening the forest is facilitated by the animals through soil disturbance and dung deposition. Young trees, such as oak or hazel, would also invade in scrub areas through introductions of seed by birds, wind, and animal dispersal. Hazel, specifically, if shaded out, and therefore not producing pollen, would not survive. By implication, woodlands would have had to be open for hazel to survive. Overall, the wildwood would have more the appearance of parkland than dense, mature forest cover.

The Vera theory also explains how species of open and 'woodland edge' habitat have been present since the Ice Age. The resulting woodland would be mixed aged woodland, with glades, rides, and a continuous succession of age classes; felling and regeneration are the most biologically productive and biodiverse woodland ecosystems. Medieval parks, forests and wood pastures are also examples of dynamic interaction between vegetation and grazing.[72]

Vera based much of his work on observing the New Forest in Hampshire, where there remain significant areas of wood pasture being grazed by animals who could find shelter beneath its trees, and which present an open canopy precisely because regeneration is restricted by grazing animals. Such landscapes are familiar from historical sources, including not only the true wood pasture of medieval commons (these generally became degraded to open areas due to poor management in historic times), but also from 'parkland' (generally a landscape created by the nobility and lesser gentry) from the medieval period onwards; these were generally not manicured areas, but private game reserves prior to the 18th century[73] A more open canopy would not only have facilitated the entry of large herbivores, but humans would have hunted them within the forests. The dichotomy between open and closed woodland canopy is interesting because of implications for contemporary forest management.

---

[72] Prehistoric vegetation from the end of the last ice age (n.d.) <http://www.bosci.net/prehistory.html> Accessed April 2020.
[73] J. Bond (1994) Forests, chases, warrens and parks in medieval Wessex, in *The medieval landscape of Wessex*, M. Ashton and C. Lewis (eds) Oxbow monograph 6, Oxbow books, Oxford, 116.

Britain at the period of 'climax' wildwood, provided Mesolithic peoples with a suitably adapted technology to live, hunt and migrate through the forest, including making temporary clearings. This is supported by some palaeo-ecologists, who think evidence points to a closed canopy on account of evidence from pollen, beetles and snails, but with open areas generated from catastrophic events such as 'wind-throw'. Large herbivores were still present.[74]

Hazel would be an important food source for Mesolithic people. Natural regeneration of any given species would occur following the death of individual trees, for example from lightning strikes, and re-colonisation by trees and shrubs after humans had produced woodland clearings. Hazel, being shade intolerant, would not thrive or fruit under dense oak canopies in any case, yet it would be capable of shading out small, regenerating oak trees. The result could be that oak would become marginalised to less productive sites, so that, in this model, oak woods would have dominated less fertile soil.

Pollen diagrams tell us nothing about the age and size structure of the woodland, making the general form of the wildwood a matter of some debate. However, George Peterken[75] helpfully identifies a continuum between 'relatively natural' and 'relatively artificial' woodland, commenting that there is a grey area where (for example) an oak stand is to be found in an area that would naturally have oak woodland as climax vegetation. Wildwood would, before human influence, have the property of 'original-naturalness'; a 'normal forest' would display a wide range of age structures.

Supporting arguments that favour the Vera ecological dynamic model include there being a place for hunter-gathering humans predating large herbivores; the model permits Neolithic farmers easy access to clearings (where they can commence planting) and is capable of supporting a range of 'historic' trees and plants, and butterflies associated with grasslands. There would have been a range of habitats for birds of open and closed surroundings, and, importantly, there are open habitats for plants in the ground flora that do not flower in the shade. However, debates around the nature of the wildwood in Wessex, as elsewhere, are set to continue.

Woods provided berries, roots, fruit, nuts, fungi, wild seeds, grains, and shellfish in middens, where taken inland from coasts. The practice of coppicing is ancient, and coppice products from hazel and oak associated with (possibly deliberately created) clearings to herd deer are identified from the British Mesolithic, although these may be a function of burning vegetation rather than deliberate woodland management.[76] Clearings could have been created as browsing for deer herds, or pigs, and deliberate planting is possible; coppicing was established by the early Neolithic.[77] Clearings would also be where people met and would have eventually taken on social and religious significance, even in areas where food gathering and grazing became more reliable. Stone Age technology could further expand the size and number of clearings by

[74] H.J. Birks (2005) Mind the gap: how open were European primeval forests? Trends in ecology and evolution, 20 (4) 154 – 156; Grazing Ecology < https://knepp.co.uk/the-inspiration > Accessed April 2020.
[75] G.F. Peterken (1981) *Woodland Conservation and Management.* Chapman and Hall London, 42-3.
[76] R.R. Bishop, M.J. Church and P.A. Rowley-Conwy (2015) *Firewood, food and niche construction: the potential role of Mesolithic hunter-gatherers in actively structuring Scotland's woodlands.* Quaternary Science Reviews 108, 51-75.
[77] B. Cunliffe (1993) *Wessex to AD 1000.* Longman, New York. Chapter 2.

felling trees with axes and honeysuckle ropes, and burning susceptible trees (birch or pine but not oak), although a more open canopy is arguably more susceptible to wind-throwing trees.[78]

On the Carboniferous limestone upland of Mendip, Mesolithic people hunted in a landscape of pine and birch trees, with open grasslands – these communities were the precursors of Cheddar man. Here are caves with subterranean rivers and gorges, including that at Cheddar, and a cave there has revealed the remains of around 50 inhumations, dated between 8200 and 8400 BC.[79] It is likely there was a change during the Early Mesolithic from cave occupation by humans to use of caves as burial sites.[80]

An interesting linking theme here is an archaeological find close to Stonehenge – beneath the modern car park to be precise. Here, close to one major cultural centre of historical Wessex, were excavated the bases of three pine timber, totem-like poles. Here was a presumed monument, likely in a wildwood clearing on Salisbury Plain, that dated to the Mesolithic, several millennia before its stone successor.[81] While there are ideas about this monument, it should be made clear that its purpose, its relationship to Stonehenge above the ground (see Chapter 2), and even the age of the different poles all remain uncertain.

1.8 Semi-natural habitats

'Natural vegetation', a term in common currency, is a difficult concept where humans are around. A climax plant community relies on an early successional community that ultimately leads to the 'climax' community, or the most stable state for the successional sequence. The vegetation will pass through a 'succession' of so-called 'seral stages'. Simpler communities may be replaced by more complex communities as environmental conditions change (notably climate), eventually resulting in a relatively stable community termed 'climax vegetation'.[82] Even without humans, such considerations are at the mercy of changing climate and sea levels (Figure 1.11). A seral community, 'sere' is an intermediate stage within an ecological succession in an ecosystem advancing towards its climax community.[83] Semi-natural habitats have more of a sense of arrival at vegetative stability, such that:

> Semi-natural habitats have ecological assemblages that have been substantially modified in their composition, balance or function by human activities. They may have evolved through traditional agricultural, pastoral or other human activities and depend on their continuation to retain their characteristic composition, structure and function. Despite not being natural, these habitats and ecosystems often have high value in terms of biodiversity and the services they provide.[84]

[78] N. Crane (2016) *The making of the British Landscape*. Weidenfeld and Nicholson, London. 51-2.

[79] N. Crane (2016) *The making of the British Landscape*. Weidenfeld and Nicholson, London. 33-4.

[80] R. Schulting (2005) '. . . pursuing a rabbit in Burrington Combe': new research on the Early Mesolithic burial cave of Aveline's Hole, *Proceedings of the University of Bristol Spelaeological Society* 23, 171-265.

[81] English Heritage (n.d.) History of Stonehenge <https://www.english-heritage.org.uk/visit/places/stonehenge/history-and-stories/history/> Accessed April 2020.

[82] J.M. Pandolfi (2008) *Succession*. *Encyclopedia of Ecology* <https://www.sciencedirect.com/topics/agricultural-and-biological-sciences/climax-communities > Accessed April 2020 >

[83] D. Blakesley and P. Buckley (2016) *Grassland Restoration and Management*. <Grassland-Restoration-Management-Conservation-Handbooks/dp/1907807802> Pelagic Publishing, Exeter, Chapter 1.

[84] European Investment Bank (2018) <https://www.eib.org/attachments/strategies/environmental_and_social_practices_handbook_en.pdf> Accessed April 2020

Concepts of 'semi-natural' move us on to other classic ecological concepts, such as 'plagio-climax'. For example, heathland presents a kind of vegetation succession that is prevented from achieving a mature vegetation cover (i.e. woodland), it is termed 'plagioclimax' because it is prevented from reaching a full 'climatic climax' by human intervention – in the case of heaths and moors, this is grazing, cutting and burning. The significance of this excursion into ecological theory is that Wessex is replete with semi-natural ecosystems as generally understood in conservation management.

1.9 Land cover and land use

Figure 1.12 shows the 'land use regions of Hampshire' that were defined in a publication of 1940.[85] The famous geographer and geologist Professor Sir L. Dudley Stamp, oversaw the production of a comprehensive survey of the land use of Britain during the years prior to the Second World War. Britain was mapped at a scale of six inches to the mile (1:10,560) – a typical scale for mapping environmental field survey, including soil and geological maps – and published at the scale 1:63,630 ('one inch to one mile'). This became known as the First Land Utilisation Survey of Great Britain. The project was only finished after the War; a second land utilisation survey in the 1960s was published at a scale of 1:25,000 but never completed.[86] The shaded areas represent detailed study areas in the original report.

In much the same way as geological and soil maps are accompanied by a memoir, so are these, produced on a county-by-county basis. In its totality, Stamp's achievement has never systematically been repeated before or since. While some have likened it to the Domesday Survey of 1086, it also resembles the usefully and spatially referenced approach of the descriptive County Reports, sponsored by the Board of Agriculture and published around 1800, a time when food security was also threatened.

Hampshire, while it contains only sedimentary rocks is – from a natural resource point of view – most diverse. While the climate is clement (the rainfall is neither high nor low by British standards, and overall temperatures are relatively warm), the geology (Figure 1.3) imparts a varied topography and soil development. The New Forest and heathland areas in the far west, which extend into Dorset, have among the poorest soils – and hence the poorest land capability in lowland Britain (see Chapter 7). Soils that developed on similar geologies in the London Basin do occur in the north, and include heaths and grazing land; most were developed on sands, gravel, and nutrient-poor clay soils.[87]

While these mostly economically unproductive soils provide a basis for woodland, commercial forestry, and heathlands (see Chapter 8), soils capable of being utilised for agriculture occur elsewhere (Figure 1.12). These include the 'New Forest fringe', the chalk downlands, including a 'central chalk arable area', identified in Stamp's day as highly suited to arable agriculture, although in the recession of the early 1930s much arable had been abandoned to scrub or put down to permanent grass.[88] East of here, the Chalk is frequently overlain by the Clay-with-

---

[85]   F.H.W. Green and H.P. Moon (1940) Hampshire, in L.D.Stamp (ed.) *The land utilisation of Britain* Geographical Publications, London.
[86]   https://discovery.nationalarchives.gov.uk/details/r/93bf9401-ab4e-44bb-8980-2600473231d7
[87]   H. Cook (2018) *New Forest: The Forging of a landscape*. Windgather Press, Oxford. Chapter 2.
[88]   F.H.W Green and H.P. Moon (1940). *The land utilisation of Britain*. Part 89, Hampshire. Ed by L.D.Stamp. Geographical Publications, London, 358.

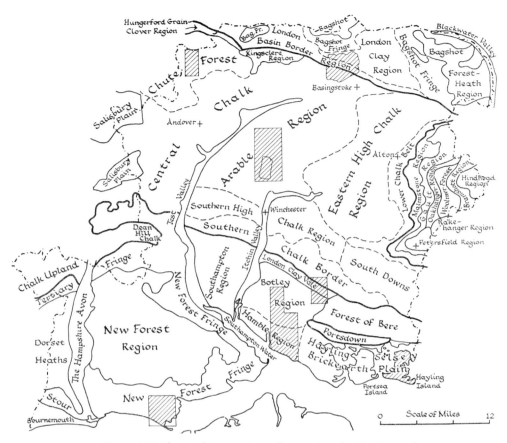

Figure 1.12. The Land Use Regions of Hampshire (public domain)

Flints formation, a weathering residual example from the Tertiary deposits that once covered the Chalk, and so is less suited to arable agriculture and today includes areas of woodland. In the extreme south-east of the region, the 'brickearth plain', which continues into West Sussex, includes some of the best soil in the country, and is famed historically for horticultural production. Between Alton and Petersfield lies the western closure of the Wealden anticline of southeast England; land use here is highly variable, reflecting a complex sequence of geological formations.

Dudley Stamp's (First) Land Utilisation Survey of Britain mobilised some 250,000 students (both school and university), and engaged local people in mapping the country's land use – something that was to prove invaluable in directing agricultural land use planning in the Second World War. His funding for this mammoth project included assistance from the Rockefeller Research Foundation and local authorities. The survey was also a foundation for Stamp's promotion of 'applied geography' in the post-war extension of urban and rural planning.[89]

---

[89] Encyclopaedia Britannica (2020) Methods in Geography <https://www.britannica.com/science/geography/

Agriculture was in recession by the mid-1930s, and much arable land had reverted to pasture. While some politicians argued against government intervention, a Food and Supply Sub-Committee was established in April 1936, which proved to be successful. At the outbreak of war, the War Agricultural Executive Committees ('War Ags') from the First World War were re-formed and given powers to order ploughing by farmers. Specifically, these groups directed agriculture by determining land usage and the types of crops to be grown. Prisoners-of-war were deployed in 'ploughing-up', and farmers were directed to grow as much food as possible to feed the blockaded British Isles. Overall, there was a reduction in in favour of arable farming, although beef and dairy stock numbers were maintained, while the numbers of other farm animals declined. Additionally, agricultural wages were controlled while product prices were fixed by negotiation with the National Farmers' Union. The formation of the 'Women's Land Army' compensated for the decline in the male rural workforce due to conscription or re-employment in other industries.[90]

Moving back across the millennia, while Mesolithic peoples had their own economy and influenced their environments, they had limited influence in land cover. The above discussion tells us that 'land use' is different from 'land cover', as it is directly concerned with economic considerations aimed at the production of food, timber, and other resources. Figure 1.12, from the 'Land Utilisation Survey', may be regionalised on a basis of agricultural production. In the centre is an 'arable region'; heathland areas are identified, with Chute Forest and New Forest also shown.

The systematic development of mapping, of surveying and the collection of soil, climate, ecological, topographical and agricultural-science information, have all led, throughout the last century, to notions of 'land evaluation', a process promoted by large, internationally or nationally significant organisations, e.g. the 'Food and Agriculture Organisation' of the United Nations or the United States Department of Agriculture. Once the characteristics of land have been established, there is a process of considering land *suitability* (for a specific purpose, including a specified crop) with land *capability*, for a range of uses, including forestry and agriculture.

Ultimately it becomes possible to recommend an *optimum land use*; although this must include socio-economic factors rather than merely the best crop to grow on a specified plot of land.[91] Into the present century, mapping has moved on in leaps and bounds, first with remote sensing data and air photographs, and now high-resolution satellites and LiDAR imagery, a laser technique that enables layers of land cover to be 'removed', revealing what may be beneath.[92] Integration and manipulation of spatial data, using Geographical Information Systems, has developed exponentially over the last four decades, making 'geographical information' a sub-discipline of modern geography and engineering and environmental sciences in general. Where once data copyright in the UK was jealously guarded by certain agencies, much information today is freely available, or charged for at modest rates.

Methods-of-geography> Accessed April 2020.
[90] National Archives (n.d.) Agriculture in the Second World War. <http://www.nationalarchives.gov.uk/cabinetpapers/themes/agriculture-second-world-war.htm> Accessed April 2020.
[91] S.G. McRae and C.P. Burnham (1981) *Land Evaluation*. Clarendon Press, Oxford. Chapter 1.
[92] LiDAR-UK.com (2020) How Lidar Works <http://www.lidar-uk.com/how-lidar-works/> accessed April 2020.

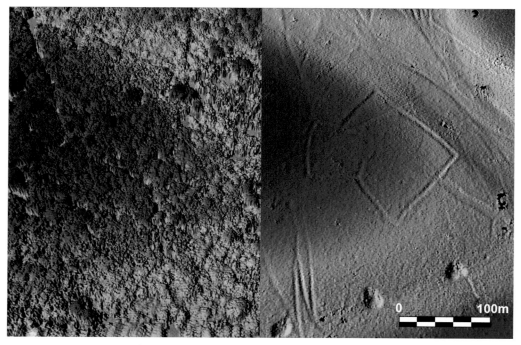

Figure 1.13. LiDAR image of an area of Savernake Forest, near Marlborough (copyright Natural England)

Geophysical information now plays a significant part when seeking evidence for, and the nature of, land use change, as well as more detailed geological, archaeological, and habitat-based land use change. In Figure 1.13, LiDAR imagery reveals what is beneath a woodland canopy in Savernake Forest. The left-hand image shows the first return of the lidar pulse, processed to show the tops of the trees: something like a conventional aerial photograph. The right-hand image shows filtered data processed to remove the vegetation. What is revealed is a late Iron Age enclosure. Therefore, the continuity of a dense tree cover before 2000 years ago may be brought into question. This part of Savernake is unlikely to be a direct descendent of the pre-Neolithic 'wildwood', even if much of it is considered semi-natural ancient woodland in the present. To be considered 'ancient', a woodland should have been in existence in AD 1600, however, even the preceding medieval Royal forest need not have been a dense tree cover.[93]

Other non-invasive ground survey geophysical techniques are extensively used in archaeology and for other purposes. These include ground penetrating radar (GPR), electrical resistivity surveys, electromagnetic ground conductivity (electromagnetic gradiometry) and gravity anomaly surveys.[94]

Figure 1.14 shows a section of a geophysical survey around Old Sarum Castle, near modern Salisbury. Today the area is under pasture, but the survey shows what are the likely buried

[93] LiDAR-UK.com (2020) How Lidar Works <http://www.lidar-uk.com/how-lidar-works/> accessed April 2020.
[94] RSK Geophysics (n.d.) A reference for geophysical techniques and applications <https://www.geos.ed.ac.uk/~whaler/environmental_geophysics_handbook_lowres.pdf > Accessed April 2020.

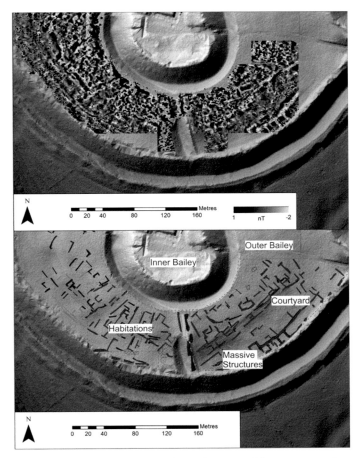

Figure 1.14. Section of a geophysical survey of the area around
Old Sarum Castle, near modern Salisbury (copyright K. Strutt and
Environment Agency)

remains of medieval trackways and buildings, indicating urban abandonment and a return ultimately to agricultural land use. We see a greyscale image of the magnetometry from the south of the outer bailey (top image), and the archaeological interpretation plot for the data overlaid on LiDAR for the area (bottom).

1.10 Multi-disciplinarity and landscape

It becomes apparent that to understand landscape requires consideration of many processes and variable, from both the natural order (including climate, geology, hydrology, soils, biological processes, and habitats) and from human activities. The latter may be carried out with a conscious purpose in mind (including woodland clearance, agricultural operations, alteration of genetic stock of plants and animals, construction of roads, watercourses, buildings, etc.), or the processes may have been unintentional (the starting of fires, soil erosion or accidental flooding, etc.).

Such processes may be described in a way that emphasises the spatial (geographical) or temporal, or, in practice, both. Figure 1.15 aims to summarise these. If there appears a degree of hair-splitting, it is because of complexity. For example, 'green history' is a relatively modern sub-discipline that implies a political dimension in writing it (but not a particular political party), and it is distinct from 'environmental history' (supposed to be more objective), which focusses on natural scientific information (such as tree rings, pollen, snail and insect remains, etc.). On the other hand, documentary history, perhaps mapped (spatial) information or records, is written with a specific purpose in mind (generally to do with land tenure or economic gain).

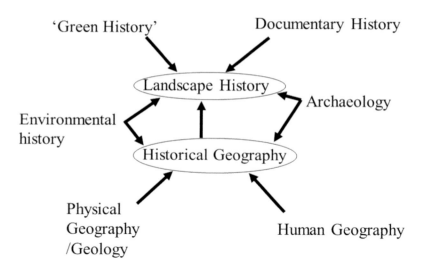

Figure 1.15. Diagrammatic representation of the disciplines involved in landscape studies (the author).

Chapter 2

Utilisation of natural resources

2.1 'A fair field full of folk'

Writing at some time between 1370 and 1390, with the landscape of England largely developed, and the countryside in some economic and social chaos, William Langland composed 'William's Vision of Piers Plowman', a Middle English allegorical narrative poem, quoted above. Rather than using rhyme, something otherwise generally regarded as 'poetic', it is expressed in unrhymed, alliterative verse so that the first consonants match. Langland's creation, unusual and even cumbersome to the modern ear, is nonetheless a vision of England long ago as a rather conservative aging man might have contemplated.

It is no exaggeration to describe the poem as a savage indictment of the Church, state and society of the day. Probably inspired by the floodplain of the River Severn viewed from the Malvern Hills in Worcestershire, Langland's Fair Field contained:

> Rich and Poor, all manner of men,
> Working and wandering, as in the world must
> Some were for ploughing, and played full seldom
> Set their seed and sowed their seed, and sweated hard
> To win what wastrels, with gluttony destroy.[1]

Langland (arguably in modern terms a dissident Catholic writer) is, unsurprisingly, greatly concerned with sin. He devotes several passages to the Seven Deadly Sins although gluttony seems pre-eminent.

The Fourteenth century was indeed calamitous with England experiencing a Great Famine (1315- 1317) which reduced the population ahead of any pestilence[2]. Subsequently, several visitations of the plague referred to as the Black Death were to create great population decline in Britain and across Europe. This catastrophe first visited England in 1348 and continued to impact the population for a long time, with notable recurrences in 1361 and 1374.[3] While Langland puts a theological interpretation on visitations of 'plague, storms and famine'[4] food production to him is precious, precarious and potentially destroyed by human greed, especially gluttony.[5] He is also concerned about idleness; to avoid this 'Each man must plough his own half-acre'. The greedy and lazy are clear targets for Langland.

---

[1] W. Langland (1912) *Piers Plowman: The vision of a Peoples Christ.* Trans. A Burrell, Everyman's Library, J.M. Dent and Sons, New York, 3-4.
[2] B. Williams (1996) Population history of England <https://urbanrim.org.uk/population.htm> Accessed April 2020.
[3] O. Benedictow (2005) The Black Death the greatest catastrophe ever. *History Today* 55(3) <https://www.historytoday.com/archive/black-death-greatest-catastrophe-ever> Accessed April 2020.
[4] W. Langland (1912) *Piers Plowman: The vision of a Peoples Christ.* Trans. A Burrell, Everyman's Library, J.M. Dent and Sons, New York, ix.
[5] The Artifice (n.d.) The Follies of Glutton <https://the-artifice.com/the-follies-of-glutton/> Accessed April 2020.

Inevitably there is a complicated economic relationship between food supply, ethics, population and politics. If the population of England fell greatly between 1347 and 1400, there were fewer mouths to feed, but there was also a labour shortage. Population reduction also meant the price of hiring a worker increased; under the 1351 Statute of Labourers there had been attempts to prevent this by re-imposing feudal bonds within society, and placing caps on the wages of labourers.

When combined with the additional provocation of a series of poll taxes, designed to reduce the tax burden on the landowning class, the result was the Peasants Revolt of 1381. However defined, English feudalism was to become a thing of the past, underlined by a steady drift to the towns.[6]

Over the next centuries the estimated population fluctuated but did not approach its 1300 level for at least 200 years. From around 1500, agriculture underwent profound change enabling a larger population to be supported as the Industrial Revolution commenced.

In recent years, covid 19 has dominated the news. Neither is climate change (the preferred modern term is 'climate crisis') ever far away from the media, and factors including drought, flooding, productivity and anticipated structural changes in agriculture are topics for serious concern. Social inequality and privilege continue and are amplified by environmental and health crises. More than 600 years later, William Langland still hits the spot, but of course such problems did not start in the 14th century.

Prehistoric agricultural societies almost certainly experienced change,[7] for example there was a period of wetter and colder weather during the Neolithic, in the second half of the 4th millennium BC as well as a deterioration in the climate towards colder, wetter conditions in the Iron Age (see Figure 1.11). Similar problems continued into post-Medieval times, when the relatively cold 'Little Ice Age' (c. AD 1440–1920) followed the Medieval Warm Period. It included a shorter period of extremely low solar activity termed the 'Maunder Minimum' (MM) c. AD 1650 – 1715 that badly impacted the climate. For the MM, climate model simulations suggest multiple factors, particularly volcanic activity, were crucial for causing the cooler temperatures in the northern hemisphere. However, a reduction in total solar irradiance likely contributed to the Little Ice Age at a level comparable to changing land use.[8]

Apart from demographic considerations, the Little Ice Age may also have been a factor in promoting structural change in English agriculture during the second half of the 17th century. It is likely that farmers invested part of their efforts improving soil fertility and this was a time of innovation in water meadows. Improvements also led to better wheat yields in the long term and a decrease in relative prices due to demographic stagnation. It seems that over time the rural sector was able to adapt to natural climate change.[9] Certainly by the 19th century,

[6] Marxists International Archive (n.d.) The Peasants' Revolt 1381 <https://www.marxists.org/history/england/peasants-revolt/story.htm> Accessed April 2020
[7] A. Bevan, S. Colledge, D. Fuller, R. Fyfe, S. Shennan, and C. Stevens (2017) Holocene population, food production, and climate. *Proceedings of the National Academy of Sciences*, 114 (49). E10524-E10531; DOI: 10.1073/pnas.1709190114
[8] M.J. Owens, M. Lockwood, E. Hawkins, I. Usoskin, G.S. Jones, L. Barnard, A. Schurer and J. Fasullo (2017) The Maunder minimum and the Little Ice Age: an update from recent reconstructions and climate simulations *Journal of Space Weather and Space Climate* 7, A33, 1-10. DOI: http://dx.doi.org/10.1051/swsc/2017034
[9] J.L. Martínes-González (2015). *Did Climate Change Influence English Agricultural Development? (1645-1740)*. Working

fertiliser was imported; guano was used from the 1840s and proved lucrative for importers. Guano derives from excrement, eggshells and carcasses of dead seabirds formed in dry, hot climates and, depending on type, can supply nitrogen, phosphorous and other essential nutrients.[10] Later, mined nitrate of soda (Chile saltpeter) and super phosphate (produced by applying sulphuric acid to bones) played an important role in raising levels of key nutrients in soils growing arable crops, and in the 20th century atmospheric nitrogen could be fixed, producing wholly artificial fertilisers through the Haber-Bosch process.[11]

Langland was a gloomy voice from a pre-scientific age. Fast forward to Thomas Malthus (1766-1834) who believed that population growth could be restricted by moral restraint (later marriage and sexual abstinence; apparently he was against contraception) at least until parents could economically support a family[12]. Malthus's theory predicts that population grows much faster than food production. Population grows geometrically (sometimes termed exponentially), while resource supply (especially food) only grows arithmetically (linearly). Growth is checked in the end by such catastrophes as famine, disease or war, a process termed the 'Malthusian Crisis'.[13] These are just the kinds of things that concerned William Langland. During Thomas Malthus's lifetime, the population of England is estimated to have increased (in a non-linear) fashion from around 6 million to 13 million.[14]

Malthus has become an influential thinker not just for other economists but also natural scientific thinkers including Charles Darwin. His popularity recovered as world population growth continued to cause concern, and we hear of 'neo-Malthusianism' as part of a strategy to promote birth control.[15] The critique of Malthusian ideas applied historically (and in England specifically) is that they underestimate the ability of agricultural technology to increase food supply.[16]

There was a disastrous famine in Ireland in 1740-41 attributed to climatic causes, which caused large-scale death by starvation; the similarly disastrous Great Irish Famine (1845–1849) was due to potato blight and there was a similar (but less fatal) event in the Scottish Highlands around the same time.[17] Otherwise there were no famines in the British Isles during the eighteenth and nineteenth centuries, just poor harvests. Food could be imported,[18] and

Paper 75, European Historical Economics Society.
[10] E. Schug, F. Jacobs and K. Stöven (2018) Guano: The White Gold of the Seabirds in H. Mikkola (ed.) *Seabirds*. Intechopen, London. DOI: 10.5772/intechopen.79501
[11] D.A. Russel and G.G, Williams (1977) History of Chemical Fertilizer Development *Soil Science Society of America Journal* 41(2) 260-265. https://doi.org/10.2136/sssaj1977.03615995004100020020x
[12] P.M. Dunn (1998) Thomas Malthus (1766–1834): population growth and birth control. *Archives of Disease in Childhood* 78(1). <https://fn.bmj.com/content/78/1/F76.info> Accessed April 2020
[13] Environmental History Resources (2020) Malthus, Population and environment, a short bibliography. < https://www.eh-resources.org/malthus-bibliography/ > Accessed April 2020
[14] S. Broadberry, B.M.S. Campbell, A. Klein, M. Overton and B. van Leeuwen (2015) *British Economic Growth, 1270–1870*. Cambridge University Press, Cambridge, 205 ; England Dep through time (n.d.) GB Historical GIS /University of Portsmouth. <http://www.visionofbritain.org.uk/unit/10061325/cube/TOT_POP> Accessed April 2020.
[15] R. Abramitzky and F. Braggion (n.d.) Malthusian and Neo-Malthusian Theories <https://ranabr.people.stanford.edu/sites/g/files/sbiybj5391/f/malthusian_and_neo_malthusian1_for_webpage_040731.pdf> Accessed April 2020.
[16] A. Howes (2020) Age of invention: Escape from Malthus <https://antonhowes.substack.com/p/age-of-invention-escape-from-malthus> accessed June 2021
[17] G. Vaughan (2015) The Irish Famine in a Scottish Perspective 1845-1851. *Cahiers du MIMMOC* 12. DOI : https://doi.org/10.4000/mimmoc.1763 Accessed May 2020.
[18] P. Sharp (n.d.) The Long American Grain Invasion of Britain: Market integration and the wheat trade between North America and Britain from the Eighteenth Century. *University of Copenhagen Department of Economics Discussion*

surplus population exported to the colonies.[19] However, such statements need to be moderated due to tariff barriers, notably between 1773 and the repeal of the Corn Laws in 1846, and fears of naval blockade during the Napoleonic wars. Eventually a combination of poor harvests and imported grain and lamb would cause agricultural recession from 1879 to 1940.[20]

Prior to the modern period, UK agriculture had been forced - or encouraged - to diversify to remain profitable. Joan Thirsk proposes three previous phases when the production of meat and grain was not the priority due to economic circumstances, prompting farmers to diversify. These were approximately 1350 to 1500 (including the impact of the Black Death), 1650 to 1750 (just prior to the Industrial Revolution) and 1879 to 1939 (when imported food kept prices down, making meat and corn production unprofitable).[21]

Humans have been changing the landscape for millennia. Today the popular notion of the 'Anthropocene' has been born, but humans have caused significant environmental change on Planet Earth for longer than the industrial age, in a physical sense by humans affecting land use cover.[22] It may be that rising greenhouse gasses also resulted from the deforestation and soil degradation associated with the advent of agriculture.[23] Natural resource exploitation is therefore far reaching, affecting both the physical and chemical nature of the Planet. The consequences are realised today in climate change and credible threats of mass extinction. Not only are there scars from the exploitation of fossil fuels, but human products, notably plastics, are widely distributed around the globe and particularly in the marine environment.[24]

2.2 Geological resources of Wessex

In Wessex, the complexity of the landscape geology (see Chapter 1) has governed the development of human economy and society as well as providing raw materials. While the mineral resources of Wessex are diverse, this does not mean industries have survived to the present day. Some extractive industries remain, (Figure 2.1) particularly the quarrying of building stone[25] and extraction of hydrocarbons,[26] while sand and gravel extraction has produced large lakes valuable for recreation and operating as nature reserves. Notably this has occurred around Blashford, near Ringwood in Hampshire, where extraction continues

Papers 8 (20) <https://core.ac.uk/download/pdf/7051639.pdf> Accessed April 2020

[19] For example, thanks to the Internet, the Author has discovered a large extended family in Australia and New Zealand as a consequence of 19th century diaspora from Somerset.

[20] H. Cook (2010) Boom, slump and intervention: changing agricultural landscapes on Romney Marsh, 1790 to 1990, in M. Waller (ed.) *Romney Marsh: persistence and change in a coastal lowland.* Romney Marsh Research Trust, Sevenoaks.

[21] J. Thirsk (1997) *Alternative Agriculture. A History.* Oxford University Press, Oxford, xi and 365.

[22] H.F. Cook (2018). River channel planform and floodplains: a study in the Wessex landscape. *Landscape History* 39, 5-25.

[23] F. He, S.J. Vavrus, J.E. Kutzbach, W.F. Ruddiman, J.O. Kaplan and K.M. Krumhardt (2014). Simulating global and local surface temperature changes due to Holocene anthropogenic land cover change. *Geophysical Research Letters* 41 (2), 623-631. DOI 10.1002/2013GL058085

[24] H. Ritchie and M. Roser (2018). Plastic Pollution. <https://ourworldindata.org/plastic-pollution> Accessed July 2021.

[25] Historic England (2017) A Building Stone Atlas Historic England for each region < https://www2.bgs.ac.uk/mineralsuk/buildingStones/StrategicStoneStudy/EH_atlases.html > Accessed April 2020; I. Geddes (2003) *Hidden Depths.* Ex-Libris Press, Bradford-on-Avon.

[26] P. Strange (2016) The Surprising Story of Oil in Dorset. <https://philipstrange.wordpress.com/2016/06/03/the-surprising-story-of-oil-in-dorset/ > Accessed April 2020.

Table 2.1. Key mineral resources in Wessex (from various sources).

Mineral Resource	Geological Age or Formation	Geographical location	Comments
Basaltic rocks, lava flows and tuffs	Silurian 'Coalbrookdale formation'	Mendip Hills	Occurrence of igneous rock, used as aggregate. Mined Moon's Hill Quarry, Stoke St Michael. Uncommon in region
Blue Lias limestones	Jurassic	Across region	Used for building and lime mortar
Brickearth	Pleistocene	Over river terraces	Brick industry important in Salisbury area
Chalk for lime	Cretaceous Chalk	Downlands / Salisbury Plain	Used for lime for cement, mortar and also liming soils
Chalk for building stone	Cretaceous Chalk	Downlands / Salisbury Plain	Hard bands within chalk for domestic buildings. 'Melbourne Rock' (Holywell Nodular Chalk Formation) and 'Chalk Rock' (Lewes Nodular Chalk Formation)
Clays	Includes Triassic, Lias, Oxford clays, Gault Clay and Coal Measures clays	Across region	Ceramics, large pipes, brick and tile making. Extraction sites are few and far between in the present
Coal	Coal measures	Bristol and Somerset Coalfield	Coal used in Roman times. Most mining occurred medieval to 1973
Flints	Cretaceous Chalk, largely in White Chalk	Downlands / Salisbury Plain	Early peak of exploitation in the Neolithic for toolmaking, later used for building, largely for ornamentation and outer covering of walls
Chert	Mesozoic, particularly Portland Limestone and upper Greensand	Upper Jurassic and lower Cretaceous outcrops	Used for aggregate, prehistoric toolmaking and limited use in buildings. Sometimes shaped into cubes
Fuller's Earth	Jurassic	Within the Great Oolite Group	Part of the process of 'fulling' cloth to eliminate oils, dirt, and other impurities and to make it thicker
Gravel	Mostly Pleistocene Gravels	River valleys below modern alluvium and on river terraces	Concrete and road construction
Hydrocarbons	Jurassic age shales (see also 'others')	Dorset coast area & Purbeck, Wytch Farm (Poole Harbour) and Kimmeridge Bay	Crude oil has been extracted 1959 (Kimmeridge Bay) and since the 1970s at Wytch Farm. Local Oil shale extracted for much longer producing both oil and gas
Iron ore	Ironstone in Devonian and Cretaceous strata	Brendon Hills, Seend area (Wilts) and Mendips	Mined from Roman times to the 19th C Lower Greensand 19th and 20th C

Mineral Resource	Geological Age or Formation	Geographical location	Comments
Limestone	Carboniferous and Jurassic formations	Mendip Hills, Cotswolds, Purbeck, Ham Hill stone (a honey-coloured bioclastic limestone)	Includes Bath Stone (Great Oolite) and aggregate limestone, Mendip Hills and Jurassic quarries on Purbeck, Portland and Chilmark stone mined at Chicksgrove, Wiltshire. Harvey Quarry Stoke Sub-Hamdon, produces Ham Hill stone
Lead	Hydrothermal veins within Carboniferous Limestone and secondary deposits of galena	Mendip Hills	Mined from Roman times to 19th C
Peat	Recent	Somerset Moors	Industry much reduced du to conservation imperatives
River alluvial (fine silty)	Recent	Somerset Levels	Bridgwater tile and brick making
Sand	Tertiary and Quaternary	Various locations	Construction industry to the present, glass-making in Bristol 17th and 18th C
Sandstone	Mostly Jurassic and Cretaceous	Various locations, used for building and construction	A mine in the upper Greensand at Hurdcott near Barford St Martin, Wilts
Sarsen stone	Tertiary weathering deposit, silica cemented sandstone resulting from weathering of sands above the Chalk.	Typically found on Marlborough Downs and Salisbury Plain	Trilithons of Stonehenge; also in vernacular building. Sarsen stone is very hard to work
Zinc ores	Hydrothermal veins within Triassic Dolomitic Conglomerate	Mendip Hills.	Less important than Lead Mining, operated 16th C to 19th C
Others	Various ages	Across region	Manganese, Barytes, Celestine and Gypsum over short timescales and the jet-like 'Blackstone' (oil shale) from the Kimmeridge Shales was used to make ornaments in prehistory

to the present. This chapter will explore these themes further by looking at the provision of geological materials, describing the water environment, outlining the development of transport and industry, and building on the material on soils in Chapter 1 to demonstrate their importance for agriculture.

Movement of stone and bulk commodities was always expensive. Due to high transport costs, brickmaking (where geology permitted) was strongly localised, and both clay deposits and

Figure 2.1. Shallow freestone quarrying Isle of Purbeck (the author).

wind-blown brickearth could be exploited.[27] Bulk commodity transport was facilitated by the introduction of canals, followed by railways in around 1840. The main mineral resources of Wessex are summarised in Table 2.1.

2.3 Early agriculture

One advantage of the adoption of agriculture was greater food production, something that would require strong social organisation. The adoption of agriculture would seem to follow on from some degree of deliberate clearing of the wildwood for purposes that included grazing, meeting and possibly rituals. As more land was cleared, so tree cover decreased, and agriculture grew in both area and output. Hunter-gatherer societies require around 150-250 ha per person, or around 0.005 persons per ha,[28] agricultural societies can support 60-100 times that density,[29] or around 0.4 persons per ha. In sustenance terms alone, agriculture is worth doing.

By around 6000 BC, during the Mesolithic, Britain was an island and new migrants came to Britain by sea. A rectangular building was constructed in the Medway valley in Kent, a construction of 7m by 18m using substantial upright posts that would have supported a

---

[27] J. Wright J (2017) *Brickmaking in Fisherton and Bemerton*. South Wiltshire Industrial Archaeological Society, Salisbury, 4-5.
[28] D. Pimental and M.H. Pimental (eds) (2007) *Food, Energy, and Society* (3rd ed.). CRC Press, Boca Raton, Chapter 6.
[29] New World encyclopaedia (n.d.) Hunter-gatherers < http://www.newworldencyclopedia.org/entry/Hunter-gatherer > Accessed April 2020.

substantial roof.[30] Usage of the site has been radiocarbon dated to around the first half of the fourth century BC. These and similar constructions have been linked to incoming agriculturists bringing domesticated animals, sheep, cattle and pigs, ceramics and polished Neolithic axes. These incomers likely arrived via the Thames estuary.[31] The expansion of agriculture has been linked with the rapid decline in the pollen record of the elm *Ulmus minor* ‹Atinia›, once thought to have been a result of selective felling, but now attributed to Dutch Elm disease spread by the beetle *Scolytus scolytus* through the fungus *Ophiostoma ulmi* which blocks water conducting tissue in elm trees[32]. The dates of this occurrence were from around 4400 to 3330 BC, more than a millennium, and its onset was rapid[33] (see also Figure 1.11). By 3900 BC it is possible that agriculture was to be found south-east of a line from Lyme bay to the Wash and there is evidence of conflicts in Wessex and elsewhere a few centuries later, in the middle part of the Neolithic, perhaps suggesting pressure on resources.[34] Hambledon Hill in Dorset is one such location.[35]

The lighter soils over chalk would have been the first to be cultivated, especially those covered by windblown loess (silt) that may once have been thick enough to produce a fertile soil, but has often been reduced to around 200mm, much diminished by soil erosion due to agriculture over around 6,000 years.[36] This produced the rendzina soils introduced in Chapter 1. On downland throughout England today, ploughing for arable agriculture may bring lumps of chalk and flint to the surface because soil erosion has brought the Chalk bedrock within plough depth. Otherwise, Clay-with-Flints soils were likely avoided as they would be too difficult to work at the time.[37] Chalk would also supply flints for tool making in Neolithic and later societies, and flint was sourced from Cranborne Chase.[38] Other lighter soils, such as those over the Jurassic limestone of the Cotswolds would also be found desirable for clearance.

The Vera hypothesis (see Chapter 1) provides ample opportunity for cycles of clearance and re-growth. Clearance by humans, on the other hand, was not always progressive; often land was abandoned for a couple of hundred years as fresh clearances was made elsewhere, while there was a general woodland regeneration during the middle Neolithic that may have been linked with agricultural decline. It may be that the episodic nature of Neolithic agriculture represents the utilisation of 'Vera cycles' to exploit the cyclically open areas for agriculture.[39]

[30] A. Barclay, A. Fitzpatrick, C. Hayden and E. Stafford (2006). *The Prehistoric Landscape at White Horse Stone, Aylesford, Kent.* Oxford Wessex Archaeology Joint Venture, Oxford. https://doi.org/10.5284/1008829

[31] N. Crane (2016) *The Making of the British Landscape.* Weidenfeld and Nicholson, London, 69-71.

[32] Forest research (2020) Dutch elm Disease <https://www.forestresearch.gov.uk/tools-and-resources/pest-and-disease-resources/dutch-elm-disease/> Accessed April 2020.

[33] A.G. Parker, A.S. Goudie, D.E. Anderson, M.A. Robinson and C. Bonsall (2002). A review of the mid-Holocene elm decline in the British Isles Progress in physical Geography. *Earth and Environment* 26 (1), 1-45 <https://doi.org/10.1191/0309133302pp323ra> Accessed April 2020.

[34] N. Crane (2016) *The Making of the British Landscape.* Weidenfeld and Nicholson, London, 79-80; J. Wilkes (2015). What is the earliest evidence for a battle in Britain? < https://www.historyrevealed.com/eras/ancient-rome/what-is-the-earliest-evidence-for-a-battle-in-britain/> Accessed April 2020.

[35] R.J. Mercer and F. Healy (2008) *Hambledon Hill, Dorset, England: Excavation and survey of a Neolithic Monument Complex and its Surrounding Landscape.* English Heritage, Swindon.

[36] J.B. Boardman, (1992) Current erosion on the South Downs: implications for the past in M. Bell and J. Boardman (eds) *Past and Present Soil Erosion.* Oxbow Books, Oxford, 9-19.

[37] Wiltshire Council (n.d.) Chute Forest Landscape Character <http://www.wiltshire.gov.uk/kennet_landscape_character_assessment_part_2_the_character_areas_-_chute_forest.pdf> Accessed April 2020.

[38] M. Green (2000) *A Landscape Revealed: 10,000 Years on a Chalkland Farm.* The History Press, Stroud, 20-21.

[39] M. Robinson (2014) The Ecodynamics of Clearance in the British Neolithic. *Environmental Archaeology* 19:3, 291-297. DOI: 10.1179/1749631414Y.0000000028

An archaeological section called Fir Tree Field Shaft at Down Farm on Cranbourne Chase in Dorset displays one of the earliest known records of woodland clearance in southern Britain, home of archaeologist Martin Green.[40] Here an original Mesolithic age vegetation mosaic comprised parkland-like land cover including open spaces. Clearance of the woods may have been for monument construction or for agriculture. Palaeo-environmental evidence from plant remains and land snails shows a vegetational relationship in continuation from the late Mesolithic through the Neolithic and into the early Bronze Age (shown by artefacts), hardly surprising if elements of woodland survived.[41]

The infill of the chalk shaft is consistent with the idea that a considerable amount of soil was detached by the plough. It is unlikely, however, that much soil, its particles thus detached, remained so close to the original location. In reality, the progressive clearance of the chalk interfluves (ridges and slopes) by early farmers must have generated a vast amount of silt that had originally been deposited as aeolian (wind transported) material during the Pleistocene and that thus became mobile downslope and added to the floodplains, and hence also had the potential for loading additional sediment into the river channels.

This archaeological section is of incredible and evocative value. The 'Neolithic Revolution' -when agriculture replaced hunting and gathering as a means of sustenance - is probably the most significant event in human history, causing extensive social and landscape changes. Even if Britain became a 'late adopter' compared with the European mainland, and prior to that the Near East, Wessex contains an exemplar of the transition.

This new agriculture likely started as small, temporary clearances over easier soils of the kind formed by pioneer farmers, akin to *Landnam*.[42] With land clearance of the wildwood there came problems, and the first was likely soil erosion, described above. As cleared slopes became longer and individual units cleared for agriculture became larger and then joined together, the disaggregation of aeolian deposited silt over the chalk was transported by water and wind to hollows in the land surface, or downslope into river valleys.[43]

2.4 River modification, deliberate or otherwise

River channel change is associated with extreme climatic fluctuation during the Pleistocene. During cold stages, the nature of rivers was largely of bedload-dominated channels either braided or wandering in planform, and this may follow an early glacial stage incision and non-depositional phase.[44] By analogy, in the modern European Arctic, single thread streams, in-channel islands and multiple channels may all be observed.[45] Following the Last Glacial

---

[40] M. Green and M. Allen (1997) An Early Prehistoric Shaft on Cranborne Chase, *Oxford Journal of Archaeology* 16 (2), 121-132.

[41] H. Cook (2018) *New Forest: The Forging of a Landscape*. Windgather Press, Oxford. Chapter 2.

[42] B. Cunliffe (1993) *Wessex to AD 1000*. Longman, New York, Chapter 2.

[43] H. Cook (2008) Evolution of a floodplain landscape: A case study of the Harnham Water Meadows at Salisbury, England. *Landscapes* 9 (1), 50-73.

[44] P. Gibbard and J. Lewin (2002) Climate and related controls on interglacial fluvial sedimentation in lowland Britain. *Sedimentary Geology* 151 , 187-210; P. Gibbard and J. Lewin (2007) Climate and related controls on interglacial river sedimentation in lowland Britain. <http://www.qpg.geog.cam.ac.uk/research/projects/interglacialrivers/> Accessed April 2020.

[45] J. E. Brittain *et al.* (2009) Arctic Rivers, in K. Tockner, U. Uehlinger and C. T. Robinson (eds) *Rivers of Europe*, Elsevier Academic Press, Cambridge Massachusetts, 337-379.

Maximum (see Chapter 1), the climate warmed over the long-term. Eventually, the sediment load of rivers became finer and stabilising vegetation developed along the banks of streams, which could be single thread or multiple channel in form.[46] The change from sinuous or meandering single thread channels to multiple-channel geometry in river channels is a response to sediment loading. During the cold periods, coarse ('bedload dominated') river channels might form classic 'braided' systems. Once the climate warmed into the Holocene, the transportation of finer material, combined with the developing vegetation, stabilised the river system. However, the invigorating loading of river channels with fine sediment caused by land clearance tended to change any conventional single-thread rivers towards another form of multiple-channel termed 'anastomosed'

The term refers to branching channels which re-join within the floodplain and which have a more irregular appearance than conventional 'braided' channels. The number of discrete anastomosing channels is typically between two and four at any given valley cross-section. Sediment loading raises the bed of the river, reducing the capacity of the channel, meaning there is an increased chance of water in the channel overtopping its banks and producing a new channel crossing the floodplain. This may become permanent once the flood subsides meaning the response has been increasing total capacity. Hence, it is flow diversions (termed 'avulsions') which create new channels on the floodplain.

There are two possible mechanisms: by formation of bypasses, while bypassed older channel-belt segments remain active; and by splitting of the diverted (avulsive) flow, leading to the contemporaneous scouring of multiple channels on the floodplain. Between avulsion events there is a presumption of enhanced stability in bank and channel through a combination of inherent coherent banks, of deliberate hydrological regulation within channels (by fixed structures) and of otherwise engineered or vegetated banks. Otherwise 'floodbasins' between channels, or between them and the valley side, will constitute a major structural element of anastomosing river floodplains as well as the alluvial, often saucer-shaped 'islands' left between the channels. These floodbasins accumulate both fine mineral sediment and organic rich deposits such as peat (Chapter 1) which condition the soils formed on the floodplain.[47] This subject, together with examples from the Wessex landscape, is further explored in Chapter 4.

To deforestation, soil erosion, the loading of river and channel streams and potentially flooding, we may add climate impacts. The interaction with climate is interesting, for climate amelioration been linked to the instigation of agriculture in northwest Europe.[48] Globally, 'Anthropogenic Land Cover Change (ALCC)' by pre-industrial societies has resulted in emissions of greenhouse gases and aerosols from biomass burning, deforestation and crop cultivation. The mechanisms are intricate and interconnected and include modification of the land-atmosphere exchange of momentum and moisture as well as radiative and heat fluxes. This process may have contributed to global climate change, on average increasing

---

[46] H. Cook (2008) Evolution of a floodplain landscape: A case study of the Harnham Water Meadows at Salisbury, England. *Landscapes* 9 (1), 50-73.

[47] H.F. Cook (2018). River channel planform and floodplains: a study in the Wessex landscape. *Landscape History* 39, 5-25.

[48] C. Bonsall, M. Macklin, D. Anderson and R. Payton (2002) Climate change and the adoption of agriculture in north-west Europe. *European Journal of Archaeology* 5(1), 9-23. doi:10.1179/eja.2002.5.1.9.

Figure 2.2. Contact between presumed downland and (in the middle distance) ploughed rendzina soils in April 2020 (the author)

temperatures by as much as 0.73°C, comparable with the 0.8°C increase associated with later industrialisation.[49]

Presuming this proposal is correct, it pushes the Anthropocene back in time to the dawn of agriculture. ALCC, river system impact and later interventional changes such as the creation of field systems, infrastructure for mills, canals and water meadows all become features of the Anthropocene impact in Wessex.[50] Figure 2.2 shows the outcome of continued ploughing of chalkland soils. Beyond the preserved downland at Laverstock, near Salisbury, spring ploughing exposes the fragmented white chalk subsoil, the once silty cover loam soil having long been eroded, generating sediment runoff and alluvium.

Figure 2.3, by a group of local historians, depicts the use of the river channels around Romsey, Hampshire, on the lower river Test and provides some context to the utilisation of rivers in south Wessex. The location of mills on the multiple-channel anastomosing river is shown; the 'barge canal' on the eastern side is entirely artificial.

---

[49] F. He, S.J. Vavrus, J.E. Kutzbach, W.F. Ruddiman, J.O. Kaplan and K.M. Krumhardt (2014). Simulating global and local surface temperature changes due to Holocene anthropogenic land cover change. *Geophysical Research Letters* 41 (2), 623-631. DOI 10.1002/2013GL058085
[50] B. Burbridge (ed.) (1998). *Romsey Mills and Waterways.* Lower Test Valley Archaeological Study Group, Romsey.

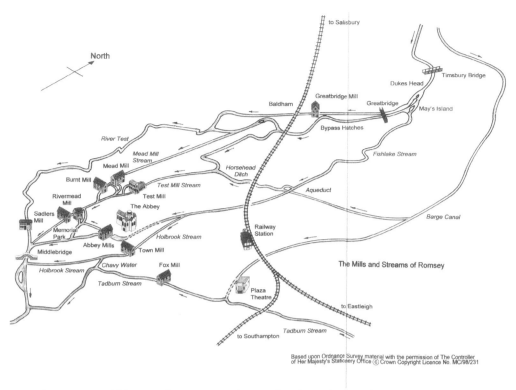

Figure 2.3. Local historians' perception of a complex channel network at Romsey, Hampshire
(Romsey Local History Society)

At the start of the Neolithic period, out of *c.* 23m ha of Britain (excluding the island of Ireland) there was an estimated 16m ha of forest and woodland; by the end of the Iron age this has been estimated at 10m ha, and woodland cover continued falling into historic times.[51] By the early Neolithic there existed a developed agrarian economy in Wessex, supported by a population large enough to farm, to mine and to create cultural and ceremonial structures of continental significance. The core was the chalk downland.[52]

Early strains of cereals such as emmer wheat (*Tricium dicoccum*), spelt wheat (*Tricium spelta*) and barley (*Hordeum vulgare*) would have been sown on the cleared land. These cereal crops had arisen from wild crosses of grass species.[53] However, it may be that early agriculturists remained semi-nomadic, perhaps practicing seasonal migration with their livestock.[54] Aurochs had been selectively bred to produce early domesticated cattle and there were pigs and early sheep, such as moufflon. These agricultural beasts would have been transported by sea to the island of Britain by Neolithic people.[55] Archaeological evidence suggests influxes

---

[51] R.H. Best (1981) *Land Use and Living Space*. Methuen, London, 11.
[52] N. Thorpe (2021) pers. comm.
[53] P.J. Reynolds (1987) *Ancient Farming*. Shire Books, Aylesbury, Chapter 3.
[54] N. Crane (2016) *The Making of the British Landscape*. Weidenfeld and Nicholson, London, 79
[55] B. Cunliffe (2013) *Britain Begins*. Oxford University Press, Oxford, 133; S. Brace *et al.* (2018). Population Replacement in Early Neolithic Britain. *bioRxiv* DOI: 10.1101/267443.

into Britain and Ireland from both (modern) northeast France and from Belgium, and also via the Atlantic seaways from western France around 4000BC. This seems to be supported by human DNA evidence that also includes a minority of genes of near eastern origin, something attributable to the introduction of agriculture into Europe.[56]

These dramatic changes may also have been accompanied by a climate change with colder winters, warmer summers and lower rainfall suitable for cereal cultivation.[57] The new communities would become multi-skilled, knowledgeable about hunting, gathering and food production for diet, while the Neolithic package was truly revolutionary: alongside dramatic landscape change, other new technologies included pottery, mining (for flint), the grinding and polishing of stone to produce attractive tools such as axe-heads, textile manufacture (replacing animal skins), building construction in wood and monument creation in stone. By the Neolithic, it should also be remembered that boat building and navigation skills were sufficient to deliver people and livestock to Britain, and indeed to Ireland.

2.5 Agricultural development

It is a shame that field systems of Neolithic age are unproven in Wessex. While it is not impossible that some prehistoric systems were in place before about 2,500 BC, we have scant evidence for ploughing and only a little for settlement, although some traces of ploughing have been found to the edge of a cultivated plot beneath South Street Long barrow, (early Neolithic) in Wiltshire. Elsewhere evidence points to progressive clearance of woodland.[58] Boundaries were likely constructed from the early Neolithic onwards: an early linear earthwork dated to around 3600 BC follows the crest of the western escarpment of Hambleton Hill, Dorset, for about 3km, an almost continuous bank and segmented ditch. Land boundaries appear in greater numbers from the middle of the Bronze Age (around 1500 BC), probably resulting from pressure on land and powerful rulers.[59] The remains of four early Neolithic houses supported by upright poles have been found at Kingsmead Quarry, Berkshire in 2008. The first house discovered measures 9.8m by 6.5m and was divided into two rooms by the partition walls.[60]

The function of such boundaries as may be found in Wessex are not altogether clear. Evidently there were cultivated fields, and linear features may have constrained livestock within large areas. Neither does the very limited number of such features as are currently known mean there were not more. These may have been lost to subsequent landscape reorganisation. Whatever the case, there are no prehistoric field systems in Britain clearly dated to the Neolithic.

---

[56] B. Cunliffe (2013) *Britain Begins*. Oxford University Press, Oxford, 177; A. Bevan, S. Colledge, D. Fuller, R. Fyfe, S. Shennan, and C. Stevens (2017) Holocene population, food production, and climate. *Proceedings of the National Academy of Sciences*, 114 (49), E10524-E10531; DOI: 10.1073/pnas.1709190114

[57] B. Cunliffe (2013) *Britain Begins*. Oxford University Press, Oxford, 136; N. Finlay, E. Anderson and P.C. Woodman (1999). *Excavations at Ferriter's Cove, 1983--95: last foragers, first farmers in the Dingle Peninsula*. Wordwell, Bray.

[58] B. Cunliffe, B. (1993) *Wessex to AD 1000*. Longman, New York, Chapter 2.

[59] Historic England (2018) *Prehistoric Linear Boundary Earthworks*. Historic England, Swindon. <https://historicengland. org.uk/images-books/publications/iha-prehist-linear-boundary-earthworks/heag219-prehistoric-linear-boundary-earthworks/> Accessed April 2020.

[60] A. Barclay, G. Chaffey and M. Symonds (2014) Horton's Neolithic houses. *Current Archaeology* 262 <https://www.archaeology.co.uk/articles/hortons-neolithic-houses.htm> Accessed April 2020.

Figure 2.4. Excavated stone wall at Céide, Co Mayo in Ireland
(the author)

Crossing the sea to Ireland, Neolithic fields at Céide in Co. Mayo were discovered during peat digging and subsequently investigated by archaeologists Probing through the accumulated peat enabled the fields beneath the peat to be mapped. This earlier landscape dates from around 3200 BC after which climatic change caused peat accumulation, smothering the earlier landscape.

At Céide (Figure 2.4) a co-axial pattern of low stone walls has been found running cross-country for distances of up to 2km with roughly parallel walls 150m to 200m apart, forming long strips sub-divided by cross-walls. Low walls up to 0.8m high resulting from field-clearance may not have been readily capable of constraining livestock, although they may delineate ownership, while pollen evidence indicates wheat and barley cultivation.[61] This part of prehistoric heritage is truly remarkable.

It is now considered likely that the people who introduced farming in Ireland came directly along the Atlantic Coast from the area of the Bay of Biscay and hence were not immigrants

---

[61] B. Cunliffe (2013) *Britain Begins*. Oxford University Press, Oxford, 147-8; Museum of Mayo (n.d) History of Céide Fields <http://www.museumsofmayo.com/ceide.htm>. Accessed April 2020.

who came via Britain. Bringing their domesticated animals, they would have made a far more perilous crossing than their supposed British counterparts arriving in the early Neolithic period and entering via the Thames estuary.[62]

In Wessex, Neolithic monuments including Stonehenge, West Kennet Long Barrow (a chambered tomb), Avebury (a henge and stone circles), Windmill Hill (a causewayed enclosure), Silbury Hill (an artificial mound), Stanton Drew (stone circles) and many more are maybe a cultural and religious dimension to Neolithic society, and suggest a food surplus and a labour force with spare time when not engaged in agriculture.

While nobody can, nor ever will be sure of the belief systems behind Neolithic monuments, we are relatively safe to presume they were intended to project social cohesion, while the astronomical alignments (of both sun and moon) reflect the pre-occupations with life, death and the festivals of an agricultural society. The ditch which pre-dates the stones around Stonehenge dates from around 3,000 BC while the presumed sanctity of the site is older,[63] not to mention the nearby Mesolithic site of Blick Mead, and the association of the 'Heel Stone' with the rising midsummer sun has long been described.[64] It may be that prehistoric festivals themselves, based in agricultural or pastoral practice, required such an alignment as a calendar.[65] Figure 2.5 shows the late Neolithic monument of Stonehenge (an icon of Wessex) with the 'Heel Stone' in the foreground, looking approximately southward. At Stonehenge on the summer solstice, the sun rises behind the Heel Stone in the north-east part of the horizon and its first rays shine into the heart of Stonehenge.[66]

Stonehenge belongs to a wider prehistoric 'ritual' landscape on Salisbury Plain, including burial mounds from the succeeding Bronze Age, suggesting not only continuity of importance but a range of significances, including the laying out of a prehistoric necropolis. Considered in the round, there was a huge construction effort over long time periods, probable astronomical significances, and considerable social value and religious significance invested. Similar things can be claimed for Avebury.

We can say with some confidence that the prehistoric society(ies) that produced Stonehenge and outlying monuments had an understanding of astronomical process governing their seasons, produced a large agricultural surplus (to feed the workers) and were hence highly organised. This was bound together by religious belief and social bonds. Archaeological investigation in the area suggests that large gatherings, including long-distance pilgrimages and excessive feasting at auspicious festivals (such as the winter solstice), were part of the story.[67] William Langland would not approve!

---

[62]   B. Cunliffe (2013) *Britain Begins*. Oxford University Press, Oxford, 131.
[63]   M. Pitts (2000) *Hengeworld*. Arrow Books, London, 106.
[64]   M. Pitts (2000) *Hengeworld*. Arrow Books, London, 230.
[65]   A. Burl (1983) *Prehistoric astronomy and ritual*. ShireBooks, Aylesbury, Chapters 7 and 8; A. Burl (1991) *Prehistoric Henges*. Shire Books, Aylesbury, Chapter 7.
[66]   English Heritage (n.d.) Solstice <https://www.english-heritage.org.uk/visit/places/stonehenge/things-to-do/solstice/> Accessed June 2020.
[67]   English Heritage (n.d.) Food and Feasting at Stonehenge <https://www.english-heritage.org.uk/visit/places/stonehenge/history-and-stories/history/food-and-feasting-at-stonehenge/> Accessed June 2020; S. Viner, J. Evans, U. Albarella and M.P Pearson (2010) Cattle mobility in prehistoric Britain: strontium isotope analysis of cattle teeth from Durrington Walls (Wiltshire, Britain). *Journal of Archaeological Science* 37 (11), 2812-2820.

Figure 2.5. Stonehenge with the 'Heel Stone' in the foreground, the arrow indicates a probable astronomical alignment (the author)

Burial customs changed throughout the Neolithic, and a widely held contention sees the early and middle period long barrow burials, including those on Mendip, as reflecting a more communal/ egalitarian society. This may have been a 'segmentary society', composed of relatively small-scale autonomous groups supporting themselves within the environments of a territory that was likely defined by the construction of monumental tombs and exemplified, in some areas, by causewayed enclosures. Individuals might have even been discouraged from 'disruptive competition for personal gain'.[68] Relations between such groups may have been maintained by the exchange of goods (for example polished stone axes) between regions, to develop a web of obligations; there was also considerable local exchange of goods. Limited evidence of warfare over the period 4500-3000 BC does however suggest mounting tension.[69]

The 'Bell Beaker Culture' arrived in Britain around after *c.*2450-2250 BC and associated with a dramatic change in material culture, notably in the adoption of large 'bell beakers' named after their similarity to inverted bells. Material and cultural changes also affected burial customs, ritual monuments, and adoption of metal working. Bell Beaker Culture demonstrably had continental links, for scientific analysis of the 'Amesbury Archer' skeleton (the man had lived *c.* 2300 BC) revealed an oxygen isotope signature from his teeth suggesting

---

[68] B. Cunliffe (1993) *Wessex to AD 1000*. Longman, New York, 73-4.
[69] B. Cunliffe (1993) *Wessex to AD 1000*. Longman, New York, Chapter 3.

he grew up in central Europe, while a presumed younger male relative buried nearby was likely raised locally.[70] When viewed overall, Beaker Culture arrived from the vicinity of the Pyrenees mountains via the western seaboard and also along the long continental interface between the Cherbourg peninsula to the Rhine Delta.[71] Recent studies suggest some mobility of individuals of 'beaker culture', both men and women, but it is proposed changes in material culture within Britain are largely due to cultural transmission causing a diffusion of a 'Beaker package', most notably involving adoption of bell beaker vessels, while human DNA evidence for dramatic change in the population, or otherwise, remains contested due to the limited size and geographic spread of samples.[72]

We may be certain that, by the succeeding Bronze Age, the whole British Isles experienced a transformed landscape: technological, economic and social. Based in agriculture and permanent settlement in order to tend the fields, there was increasing population and with it, other associated pressures that would sustain changes. Social change into the Bronze Age is suggested by a change in burial customs towards individual burials in (round) barrows or cemeteries. Grave goods are common and serve to display the status of an individual. This may be linked to an increasingly territorial society dominated by chiefdoms, and this period saw the zenith of ritual henge monuments (including Avebury), which may have been created to display authority. The main period of prestige goods, approximately 2000 BC- 1400 BC, suggests extensive trade outside Britain.[73]

Such circumstantial evidence likely indicates the emergence of powerful elites during the later Neolithic. While evidence for Neolithic land division is sparse in Wessex, this is certainly not the case for the succeeding Bronze Age. Outside our region, the 'reave systems' developed on Bronze Age southern Dartmoor (similar in pattern to the Céide field system) are a famous example of planned countryside which are clear on the ground today and survive largely because by the Iron Age, excluding certain hillforts that were constructed, climatic deterioration and soil degradation had probably resulted in the abandonment of Dartmoor.[74] Although the degree of deliberate landscape planning has been challenged on Dartmoor,[75] it has also been suggested that something similar to Bronze Age reave field systems operated on Exmoor.[76]

These field systems are based in co-axial lines, comprising parallel largely topographically blind walls spaced between about 300m and 500m apart and including paddocks and dwelling huts. Similarly patterned field systems have been identified in East Anglia, near to Bungay in Suffolk and in Essex. This would make land allocation relatively easy as well as enabling

[70] Wessex Archaeology (n.d.) The Amesbury Archer: Pilgrim or Magician? <http://www.wessexarch.co.uk/book/export/html/5> Accessed April 2020

[71] B. Cunliffe (2013) *Britain Begins*. Oxford University Press, Oxford, Chapter 6.

[72] A. Gibson (2020), Beakers in Britain. The Beaker package reviewed, *Préhistoires Méditerranéennes* 8. <http://journals.openedition.org/pm/2286> Accessed July 2021.

[73] B. Cunliffe (2013) *Britain Begins*. Oxford University Press, Oxford, Chapter 3.

[74] Dartmoor National Park (2017). Bronze Age to Roman <http://www.dartmoor.gov.uk/wildlife-and-heritage/heritage/iron-age> Accessed April 2020; V. Straker, A. Brown, R. Fyfe, J. Jones and K. Wilkinson (2007) Later Bronze Age and Iron Age Environmental Background, in C. Webster (ed.) *The Archaeology of South West England*. Somerset County Council, Taunton, 103-116.

[75] R. Johnson (2005) Pattern without a Plan: Rethinking the Bronze Age Coaxial Field Systems on Dartmoor, South-West England. *Oxford Journal of Archaeology* 24, 1-21. DOI: 10.1111/j.1468-0092.2005.00222.x.

[76] H. Riley and R. Wilson-North (2001). *The Field Archaeology of Exmoor*. English Heritage, Swindon, Chapter 2.

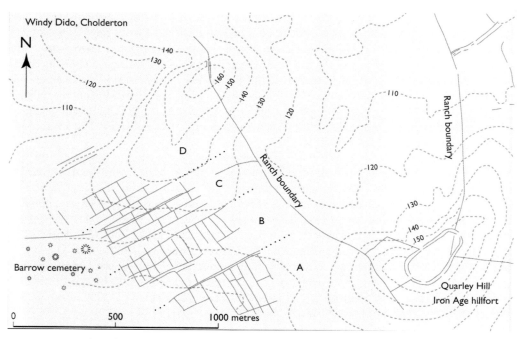

Figure 2.6. Windy Dido in western Hampshire, near West Cholderton. Four blocks of regularly laid out fields spreading downslope from a ranch boundary, probably late second millennium BC. Quarterly Hill has an Iron Age Hillfort (after B. Cunliffe, 1993. *Wessex to AD 1000* Longman, New York p.255)

the distribution of land of different capabilities.[77] Overall, large-scale dated field systems in southern England would seem to fall between 1700 and 1100 BC, although there are earlier examples.[78]

On Cranborne Chase in Dorset, the dates of such systems would seem to fall in the centuries around 1000 BC, implying a large degree of reorganisation during the middle part of the Bronze Age, and it would follow that this reduced soil erosion by creating field boundaries that would become lynchet features. Into the Iron Age, a more settled agricultural economy became established including ranching over the chalklands.[79]

Linear features, often stretching for many kilometres across the landscape, may be former ranch or estate boundaries. This implies the existence of ruling elites who were powerful enough to not only plan the countryside on a large scale, but also to enforce some form of land tenure system.[80] And there is evidence of conflict leading to murder, with bodies dumped in a

[77] H. Riley and R. Wilson-North (2001). *The Field Archaeology of Exmoor.* English Heritage, Swindon, Chapter 8; O. Rackham (1987) *The History of the countryside.* J.M. Dent, London, Chapter 8.
[78] Historic England (2018) Field Systems: Introductions to Heritage Assets. Historic England. Swindon <HistoricEngland.org.uk/ listing/selection-criteria/scheduling-selection/ihas-archaeology/> Accessed April 2020.
[79] M. Allen (2000) Soils, pollen and lots of snails, in M. Green (2000) *A Landscape Revealed: 10,000 Years on a Chalkland Farm.* The History Press, Stroud, 36-49.
[80] B. Cunliffe (2013) *Britain Begins.* Oxford University Press, Oxford, 253-259, 271-272.

ditch at Tormarton in south Gloucestershire during the Bronze Age. It is possible this related to a boundary dispute.[81]

Figure 2.6 shows a 'reave-like' co-axial field system on the chalk downland of Hampshire. Linear earthwork features (such as the 'Ranch Boundary') are frequently associated with Bronze Age and early Iron Age field systems. These may have had a boundary function or were related to transport of people and animals.[82]

The Iron Age (c.750 BC – c.AD 40) landscape reflects the multitude of changes that can be deduced from the archaeological evidence. In addition, the large-scale deterioration in climate and soil conditions evident from Dartmoor would have implications across southern England; Figure 1.11 indicates a change to cooler and wetter conditions. The emergence of elites during the Neolithic, added to population pressure on resources would further foster the rise of tribal societies.

From the late Bronze Age, new hilltop structures emerge. These were the 'hillforts' and were different in structure to the earlier causewayed enclosures, where the ditch and banks were interrupted by causeways of intact ground. While many smaller examples have uncertain purposes and may have had functions that were other than defensive (in the military sense), including simply corralling livestock,[83] other 'hillforts' were of incredible significance in marking the prestige of rulers. In an economic sense, they would also come to introduce elements of proto-urbanisation as central places within a landscape hitherto dominated by dispersed settlement patterns.

Through the late Bronze Age and Iron Age, hillforts evolved (for example at Woolbury near to Stockbridge and at Windy Dido, Figure 2.6) from fairly simple structures set amongst Bronze Age fields to late Iron Age complex[84] 'developed hillforts' typified by Maiden Castle near Dorchester, Danebury, Egdon Hill, or Hambledon and Hod Hill near Blandford Forum in Dorset.[85] While the Roman general Vespasian would have encountered resistance during his progress through Wessex in AD 43, it is likely that by that time certain classic hillforts has been abandoned or were re-fortified with the arrival of Roman forces.[86]

Oppida were Iron Age fortified towns with strong central place functions, particularly as commercial centres with a distinctive hierarchy of building types, which often developed into Roman towns. In Wessex was Winchester (*Venta Belgarum*),[87] the principal tribal confederation centre of the Belgae, and Silchester (*Calleva Atrebatum*). Major hillforts such as Maiden Castle (the enormous hillfort by Dorchester), which was probably a tribal confederation centre for the Durotriges, and Danebury Hillfort in Hampshire may comprise aspects of planning of

---

[81]   N, Thorpe (2019) Scales of Conflict in Bronze Age to Iron Age Britain: Enemies both Outside and Within, in S. Hansen and R, Krausepp (eds) *Materialisation of Conflicts*. Habelt Verlag, Bonn, 259-276.

[82]   A. Tullett (2010) Information Highways – Wessex Linear Ditches and the Transmission of Community, in M. Sterry, A. Tullett and N. Ray (eds) *In Search of the Iron Age Proceedings of the Iron Age Research Student Seminar 2008*. Leicester Archaeology Monographs 18. University of Leicester Press, 111– 126.

[83]   H. Cook (2018) *New Forest: The Forging of a Landscape*. Windgather Press, Oxford, 54.

[84]   B. Cunliffe (1993) *Wessex to AD 1000*. Longman, New York, 144.

[85]   B. Cunliffe (1993) *Wessex to AD 1000*. Longman, New York, 167.

[86]   B. Cunliffe (1993) *Wessex to AD 1000*. Longman, New York, 215-216.

[87]   English Heritage (2018) *Oppida* <https://historicengland.org.uk/images-books/publications/iha-oppida/heag213-oppida/> Accessed April 2020.

streets and houses, but they are generally regarded as major hillforts.[88] The exact definition of the term *oppidum,* as applied in Wessex at least, remains a matter of contention.

By the Iron Age, the Wessex landscape comprised large, open settlements, those within large (potentially defendable) ditched enclosures, which might potentially develop into larger and more impressive hillforts, and smaller 'banjo enclosures' approached by narrow roads, flanked by ditches that are continuous with the enclosure ditches; these may be livestock-related but their exact purpose is uncertain. Both may be found across the chalklands. The dominant type of dwelling was a roundhouse, with farmsteads typically tending a patchwork of small rectangular fields, especially on the downs; granaries also seem to have been commonplace.

The land would have been extensively farmed and under the control of powerful rulers of tribes, or tribal confederations. Iron tipped ploughs enabled expansion of arable land to heavier (clay-rich) soils; it is likely the downland soils were suffering from degradation. Wessex maintained a balanced rural economy with a livestock sector comprising cattle, sheep, pigs and horses.[89] Specialism may have been local, but it is likely that across Britain there had emerged a tendency for a pastoral economy in the west with more arable land in the south and east in the Iron Age,[90] a situation analogous to the present because it is largely determined by soils, topography and climate.

Surviving excerpts from Pytheas of Marsailles within classical documents by other writers relate to a lost description of voyage(s) to Britain as early as *c.* 320 BC. According to Pytheas, the island of Britain may have been called 'Pretannia' and tin was found there. Britain was apparently thickly populated with a cold climate. The *Pretanni* were a tribal people, ruled by kings and aristocrats, who lived in houses of reeds or timber and engaged in agriculture; they cut off only the heads of grain and stored them in roofed buildings. Each day they selected the ripened heads and grind them, in this manner getting their food.[91]

After the Roman expeditions of 55 BC and 54 BC, Julius Caesar could write, most likely about the same part of Britain as did Pytheas:

> The most civilized of all these nations are they who inhabit Kent, which is entirely a maritime district, nor do they differ much from the Gallic customs. Most of the inland inhabitants do not sow corn, but live on milk and flesh, and are clad with skins. All the Britains, indeed, dye themselves with woad, which occasions a bluish colour, and thereby have a more terrible appearance in fight. They wear their hair long, and have every part of their body shaved except their head and upper lip....[92]

---

[88] M. Papworth (2011) *In Search of the Durotriges*. The History Press, Stroud, Chapter 1.
[89] B. Cunliffe (1993) *Wessex to AD 1000*. Longman, New York, 180-193.
[90] H.M. Jewell (1994) *The North-South Divide*. Manchester University Press, Manchester, 8-9.
[91] T.S. Garlinghouse (2017) On the Ocean: The Famous Voyage of Pytheas. *Ancient History Encyclopedia*. <https://www.ancient.eu/article/1078/on-the-ocean-the-famous-voyage-of-pytheas/> Accessed April 2020; B. Cunliffe (2003) *The Extraordinary Voyage of Pytheas the Greek*. Penguin, London.
[92] Caesar, *Gallic War* 5.14 <http://www.perseus.tufts.edu/hopper/text?doc=Perseus%3Atext%3A1999.02.0001%3Abook%3D5%3Achapter%3D14> Accessed July 2021.

While this undoubtedly led to, or fed off, stereotypes about 'ancient Celts', Caesar was commenting (inaccurately because cereals and textiles were clearly important) on the agricultural economy. According to Strabo (*c.* 64 BC to c. AD 24) Britain:

> produces corn (cereals), cattle, gold, silver and iron. These things are exported, along with hides, slaves, and dogs suitable for hunting. [93]

To this list may be added wool, leather, zinc and lead. Whatever the case, the economy was rich enough to attract interest from the Romans. It was not until AD 43 that the Emperor Claudius systematically and successfully invaded Britain. The lowlands were relatively rapidly subdued; more peripheral areas such as Cornwall, modern Wales and Caledonia (modern Scotland) were a different story. While Roman influence was felt in Ireland, they never conquered that country.

Julius Cesar's arrogant quote relating to his conquest of Gaul, 'I came, I saw, I conquered' might, by an economic historian be rendered 'I came, I developed, I taxed'. This however would be a long-term view. Britain was a going concern by the end of the Iron Age and the economy was already partly integrated with the Roman Empire through trade. This trade had strengthened since Caesar's conquest of Gaul, switching the emphasis towards the Thames estuary as an entrepôt, largely replacing Hengistbury Head. [94]

An immediate pretext for the Claudian invasion was civil war between the tribes of the southeast that now threatened to disrupt trade with Rome. An internal (Roman) political reason was that Claudius needed to show himself capable of restoring internal peace after the chaotic and bloodthirsty rule of Caligula. The plunder and later tax won from conquest would bolster his prestige.

There was already a strong relationship between the tribes of southern Britain and Rome, and client tribal leaders may have been receptive to Roman rule in return for confirmation of their status. One likely contender may have been *Cogidubnus,* possibly the builder of the Roman palace at Fishbourne near Chichester. There is also a claim for a late Iron Age (pre-conquest) base here including imported goods.[95] Others may have been rulers of the Atrebates and Dobunni who collaborated or sought the patronage of Claudius.[96] Vespasian (later emperor) fought his way across southern Britain suppressing resisting hill forts, particularly those of the Durotriges where action was required for conquest.[97]

All things being equal, the Romans had little motive for invasion while trade continued uninterrupted, but the tribes were constantly at war with each other making 'divide and rule' a desirable political option, and a common recipe for imperial expansion. The Romans could take advantage of tribal battles to gain allies helping to control opponents[98]. Once within the Empire, Roman public contractors called *publicani* (referred to as 'tax farmers') were

---

[93]   Strabo, *Geographica* 4.5.2 <http://www.perseus.tufts.edu/hopper/text?doc=Perseus%3Atext%3A1999.01.0239%3A book%3D4%3Achapter%3D5%3Asection%3D2> Accessed July 2021.
[94]   B. Cunliffe (2013) *Britain Begins.* Oxford University Press, Oxford, 361.
[95]   J. Manley and D. Rudkin (2005) A Pre-A.D. 43 Ditch at Fishbourne Roman Palace, Chichester. *Britannia*, 36, 55–99.
[96]   B. Cunliffe (1993) *Wessex to AD 1000*. Longman, New York, 213.
[97]   B. Cunliffe (1993) *Wessex to AD 1000*. Longman, New York, 215-6.
[98]   H. Wake (2006) Why Claudius invaded Britain <http://romans.etrusia.co.uk/whyinvade.php> Accessed April 2020.

resented for collecting taxes as tithes and customs;[99] this situation may in the long-term have contributed to the end of the empire. Looking forward to post-Roman Britain after AD 410, the medieval economy, its governance and enclosure are described in Chapter 3.

2.6 Trends in settlement, manufacture, and urbanisation

It has been inferred that permanent settlement was established with the onset of agriculture. Earlier people probably had seasonal encampments which might be returned to year on year. Settlements that involved clustering together of individuals (rather than houses that were dispersed among cultivated or grazed fields) may have come about for reasons of mutual safety, with the added advantage that goods and services could be exchanged. This 'proto-urbanisation' certainly fits the bill for the major hillforts where archaeology demonstrates that in many cases, a population dwelled within.

Once pacified, the new province of Britannia had to be developed further in order to exploit its economy. Romans were urban-minded and towns, sometimes on the site of significant Iron Age settlements, were established to act as markets, as economic and religious centres, to administer justice and organise taxation. They also produced a planned system of roads for the movement of troops and goods which linked important settlements. Generally, major towns were based on British tribal confederations and the aristocracy were involved in their administration, becoming 'Romanised' in the process. Winchester (*Venta Belgarum*) and Silchester (*Calleva Atrebatum*) are examples where an *Oppidum* developed into a Roman town; elsewhere, in the western part of Wessex, new settlements were established near to major hillforts, such as Dorchester (*Durnovaria*) replacing nearby Maiden Castle.[100]

Other significant Roman centres in and around Wessex were at Chichester (*Noviomagus Reginorum*), Exeter (*Isca Dumnoniorum*), Gloucester (*Glevum*), Cirencester (*Corinium Dobunnorum*) and Bath (*Aquae Sulis*). Additionally, there were smaller towns and villages as well as the 'villa system', which would have dominated the countryside, but it is unlikely these had any significant administrative role (Chapter 3). Places described as *civitas* in Latin were in effect tribal confederation centres and included *Venta Belgarum* for the Belgae and *Durnovaria* for the Durotriges.

The end of Roman Britain is conventionally dated to AD 410, when the Roman army was supposedly recalled following the sack of Rome by the Goths, although the army in fact left in stages in the preceding years to counter invasions from across the Rhine.

Roman control never returned to Britain and the speed of administrative collapse appears to have been rapid. Perhaps it was not cost-effective to return to administering the former province. However, long-term the cultural significance of the old Empire would continue in Britain as elsewhere. In the ensuing power vacuum new kingdoms were to emerge in what was to become England under Germanic control, of which Wessex became arguably the most significant (see the Introduction).

---

[99] UNRV (2019) Roman taxes <https://www.unrv.com/economy/roman-taxes.php> Accessed April 2020
[100] Historic England (n.d.) *Oppida* <https://historicengland.org.uk/images-books/publications/iha-oppida/heag213-oppida/> Accessed April 2020.

The relationship between existing towns and the emerging Anglo-Saxon societies in earlier times remains unclear. The collapse of Roman rule brought an end to organised urban living in the 5th century. However, in the long-term the occupants of the new kingdoms were to make many Roman centres their own. *Aquae Sulis* would aptly become Bath, *Venta Belgarum* Winchester, *Durnovaria* Dorchester and so on; the suffix '-chester' recalling the Latin *castrum* for a military camp, even where these evolved into proper towns. The Anglo-Saxons (or perhaps, following Bede, Jutes in southern Hampshire) provided for some continuation with *Hamwic* (or *Hamtun*) founded c.700, near to the Roman town of *Clausentum*. It became an administrative centre and a significant port for Wessex, eventually being known as Southampton[101.] *Calleva Atrebatum* (the 'Atrebatean Place in the Woods') was built on an Iron Age predecessor, while the suffix rule '-chester' was followed in its renaming as Silchester by the Anglo-Saxons, although no significant settlement emerged after that time.[102]

The Anglo-Saxons would eventually organise around manor-houses providing rural nucleated settlements. Churches added further opportunities for stable settlement; early large missionary or 'minster' parishes would be sub-divided into parishes while the land would be divided administratively into shires and hundreds.[103] Eventually in Wessex, in the late 9th century, around the time of Alfred the Great (who reigned 871-99), a defensive system of planned 'burhs' developed during the Danish attacks; these also functioned as market towns giving rise to full urban life once more (Chapter 3).

By the time of the Domesday Book (1086) Wessex was an area of 'large, consolidated manors with many villages belonging to a single lord'.[104] It has been noted that this sense of feudal power is further re-enforced by the relatively high proportion of serfs compared with elsewhere in England and suggests a downgrading of the peasantry post-Norman invasion.[105] Towns were clearly established by this time with 31 boroughs[106] recorded in Domesday Book, including Wareham, Reading, Wilton, Old Sarum, Bristol, Bath, Winchester, Southampton, Wallingford, Shaftsbury and Dorchester. Populations seem to have varied between about 500 and 3,000. Markets were commonplace providing outlets for agricultural products from the outlying villages. The Norman invasion produced considerable urban change: redevelopment through the building of castles affected Winchester, Wareham and Wallingford) and destruction by the passage or armies damaged Dorchester, Bridport and Shaftsbury, while Exeter was besieged in 1068.[107]

While the governance of medieval towns was complex and changed over time, there were interests represented by the aristocracy, bishops, monasteries, merchants, trades (through

---

[101] Hampshire History (2012) < http://www.hampshire-history.com/hamwic-hampshires-anglo-saxon-port/ > Accessed April 2020.
[102] English Heritage (2018) Calleva Atrebatum< https://www.english-heritage.org.uk/visit/places/silchester-roman-city-walls-and-amphitheatre/history/ > Accessed April 2020; B. Johnson Silchester Roman Town (Calleva Atrebatum) <https://www.historic-uk.com/HistoryMagazine/DestinationsUK/Calleva-Atrebatum-Silchester-Roman-Town/> Accessed April 2020.
[103] Hampshire History (2012) < http://www.hampshire-history.com/hamwic-hampshires-anglo-saxon-port/ > Accessed April 2020.
[104] J.H. Bettey (1986) *Wessex from AD 1000*. Longman, New York, 19.
[105] J.H. Bettey (1986) *Wessex from AD 1000*. Longman, New York, 19-20.
[106] The term 'Borough' is taken to refer to a settlement (a town but not a city) with a high degree of self-government through an elected council or 'corporation'. It may have privileges (for example a market) granted by a royal charter.
[107] J.H. Bettey (1986) *Wessex from AD 1000*. Longman, New York, 23-25.

guilds) and others. The tendency away from a pure form of feudalism towards more market-driven economies in the towns, was driven by famine and increasing population. Visitations of the Black Death, notably in the second half of the 14th century, broke the status quo for social relations in the countryside and, apart from creating overall depopulation in England, allowed for substantial urban growth by migration to the towns which grew in population, wealth and influence.

One indication of the rising importance of towns around the start of the 13th century was the grant of a Papal bull to relocate Salisbury Cathedral in 1219,[108] from the nearby site of Old Sarum (the former site of an Iron Age hillfort) to the plain below. The new town of Salisbury was laid out on a grid system around the Cathedral and there was confirmation that the Bishop could hold a weekly market in its Royal Charter of 1227.[109]

By the 1370s, Salisbury ranked as England's seventh city.[110] Other new towns dating from the 13th century were at Chipping Sodbury (the name 'Chipping' refers to a market), Marshfield, Sherston, while new markets were established at Warminster and elsewhere. On the coast, Portsmouth was founded as a deep-water port in 1194 because the former Roman port of Portchester could not accommodate larger ships.[111]

Wessex overall was relatively prosperous when compared with elsewhere in Britain and Ireland. During the period 1500 to 1660 economic activity and overall population growth began to pick up across England after a sustained contraction following the Black Death.[112] In Wessex, urban growth could likewise be considerable with manufacturing strong. Industry was heavily based in textiles (Taunton, Devizes, Newbury, Frome, Bradford-on-Avon and Trowbridge) and Bristol exported cloth from Wessex and the midland counties via the River Severn and from South Wales. Other towns (Basingstoke, Abington, Wantage, Shaftesbury, Warminster and Andover) benefitted from trade in cattle, corn and cheese supplying London and elsewhere. London was furthermore to stimulate English overseas trade in the post-Medieval Period. A further boost to regional trade arose as adventurers explored and established colonies in Newfoundland and New England.[113]

Ironically, this activity was not reflected in Salisbury's economic fortunes after about 1600 and the population of the city declined slightly in population until 1800,[114] as did manufacturing. This was a fall from its status as a 'boom town' in the 15th century when merchants such as the notorious rivals John Halle and William Swayne made a fortune trading wool and other commodities, and Salisbury produced finished cloth, the striped Salisbury 'ray'.[115] With the emergence of a new class of capitalist clothiers in west Wiltshire, the locus of the weaving

---

[108] E. Crittall (1956) The cathedral of Salisbury: From the foundation to the fifteenth century, in R.B. Pugh and E. Crittall (eds) *A History of the County of Wiltshire Volume 3*, Victoria County History, London, 156-183. <http://www.british-history.ac.uk/vch/wilts/vol3/> Accessed April 2020.

[109] E. Crittall (1962) Salisbury: Markets and fairs, in E. Crittall (ed.) *A History of the County of Wiltshire Volume 6*. Victoria County History, London, 138-141. <http://www.british-history.ac.uk/vch/wilts/vol6/> Accessed April 2020.

[110] J. Chandler (2001) *Endless Street*. Hobnob Press, Salisbury, 42.

[111] J.H. Bettey (1986) *Wessex from AD 1000*. Longman, New York, 56-58.

[112] J. Chandler (2001) *Endless Street*. Hobnob Press, Salisbury, 43.

[113] J.H. Bettey (1986) *Wessex from AD 1000*. Longman, New York, 142-143.

[114] J. Chandler (2001) *Endless Street*. Hobnob Press, Salisbury, 42.

[115] E. Crittall (1959) The woollen industry before 1550, in E. Crittall (ed.) *A History of the County of Wiltshire 4*, Victoria County History, London, 115-147. <http://www.british-history.ac.uk/vch/wilts/vol4/> Accessed April 2020.

Figure 2.7. Rev. Peter Hall's etching of Harnham Mill in 1834. The Tudor building is the long, low chequerboard factory, the taller building on the left is Georgian and described as a yarn factory in 1810 (public domain).

industry moved elsewhere, notably Bradford-on-Avon[116] and Trowbridge. Southampton also lost its pre-eminence as trade with Italy declined and the demand for striped kersey-like cloth declined in favour of undyed broadcloth.[117]

Manufacturing in Salisbury had been important for centuries. For example, an unknown merchant built the Harnham Mill (the 'Old Mill') at West Harnham at some time around or after 1500, although this is not the original mill building on the site (Figure 2.7). From even before 1300, the mill buildings seem to have been specifically industrial (frequently employed in manufacturing) rather than used for flour milling. Their uses included fulling of cloth, yarn manufacture, bone crushing (for fertiliser), tallow (for candle making), parchment - and importantly paper. Tallow and parchment manufacture would not require waterpower in themselves, unlike bone crushing, but it would be necessary in obtaining their raw materials, which are rendered from animal carcasses. Although waterpower would have been used for fulling and was probably the first industrial activity on the site, Harnham Mill may be the oldest extant paper mill in England.[118] Mills should be seen as 'fixed assets' economically speaking which could change function according to market demand. It is likely that as local

[116] Wiltshire Community History (2011) Bradford on Avon. <https://history.wiltshire.gov.uk/community/getcom2.php?id=26> Accessed April 2020.
[117] J. Chandler (2001) *Endless Street*. Hobnob Press, Salisbury, Chapter 3.
[118] J. Chandler (2001) *Endless Street*. Hobnob Press, Salisbury, 80; M. Cowan (2008) *Hanham Mill Salisbury*. Sarum Studies 2. Hobnob Press, Salisbury.

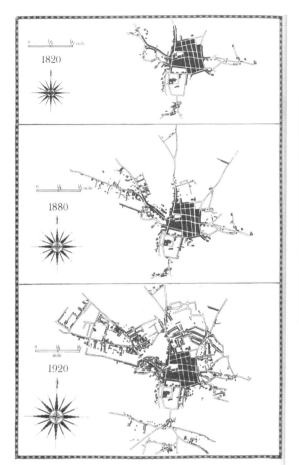
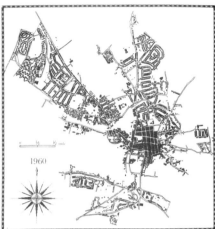

Figure 2.8. The development of Salisbury 1820-1960. The Cathedral Close is clearly identifiable on the south side (after J. Chandler (2001) *Endless Street*. Hobnob Press)

cloth manufacture declined in the long run, increasing literacy and the added value of a manufactured good drove construction of a paper mill. Unfortunately, this is conjecture based upon the structure of the building and use as a paper mill is not documented.

The development of Salisbury in the nineteenth and twentieth centuries (Figure 2.8) was dramatic and consistent with an increase in population from around 8,000 to 33,000 citizens. The increase during this time was actually exponential following the long period of gradual decline (roughly 1500-1800), but although by its own standards growth was dramatic, in absolute population terms Salisbury remained behind many established centres of population.[119] In 1820, the City had barely outgrown its medieval extent, indicated by the grid-like plan of the street pattern forming 'The Chequers'. Even by 1880, the arrival of the

---

[119] J. Chandler (2001) *Endless Street*. Hobnob Press, Salisbury, 42.

railway seems to have made only a modest impression, although 19th century building was commonplace.

By 1920 there was extensive growth to the west of the city, on the river terraces of Bemerton Heath; this continued into mid-century with encroachment on the downs and higher river terraces to the north. Bricks tend to be produced as locally as possible, largely due to the high cost of transportation. The 19th century westward growth included development on the brickfields,[120] especially at Fisherton and Bemerton (i.e. around the railway station and along the Devizes road),[121] where the brickearth also provided valuable land for growing that was close to markets for vegetables. Indeed, historic mapping shows that market gardening was commonplace around the City and located on better soils. Building, on the other hand, often (but not always) avoided the active floodplain including water meadow areas (see Chapter 1). One way of seeing the 19th and 20th century development of Salisbury is that space was contested between brick making, horticulture and urban development. The latter would win out.

The revival of Salisbury's fortunes was not spectacular, when compared to the growth of Bristol in trade and manufactures, or the long-term recovery of Southampton. Between 1780 and 1810, there was some revival of the textile industry[122] (as witnessed by the construction of the yarn factory at Harnham Mill), but the arrival of the railway in 1856 was late by regional standards, and the city would never compete with the developed industrialised manufacturing of the north of England and elsewhere, even within Wessex.

While industrialisation never reached the degree of that in South Wales, the Midlands, or North of England, nonetheless the growth of towns in Wessex was impressive. Bristol's population increased from c. 27,000 in 1701 to c. 50,000 in 1751, and to 142,825 in 1831. The populations of all the region's counties increased between 1600 and 1800, and between 1801 and 1901 Bristol, Portsmouth and Reading all experienced population increases at or above 395%; Southampton experienced growth of over 1,200% and Swindon an astounding 3,650%. The development of the holiday resort caused impressive growth of Bournemouth and Weston-super-Mare. These figures are, however, still far below rates of growth in the Midlands and North[123]. In 1991 the population of Wessex was about 3.19m, around half of which was in Hampshire. In 2017 the estimated population of southwest England was 5.56 m.[124]

2.7 Tracks old, straight, or otherwise

Human tracks are preserved in the alluvial deposits of the Severn Estuary at Goldcliff, some nine km south-east of Newport, Monmouthshire. Footprints of both adults and children are preserved in the sediment, from people who exploited an oak forest associated with an island preserved in the estuarine sediments. Because sea level was lower, silts and peats became

---

[120] P.M. Hopson, A.R. Farrant, A.J. Newell. R.J. Marks, K.A. Booth, L.B. Bateson, M.A. Woods, I.P. Wilkinson, J. Brayson and D.J Evans (2007) *Geology of the Salisbury District*. Salisbury, Keyworth, Sheet explanation of the BGS 1:50,000 Sheet 298.
[121] J. Wright (2017) *Brickmaking in Fisherton and Bemerton*. South Wiltshire Industrial Archaeology Society, Salisbury.
[122] J Chandler (2001) *Endless Street*. Hobnob Press, Salisbury, 89.
[123] J.H. Bettey (1986) *Wessex from AD 1000*. Longman, New York, Chapters 6 and 7.
[124] Statistica (2019) <https://www.statista.com/statistics/294681/population-england-united-kingdom-uk-regional/> Accessed April 2020.

buried. The basal sequence comprises remains of a substantial oak forest which existed between about 6000 BC with occupation continuing until c 4700 BC.[125] From pre-Neolithic times settlement sites could be re-occupied time and again by transhumant and hunter gathering people. These observations open the possibility of tracks being established that continued into times of settled agriculture.[126]

Upper Paleolithic and Mesolithic peoples were also associated with upland areas, for example their presence between Exmoor, around the Somerset Levels and on Mendip is well attested from artefacts.[127] Human bones from the Mesolithic have been found from Aveline's Hole near Burrington Combe that suggest funerary rights were practiced in the early Mesolithic.[128] It is probable that upland areas were hunted during the summer months, while wooded lowland areas were settled during the winter.[129] Mesolithic settlement has in fact been identified across Wessex (including notably Cranborne Chase) with artefacts made from Portland type chert, slate from Cornwall and north Devon and 'pebbles of South West origin' of the time clearly exchanged or carried and deposited over wide areas across southern England.[130]

No academic discipline is exempt from 'unifying theories' and landscape studies are no exemption. 'Ley-lines' are no longer considered a serious proposition, aside from those of a more esoteric inclination. However, if ideas involving the deliberate placement of landscape features along lines reaching over considerable distances have been marginalised, it is realised that ley line theory has some sort of rational basis, even if the outcomes lack credibility in the eyes of modern researchers. The instigator was Alfred Watkins (1855 – 1935), who in 1921 could write:

> Presume a primitive people, with few or no enclosures, wanting a few necessities (as salt, flint flakes, and, later on, metals) only to be had from a distance. The shortest way to such a distant point was a straight line, the human way of attaining a straight line is by sighting, and accordingly all these early trackways were straight, and laid out in much the same way that a marksman gets the back and fore sights of his rifle in line with the target.[131]

Watkins' ideas rely upon early peoples on unenclosed landscapes sighting on hills and filling in the way with 'secondary markers', typically on lower land areas. These became trackways that determined the location of archaeological monuments and, later, churches, but given time tracks deviate from straight lines. In a landscape more wooded than the present-day, however, there may have been sighting problems and millennia later, it took the surveying

---

[125] Severn Estuary Levels Research Committee (n.d.) Goldcliff Mesolithic Archaeology <http://www.selrc.org.uk/maplocation.php?location_id=38 > Accessed April 2020; M.G. Bell (2013) *Prehistoric Coastal Communities: The Mesolithic of Western Britain.* Council for British Archaeology, York, 153.
[126] P.R. Preston and T. Kador (2000) Approaches to Interpreting Mesolithic Mobility and Settlement in Britain and Ireland. *Journal of World Prehistory* 31, 321–345; M. Green (2000) *A Landscape Revealed: 10,000 Years on a Chalkland Farm.* The History Press, Stroud, Chapter 1; R. Peberdy (2020) Paths and Tracks <https://www.victoriacountyhistory.ac.uk/explore/themes/transport-and-communications/paths-and-tracks> Accessed April 2020.
[127] B. Cunliffe (1993) *Wessex to AD 1000.* Longman, New York, Chapter 1.
[128] B. Cunliffe B (2013) *Britain Begins.* Oxford University Press, Oxford, 125.
[129] *Britain Begins.* Oxford University Press, Oxford, 106.
[130] B. Cunliffe (1993) *Wessex to AD 1000.* Longman, New York.
[131] A. Watkins (1922) Early British Trackways, Moats, Mounds, Camps, and Sites. A Lecture given to the Woolhope Naturalists' Field Club, at Hereford, September, 1921, < http://www.ancient-wisdom.com/alfredwatkinslecture.pdf > Accessed April 2020.

skills of the Roman Army to achieve something almost like the 'straight tracks' envisaged by Watkins. There are, by contrast, few controversies around Roman roads!

In effect, Watkins' ideas allowed for ley-lines at different scales, from almost a field-scale to the regional. Wessex is said to be traversed by the St Michael's ley-line that runs from somewhere near to Great Yarmouth to Penzance, taking in Glastonbury and Avebury, tracing the path of the Sun on 8th May (the spring festival of St. Michael). Stonehenge, apparently has its own separate but not unrelated arrangements.[132]

There is little support for the idea of ley-lines among the archaeological or landscape history communities. A classic piece of research convincingly debunked the idea based on detailed observation of mapping, temporal considerations and simulation exercises.[133] Detractors furthermore decry the association of the ley-line hypothesis with more esoteric ideas including astronomical alignments over long distances, which developed as ideas originating from Watkins diverged ever further from the mainstream.

To be fair, Watkins did generate research questions relating to the origins of track networks and the citing of prehistoric and later monuments. While astronomical alignments are important in prehistory in relation to specific monuments and indeed their siting, the origins of trackways lie in less tidy, but more pragmatic considerations based ultimately in geology, hydrology, soils and places where early societies chose to congregate, linked by migration, transhumance and trade. The answers (if that is the appropriate word) lie somewhere between natural topography and G.K. Chesterton's 'rolling English drunkard',[134] or at least his forebears. Maybe the idea of ley lines arose from a backwards projection of the achievements of Roman surveyors into prehistory?

Yet people do exchange goods, have social and religious gatherings, and migrate – generally on a seasonal basis – and it is likely meetings were multi-purpose. A territoriality based in river basins has been proposed, each perhaps involving a few hundred individuals.[135] It is probable that some form of territoriality was established by the end of the Mesolithic. The locations of early Neolithic long barrows in Wessex suggest that these were communal burial sites for that reflected a need to identify territoriality between differing agricultural communities.[136]

Figure 2.9 shows the principal prehistoric 'ridgeways' as they might have been in the Neolithic. The pioneering aerial archaeologist O.G.S. Crawford identified four 'ancient' trackways, three mostly along chalk escarpments (the North Downs (including the 'Hogs Back'), South Downs, and the chalk escarpment which includes the Chilterns), and one along Jurassic limestones.[137] This simple model of prehistoric communications has a number of points in its favour: it is topographical and geological, it seems to allow for convergence on Stonehenge (and link with

---

[132] Ancient Wisdom (n.d.) St Michael <http://www.ancient-wisdom.com/stmichael.htm> Accessed April 2020.

[133] T. Williamson and L. Bellamy: *Ley lines in Question*. Tadworth: World's Work, 1983.

[134] 'Before the Roman came to Rye or out to Severn strode, The rolling English drunkard made the rolling English road. A reeling road, a rolling road, that rambles round the shire, And after him the parson ran, the sexton and the squire; A merry road, a mazy road, and such as we did tread The night we went to Birmingham by way of Beachy Head.' Chesterton, G. K. (1927) *The Rolling English Road*.

[135] P.R. Preston and T. Kador(2000) Approaches to Interpreting Mesolithic Mobility and Settlement in Britain and Ireland. *Journal of World Prehistory* 31, 321–345.

[136] P. Ashbee (1970) *The Earthen Long Barrow in Britain*. J.M. Dent, London.

[137] O.G.S. Crawford (1918) *Archaeology in the Field*. Phoenix House London, 77-78.

the ancient port at Hengistbury Head, Christchurch) and would allow for long-distance trade of flint artefacts. The Icknield Way passes many monuments from the Neolithic, Bronze Age and Iron Age. Furthermore, the Jurassic thoroughfare from the Cotswold Hills up towards modern Lincoln and beyond presents a northern link. It is frustrating there are no mapped ancient trackways west of Bath and Salisbury. Nonetheless, there are real problems of dating when considering prehistoric trackways, meaning that any 'developed' model of communication networks may mask earlier variation and present problems for establishing their evolution.[138]

Land clearance at the Mesolithic/Neolithic boundary likely occurred on higher ground, especially interfluves and the crests of escarpments. Early farmers would have been attracted to cultivate the loess-covered aeolian deposits that capped the chalk in most areas, while on or below the crest of the downs many Neolithic monuments may be found.[139] Claims that the Icknield Way is the oldest track in Britain seem at first plausible, although the great antiquity of this and maybe other tracks has been challenged on archaeological and topographic grounds.[140] Without necessarily always having a line of sight, prehistoric peoples would have found these escarpments convenient and, unlike river valleys or wetlands, for example, would generally enjoy year-round passage. Apparent coincidences are not ruled out: the path of the Icknield way from Norfolk follows the chalk escarpment north of London for much of its way then bears southwest toward Salisbury Plain on an overall bearing similar to that of the 'St Michael's Ley line'. There is no need for either ley-lines or astronomical alignments over long distances.

Discovered in 1966, during the extension of what is now the old carpark at Stonehenge, is a Mesolithic site comprising three postholes;[141] since then, a fourth post hole has been identified at Stonehenge. Considerable additional Mesolithic activity has been identified at nearby Blick Mead,[142] dated to the 9th-7th millennia BC, with Mesolithic use of the area continuing into the 5th millennium BC and the dawn of the Neolithic period. It has been speculated that the postholes had some function that was perhaps analogous with totem poles.[143]

If such speculation is correct, it suggests there was not only considerable activity in the area throughout the Mesolithic but also introduces the possibility of pre-Neolithic meeting places and even continuity of some religious or other communal activity at the site. Recent research at Durrington Walls (a large henge monument 3.2km northeast of Stonehenge),[144] suggests that Neolithic People migrated great distances to the Stonehenge area with their livestock,

---

[138] M. Bell and J. Leary (2020) Pathways to past ways : A positive approach to routeways and mobility. *Antiquity* 94 (377), 1349-1359.

[139] South Downs National Park (2005) South Downs Final Report (updated 2011) Integrated Landscape Character Assessment (Updated) 95 Part 2: Character of the South Downs Landscape <https://www.southdowns.gov.uk/wp-content/uploads/2015/03/ILCA-Appendix-A-Open-Downland.pdf> Accessed April 2020.

[140] K. Fitzpatrick-Matthews (2016) North Herts Museum update: how old is the Icknield Way? <https://northhertsmuseum.org/north-herts-museum-update-how-old-is-the-icknield-way/> accessed Aug 2021.

[141] G. Vatcher and F. de M. Vatcher (1973) Excavation of three post-holes in Stonehenge car park, *Wiltshire Archaeological and History Magazine* 68, 57-63.

[142] D. Jacques and T. Philips (2014) Mesolithic settlement near Stonehenge: excavations at Blick Mead, Vespasian's Camp, *Wiltshire Archaeological and Natural History Magazine* 107, 7-27.

[143] English Heritage Research Records (n.d) Monument Number 219856. https://www.heritagegateway.org.uk/Gateway/Results_Single.aspx?uid=219856&resourceID=19191> Accessed July 2021.

[144] English Heritage (n.d.) Food and feasting at Stonehenge < https://www.english-heritage.org.uk/visit/places/stonehenge/history-and-stories/history/food-and-feasting-at-stonehenge/ > Accessed April 2020.

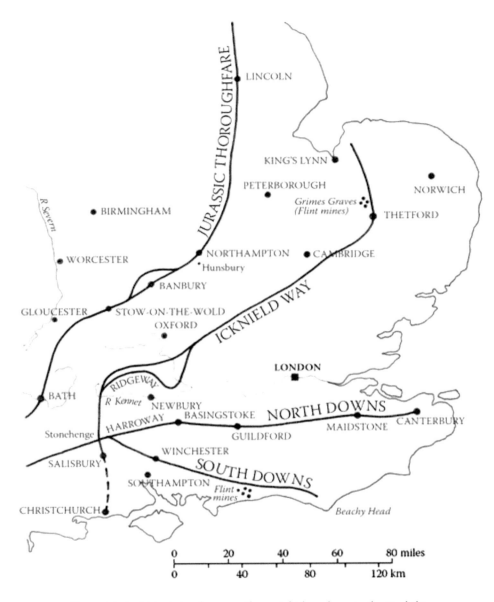

Figure 2.9. Prehistoric trackways as they might have been in the Neolithic
(O.G.S. Crawford, public domain)

from (modern day) Wales, northern England and Scotland. It is now widely accepted that there was a significant social element in this, including feasting to mark the agricultural year (most likely the Winter Solstice), and we imagine that rituals encouraged social cohesion. It is evident these seasonal events involved excessive feasting on meat (mostly piglets but also beef) and dairy products. While it is likely, there is no conclusive evidence for production of beer or mead during the Neolithic.[145]

Wessex during the Neolithic was also externally linked via rivers such as the Hampshire Avon to the sea. This is demonstrated by imported jadeite axe-heads of Neolithic age found at Breamore in Hampshire, and elsewhere in southern England that originated in the Alpine region.[146] Hengistbury head provides shelter for the modern Christchurch Harbour; Poole Harbour is similarly sheltered. Economic activity has been established at both sites since the Bronze Age, when there was development elsewhere along the southern coast, enabling the possibility of a contact zone with the continent. On the Severn Estuary, economic development is demonstrated by the construction of wooden trackways across the waterlogged soils of the Somerset Levels and Moors.[147] This process started in the Neolithic with construction of the 'Sweet Track' but seems to have ceased by around 800 BC due to inundation.[148] These tracks enabled linkages between pastureland and the arable of former islands with upland areas where settlements were to be found.

2.8 Attitudes, economy and engineering

By the end of the Iron Age, a complex of routeways criss-crossed Wessex and linked the region not only to elsewhere in Britain, but also further afield by sea via the Severn Estuary and English Channel. There was migration from the Continent in the form of the Belgae tribal confederation, and this is a reflection of an outward-looking economy. The rise of hillforts (later *oppida*) suggests an elite was in command who chose to display their importance and is also linked with urban development. These considerations start to explain why, by the mid first-century BC, Rome was showing a real interest in Britain.[149] Iron Age Britain was not an undeveloped island populated by fierce unsophisticated tribes.

To supplement the existing, if limited, road network, the Romans rapidly developed their own roads and this in turn would support a mature urban infrastructure. It should be stressed that any names for roads referred to here were given during the Anglo-Saxon period; the original names (unlike the names of significant settlements) are lost, but they are largely physically traceable and include stretches of modern roads. The Roman imperative for road construction was originally military and the regular army were responsible for their building, although both their function and maintenance increasingly moved towards civilian and commercial roles during the Roman occupation.[150]

Neither were they always straight; rather Roman military surveyors would have adapted the roads to gradients and other topographic factors, including the construction of terraceways

---

[145] M. Dineley (2018) What did Neolithic People drink at Feasts? English Heritage < http://blog.english-heritage.org.uk/neolithic-drink-at-feasts/ > Accessed April 2020.

[146] A. Sheridan, D. Field, Y. Pailler, P. Pétrequin, M. Errera and S. Cassen (2010) The Breamore jadeitite axehead and other Neolithic axeheads of Alpine rock from central southern England. *Wiltshire Archaeological and Natural History Magazine* 103, 16-34.

[147] B. Cunliffe (1993) *Wessex to AD 1000*. Longman, New York, 146-8.

[148] B. Cunliffe (1993) *Wessex to AD 1000*. Longman, New York; Avalon Marshes Somerset (n.d.) The Sweet Track and Other trackways <http://avalonmarshes.org/the-avalon-marshes/heritage/sweet-track/> Accessed April 2020.

[149] Cunliffe B (2013) *Britain Begins*. Oxford University Press, Oxford, 334.

[150] R.W. Bagshawe (1979) *Roman Roads*. Shire Books, Aylesbury, 7.

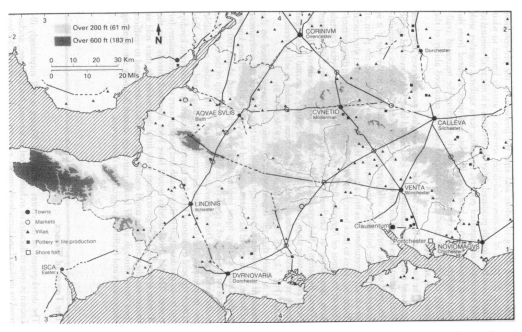

Figure 2.10. Roman Wessex (after B. Cunliffe, 1993. *Wessex to AD 1000* Longman, New York p.242)

and zig-zags down across steep hillsides.[151]An interesting illustration of adaptability is the curving of a Roman Road around the enigmatic Neolithic mound of Silbury Hill near Avebury, including a part of the modern A4. The Roman surveyors setting out the road eastwards from *Aquae Sulis* (modern Bath) evidently used it as a sighting point, and recently a substantial Roman Settlement has been discovered at the foot of the mound.[152] Figure 2.10 shows the road system and principal towns of Roman Wessex.

Hillforts soon lost their function during the occupation or even beforehand, although the *oppida* represent an element of continuity from the late Iron Age. Such were at Winchester and Silchester (now vanished as a settlement); other important Roman towns, the forerunners of Bath, Chichester and Dorchester, were developed by the end of the first century AD. During the fifth century AD the Roman transport network (Figure 2.10) probably fell into disrepair and it is likely progress by road became restricted, by design or otherwise, although many Roman roads would have continued to be used.

Into the late Anglo-Saxon period, the Ridgeway, a route from Ilchester to London including part of the North Downs Way, and a route from Ilchester to Bath becoming the Fosse Way, likely remained significant roads.[153] Saxon charters indicate that both prehistoric trackways

---

[151] R.W. Bagshawe (1979) *Roman Roads*. Shire Books, Aylesbury, Chapter 2.
[152] English Heritage (n.d.) History of Silbury Hill https://www.english-heritage.org.uk/visit/places/silbury-hill/history/ Accessed April 2020.
[153] J.E. MacDonald (2001) Travel and the Communications Network in Late Saxon Wessex: a Review of the Evidence. DPhil University of York, 55 and Figure 2. <http://etheses.whiterose.ac.uk/9828/1/369323.pdf> Accessed April 2020.

along watershed lines and Roman roads continued in use after the Roman evacuation.[154] For example, out of 79 Wiltshire roads mentioned in the charters only 38 were almost certainly of Saxon origin, and well over half of these were purely local in extent. However, the Saxon period did see the growth of roads along the river valleys. For example, a road between Burbage to Pewsey and Manningford Bruce probably continued south along the Avon valley to Amesbury and Old Sarum.[155]

In a recent study of travel and communications in early Medieval Wessex, Alex Langlands comments in detail on the importance and distribution of Wessex's roads of that period, stressing their importance in the everyday life for all conditions of people, be they peasants, pilgrims, drovers, traders, warriors, bishops or royalty. The general Anglo-Saxon term for a road is *weg*, but *herepath* refers to a significant route, likely has its origins in roads used for both defensive or offensive purposes, and in later for royal progress. The term may have been associated with fear, for the *herepath* represents a route for raiding armies. However, road networks were also inherited from earlier periods and hence display a complex relationship with the historic landscape.[156]

Into the medieval period it becomes important to revise certain concepts of road networks. The period approximately AD 1000 to AD 1550 was one of turbulence caused by warfare, famine, plague and economic disruption and as a result both population levels and the economy fluctuated. During the early part of this period, the royal court was itinerant, coming eventually to rest at Westminster, while producers and traders required access to markets for agricultural products, wool, minerals, and finished goods. While the lack of evidence precludes any statement ruling out new road construction, it is reasonable to assume near total reliance on the existing network of prehistoric and Roman roads. A few roads were constructed or improved for military purposes or crossing problematic areas prone to flooding. It is likely that, to the medieval mind, the term 'road' was more akin to the modern 'right of way'.

For example, trackways could break up into multiple paths as travellers selected the easiest rout up and down slopes or avoiding 'foundrous' or impassable stretches in wet weather; there was a right to deviate from the track, even if this involved trampling crops! The four great highways in England, Watling Street, Ermine Street, Fosse Way and Icknield Way (three of Roman origin) were maintained or improved by statute.[157] Leaving Wessex, the Fosse way and certain northern parts of Ermine Street (the Great North Road) take advantage of Jurassic ridges (Figure 2.9).

Problems with road travel in Wessex are recorded in the 17th century. Samuel Pepys complained about a journey he made with his wife and servants between Oxford and Salisbury via Hungerford in 1668. The roads were bad, signposting was poor, the inns expensive and inadequate and his bed lousy. Celia Fiennes in 1687 found the road so waterlogged between Bath and Salisbury via Warminster that her coach got stuck fast in the mud: 'severall men were

---

[154] A. Langlands (2020). Ceapmenn and Portmenn: Trade, Exchange and the Landscape of Early Medieval Wessex, in A. Langlands and R. Lavelle (eds) *The Land of the English Kin: Studies in Wessex and Anglo-Saxon England in honour of Professor Barbara Yorke*. Brill, Leiden, 294-311. DOI: 10.1163/9789004421899_016

[155] E. Crittall (1959) Roads, in E. Crittall (ed.) *A History of the County of Wiltshire Volume 4*. Victoria County History, London, 254-271.

[156] A. Langlands (2019) *The Ancient ways of Wessex*. Windgather Press, Oxford, Chapter 1.

[157] P. Hindle (1982) *Medieval Roads*. Shire Books, Aylesbury, 5-7.

forced to lift us out; its made only for packhorses which is the way of carriage in those parts'. She declared the road only fit for packhorses and it was used to bring coal from Kingswood to Bristol. In the early 18th century, Daniel Defoe was likewise unimpressed.[158]

Located at the western end of the Weald, at the closure of the Wealden anticline and around where the grey chalk, Gault clay and Upper Greensand terrace meet, is the village of Selborne. It has become a place of pilgrimage to the home of the pioneering English naturalist and ornithologist, the Revd. Gilbert White FRS (1720 – 1793). White regularly made the journey to his old college, Oriel, in Oxford, regularly, battling out through the lanes which in winter were often so flooded.[159]

By the late 17th century it was apparent that 'something had to be done' about the road network. The Romans, like the designers of modern motorway networks, understood about planning road systems. On the other hand, history had produced what it had, and by the 18th century a piecemeal approach was required. Some historic routes fell out of use, while others saw heavy traffic, and as stagecoaches developed for rapid journeying, there emerged a crisis for efficient road transport. Parishes had become responsible for road improvements and funding from other sources including the Crown was inadequate at best.[160]

Parliament increasingly took over responsibility for repairing and maintaining roads from the parishes from the late 17th century. Turnpike Acts authorised a trust to levy tolls on those using the road and to use that income to repair and improve the road and its signage, including the erection of milestones. They had powers to purchase property to widen or divert existing roads. The trusts were not-for-profit and maximum tolls were set. The 'turnpike' was the gate which blocked the road until the toll was paid.[161]

Creating a fit-for-purpose network of roads was essential to an expanding economy that would eventually industrialise. Improved highways allowed for efficient transportation of goods and travellers. The effect was to free markets, reduce costs, connect growing towns, and integrate the economy of England. Pressure from the centre to improve transport was considerable. Bodies of local trustees were given powers to levy tolls on the users of a specified stretch of road, generally around 20 miles [32km] long. Turnpike Acts were limited to a period of 21 years although continuation of the legislation meant trusts remained responsible for most English trunk roads until the 1870s.[162] By 1836, there had been 942 Acts for new turnpike trusts in England and Wales, and turnpikes covered around 22,000 miles [35,000km] of road, about a fifth of the entire network of the time.[163]

---

[158] J.H. Bettey (1986) *Wessex from AD 1000*. Longman, New York, 195-197.
[159] Hidden Europe (2015) Selbourne, naturally <https://www.hiddeneurope.eu/selborne-naturally> Accessed July 2021.
[160] D. Bogart (n.d.) The Turnpike Roads of England and Wales <https://www.campop.geog.cam.ac.uk/research/projects/transport/onlineatlas/britishturnpiketrusts.pdf> Accessed April 2020.
[161] Roads and Railways (n.d.) Turnpikes and Tolls <https://www.parliament.uk/about/living-heritage/transformingsociety/transportcomms/roadsrail/overview/turnpikestolls/> Accessed April 2020.
[162] A. Rosevear Turnpike roads in England (2008) <http://www.turnpikes.org.uk/The%20Turnpike%20Roads.htm> Accessed April 2020.
[163] Roads and Railways (n.d.) Turnpikes and Tolls <https://www.parliament.uk/about/living-heritage/transformingsociety/transportcomms/roadsrail/overview/turnpikestolls/> Accessed April 2020.

Over the succeeding 150 years or so, road networks continued to develop with public bodies at both local and national designing and funding new projects. During the 20th century one solution to congestion in expanding towns has been the bypass, while A roads (including trunk roads) have been upgraded.[164] The motorway network became the first fully planned national system, although it often joined up pre-existing dual carriageway bypasses. One example is the Exeter bypass. Constructed before the Second World War, it soon itself became congested. In the 1970s a new bypass on the M5 was built.[165] Significant bypasses in the region have been constructed at Winchester, Basingstoke and Newbury (in the 1990s), while it was hoped that Bristol's traffic problems would be alleviated by making it a nexus in the post-War Motorway system where the London-South Wales M4 and Birmingham-Exeter M5 meet. Road system development continues, as do controversies around them.

2.9 Canals in Wessex

The canal age, 1760 to 1835, was entirely 'Georgian' and reflected the imperatives of the 18th century. In Wessex major canal building commenced in 1789, somewhat later than in the north of England. However, lowland waterways within the low-lying parts of catchments of the Brue, Tone, Yeo and Parrett including the Somerset Levels that had been in use since the Roman period were improved and extended. The objectives were the same: Efficient and fast transport of bulk commodities from place of production to markets in expanding towns and cities, generally where it had only previously been possible to use packhorses. Typical commodities were agricultural products (notably corn), coal, iron ore, stone, salt and finished items. High value goods, typically going inland from Bristol, included sugar, wine, timber and leather goods. The 'Golden Age' of canals continued until about 1830 when competition from the railways caused steady decline; by 1880 canal tolls were seriously reduced.[166]

Canals raised the profile of engineers such as James Brindley and John Rennie (the Elder) who worked on the Kennet and Avon Canal, including associated features the Dundas Aqueduct, Avoncliffe Aqueduct, Caen Hill Locks and the Crofton Pumping station. In the development of the canal system, we see the first application of steam technology to transport. The pioneering geologist William Smith is strongly associated with the region, especially the Somerset Coalfield, and he supervised the construction of the Somerset Coal Canal.[167] Smith's observation of fossil assemblages enabled equivalent strata to be identified, founding the geological sub-discipline of stratigraphy and enabling publication of the first geological map of Britain in 1815 (Chapter 9).

Canals and improvements to river navigation in Wessex or nearby pre-date 1760. For example, the Exeter Ship Canal, completed in 1566, was built to bypass weirs on the River Exe and thus enable goods to reach the port of Exeter, first in barges, later in ocean-going ships.[168] In an attempt to link Salisbury with the Sea at Christchurch, there was an effort to improve

[164] Highways England (2014) Map of roads managed by Highways England. < https://www.gov.uk/government/publications/roads-managed-by-highways-england> Accessed April 2020.
[165] https://www.roads.org.uk/motorway/a30
[166] J. H. Bettey (1986) *Wessex from AD 1000*. Longman, New York, 235-238.
[167] Yorkshire Philosophical Society (n.d.) William Smith (1769-1839) – The Father of English Geology. < https://www.ypsyork.org/resources/yorkshire-scientists-and-innovators/william-smith/ > Accessed April 2020.
[168] Friends of Exeter Ship Canal (n.d.) Britain's oldest pound-lock canal <https://www.friendsofexetershipcanal.co.uk/history> Accessed April 2020.

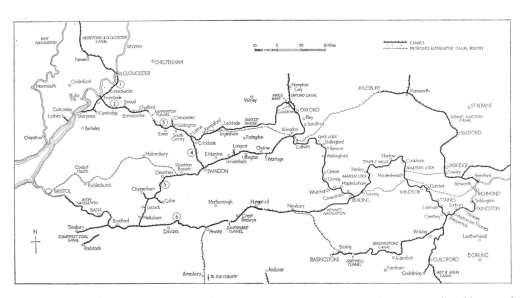

Figure 2.11. The Thames and its links with the Rivers Severn and Avon. 1. Gloucester and Berkley Canal. 2. Stroudwater Navigation 3. Thames and Severn Canal. 4.North Wilts Canal. 5. Wilts and Berks Canal. 6. Kennet and Avon Canal. Proposed routes are shown with thinner lines (C. Hadfield, 1969. *The Canals of South and South East England*. David and Charles Newton Abbot, p.196, public domain)

navigation in the lower Avon, largely by bypassing mills and weirs, which were often associated with water meadows. An enabling Act of 1664-5 authorised the Earl of Clarendon to make the Avon navigable from Christchurch to Salisbury and the Earl of Pembroke and Montgomery to do the same for the Wyle Between Salisbury and nearby Wilton. Timber and ironstone were envisaged as cargoes. Work was undertaken, financed by the Salisbury Corporation. Unfortunately, the project was dead by 1730 when the Salisbury Corporation tried to revive the idea of the canal, but this was opposed by owners of water meadows and mills.[169]

These plans for canals show the importance of aristocratic patronage before 1760. It is also worth noting that a second attempt to link Salisbury with the sea, this time via West Dean (with a wharf and lock east of Salisbury) and Lockerly to Redbridge at the head of the Test Estuary near Southampton; this time an engineer, Robert Whitworth, an assistant to James Brindley, was involved in the survey.[170] Construction of this canal commenced in 1795 but was abandoned in 1808. There was a terminal wharf at Alderbury that necessitated carting goods on to Salisbury to the southeast. The detail is complicated, but the project went bankrupt before the railway age, apparently due to the additional cost of carting and an inability to raise the capital to complete the waterway back to Salisbury.

Figure 2.11 shows the canals of the Thames with their links with Severn and Bristol Avon. Note the proposed alternative canal routes (thin lines) that were not completed. Historically there

---

[169] C. Hadfield (1969) *The Canals of South and South East England*. David and Charles, Newton Abbot, 166-169, 178-187.
[170] H. Braun (1962) The Salisbury Canal- a Georgian Misadventure. *Wiltshire Archaeological and Natural History Magazine* 58, No. 210, 171-180.

were operational links between the Thames and Basingstoke and Guildford (and across the Weald) as well as north beyond Oxford. The main regional canals are the Kennet and Avon to Bath and Bristol and a series of canals between Lechlade and Gloucester linking the Thames with the Severn as well as Abingdon to Swindon and through Wiltshire to Melksham to join the Kennet and Avon.

The Somerset Coal Canal (SCC) was constructed between 1794 and 1805 (and closed in 1900). It carried coal from the Paulton/Timsbury and Radstock coalfields to the Kennet and Avon Canal (6 in Figure 2.11). It comprised two branches, one from Paulton and one from Radstock, which converged at Midford and then connected to the Kennet and Avon at Dundas. The SCC was built to connect the mines with the major towns and was a very profitable venture; although there were problems with poor roads in the surrounding area, increasing the price of coal, its decline appears to have resulted from the railways and the working out of coal deposits. In 1904, the abandoned canal was sold to the Great Western Railway who built the Camerton to Limpley Stoke railway line over much of its course (1907-10).[171]

The Kennet and Avon Canal (K&A) comprises three different sections: two improved navigable rivers and one canal. The canal starts in Reading and ends at Bristol. The river sections were in use by the early 18th century, while the construction of the canal section took place between 1794 and 1810. The canal re-opened in1990 following restoration, mostly by volunteer labour. The K&A is almost 87 miles [140km] long, includes 105 locks and is 13 ft 8 in [14.16m] wide with a draft of 4ft 1in [1.24m].[172]

Between Bradford-on-Avon and Bath are the magnificent Avoncliffe and Dundas Aqueducts on the K&A as well as the Caen Hill Locks at Devizes. The locks comprise two flights of 29 locks have a rise 72m in 3.2 km or a 1:44 gradient to take the canal over the Upper Greensand Formation.[173] Another spectacular feature of the K&A is the Crofton pumping station. This was completed in the 1800s to supply water to the western end of the K&A which represented a major challenge to John Rennie. The Crofton Beam Engines at Crofton Locks Summit use two steam pumps to lift water 40 feet [12.2m] to the canal. These engines, built in 1812 and 1845, are among the world's oldest working steam beam engines.[174] Despite all the investment, however, the 'Golden Age' of canals was ultimately doomed due to the development of the railway network.

2.10 Railways

The next serious impact on the landscape was to be the railways; their development was complex and would affect every aspect of urban and rural life, often reaching quite remote settlements. The railway from London reached Southampton in 1840 and Brunel's broad gauge reached Bristol a year later. Portsmouth and Salisbury were reached by 1847. During

---

[171] The Somersetshire Coal Canal Society (n.d.) <http://www.coalcanal.org/> Accessed April 2020.
[172] UK Waterways Guide (2019) A brief history of the Kennet & Avon Canal < https://www.ukwaterwaysguide.co.uk/map/kennet-avon-canal/main-canal> Accessed April 2020.
[173] Canal and Rivers Trust (2017) An uphill struggle - Caen Hill's history. < https://canalrivertrust.org.uk/places-to-visit/caen-hill-locks/an-uphill-struggle-caen-hills-history > Accessed April 2020.
[174] Canal and Rivers Trust (n.d) Crofton <https://canalrivertrust.org.uk/places-to-visit/crofton> Accessed April 2020.

Figure 2.12. The network of the Great Western Railway in England and Wales in the 1930s. (https://commons.wikimedia.org/wiki/File:GWR_map.jpg)

this rapid expansion, the Salisbury line expanded to Gillingham, Crewkerne, Chard and Exeter in 1860, and Minehead in 1874, while the Somerset coalfield was linked to Bristol in 1873.[175]

Economically the impact on agriculture was dramatic. New markets opened because goods including milk, livestock and equipment could easily be moved. Indeed, the London market was opened to the West of England. The 'Cheese Country' of Wessex, providing dairy products, would especially benefit (see Chapter 5). Other industries were able to achieve national rather than local markets: textile production, coal, stone-quarrying, brewing, brick and lime making, and the manufacture of agricultural implements and steam engines. Swindon, a market town, was reborn as a railway town on the Great Western Railway with workshops operating from 1843.[176]

Figure 2.12 shows the network of the Great Western Railway in the 1930s.

---

[175] J.H. Bettey (1986) *Wessex from AD 1000*. Longman, New York, 238-243.
[176] J.H. Bettey (1986) *Wessex from AD 1000*. Longman, New York, 238-243.

Wessex was linked with London to the east, Wales to the west and parts of Devon and Cornwall to the southwest. During the inter-war period, the area of South Wessex and the Isle of Wight was served by the London and South Western Railway (1838 to 1922); after a re-grouping the Southern Railway served this area, although this new company still reached as far west as Plymouth.

Subsequent major events in the history of the railway system included its reorganisation into four companies in the 1920s, its nationalisation to create British Railways in 1948 and the closures following the 'The Reshaping of British Railways' report, published in 1963, when 2363 stations and 5,000 miles [8,000km] of track were earmarked for closure.

Closures of parts of the rail network, however, dated from before the Second World War, and the Report, controversial though it remains even to today, was an attempt to make the railway system economically viable in an age when road transport was in the ascendant.[177] One victim was the Salisbury and Dorset Junction Railway, a linking branch line between Salisbury and Bournemouth via West Moors and Wimborne.[178] While this has sadly disappeared, another victim of closure, the West Somerset Railway (1862 from Taunton to Watchet, extended to Minehead in 1874) was closed in 1971 and re-opened as a preserved railway from 1976. Today it runs between Bishops Lydeard and Minehead.[179] In a strange twist of fate, the preserved Swanage railway that links this town with Corfe Castle and Wareham may become re-connected with the national rail network. [180]

---

[177] Network Rail (2020) Dr Beeching's Axe <https://www.networkrail.co.uk/who-we-are/our-history/making-the-connection/dr-beechings-axe/ > Accessed April 2020.
[178] J. Speller (n.d.) Salisbury & Dorset Junction Railway <https://spellerweb.net/rhindex/UKRH/LSWR/SarumDor.html> Accessed April 2020.
[179] West Somerset Railway (2020) <https://www.west-somerset-railway.co.uk/railway/stations> Accessed April 2020.
[180] https://www.visit-dorset.com/listing/wareham-to-swanage-by-train/326941301/ accessed Dec 2023

Chapter 3

Environmental governance and change

> The law locks up the man or woman
> Who steals the goose off the common
> But leaves the greater villain loose
> Who steals the common from the goose.
>
> The law demands that we atone
> When we take things we do not own
> But leaves the lords and ladies fine
> Who takes things that are yours and mine.
>
> The poor and wretched don't escape
> If they conspire the law to break
> This must be so but they endure
> Those who conspire to make the law.
>
> The law locks up the man or woman
> Who steals the goose from off the common
> And geese will still a common lack
> Till they go and steal it back.
>
> A 17th century anonymous folk poem railing against the
> enclosure movement and the hypocrisy of the landed classes.[1]

3.1 Broad landscape types for Wessex

This chapter outlines key aspects of historic environmental management because institutional arrangements and organisational behaviours are key to understanding landscape change. Widespread conscious intervention sustained, changed and continues to change the Wessex landscape, often in dramatic ways, responding to economic and political forces affecting both form and function. Major examples include the manorial system, royal hunting forests, church land, enclosure, internal drainage boards and the rise of governmental involvement. These are all enormous topics and worthy of description. A brief overview of technological innovation in agriculture and land use change is included.

The structure of the medieval landscape of Wessex was not simple and underlies what may be seen today. The essential division is (a) upland regions, (b) woodland regions (sometimes 'ancient countryside') and (c) open field 'champion' regions (sometimes 'planned countryside'). Upland landscapes included infield-outfield systems, where the infield was located near to the settlement and was manured and cropped continuously. The much larger outfield was

---

[1] Anon (n.d.) *Stealing the Goose from the Commons.* <http://www.onthecommons.org/sites/default/files/celebrating-the-commons.pdf> Accessed April 2020.

subdivided into areas that were cropped at intervals and otherwise left allowing fertility to recover.

Woodland landscapes (areas that could yield considerable amounts of wood from hedges) were dominated by ancient hedges defining irregular 'closes' with small woods frequently resulting from assarts, which often date from prehistory, although medieval and later clearance is well-known. Assarting in its widest sense, is the clearance of waste (rough grazing), scrub or woodland for more productive agricultural activities.

Within the woodland landscape, there was often parkland grazed by animals for hunting, especially in deer parks maintained for elites. Open fields did exist within woodland manors but were largely enclosed by 1700. Heaths were common, as were ponds.

Champion landscapes were planned and are typified by Open Field Agriculture (OFA). They most likely date from the centuries pre-Norman Conquest but are unlikely to be Roman or early Anglo-Saxon. Where most hedges form regular and typically rectangular fields, they date from the historic period and are often post-medieval. Woods were rare, as were large heaths and there were few ponds.[2]

This tripartite division applied to the wider Wessex landscape includes legal concepts such as 'common land' that are found in all manors. Common land is land subject to rights enjoyed by one or more persons to take or use part of a piece of land or of the produce of a piece of land which is owned by someone else – these rights are referred to as 'rights of common'. Those entitled to exercise such rights were called commoners.[3]

Upland landscapes, common throughout northern and western Britain, are really only represented in west Somerset, typically by Exmoor. Woodland regions dominate much of Somerset, Hampshire and Berkshire. In between these counties, much of Dorset and Gloucestershire and virtually all of Wiltshire were originally champion landscapes that become enclosed in post medieval times.[4] Similar divisions are found elsewhere in Europe. The enclosed wooded *bocage* country of Normandy and Brittany, and what is termed *boulonnais* inland from Boulogne, comprise small woods and pasture with winding trackways. This contrasts with the open *champagne* country of France.[5]

The term 'governance' is applicable (for example) to manors and hunting forests refers to 'how things get done'.[6] Wherever there is some form of societal organisation, governance is present either implicitly (because people operate together) or explicitly, generally in the

---

[2]   O. Rackham (1987) *The history of the Countryside*. J.M. Dent, London, Chapter 1. These definitions exclude common land, which was generally owned by the lord of the manor, although managed for the benefit of tenants with rights of common.

[3]   National Archives (n.d.) Land ownership, use and rights: common lands <https://www.nationalarchives.gov.uk/help-with-your-research/research-guides/common-lands/> Accessed April 2020.

[4]   T. Williamson and L. Bellemy (1987) *Property and Landscape : A Social History of Land Ownership and the English Countryside*. George Philip, London, Chapter 1.

[5]   O. Rackham (1987) *The history of the Countryside*. J.M. Dent, London, 181.

[6]   The origins of 'governance' lies in ancient Greek and relates to the process of 'steering'. Modern conceptions date from the old French 'gouvernance' and subsequently (in English) the term has been taken to mean a process of steering society (D. Benson 2020 pers. comm.).

form of institutions. Governance is the outcome of hierarchies (linked to government at some scale), networks (people with common interests), markets and communities.[7]

3.2 To the manor born

The term 'manor' is derived from the Latin *manerium* (a residence). The institution was the dominant form of tenure in the medieval period, continuing into later times. Originally 'manor' referred to a territorial unit held by an overlord. Administration was concerned with land, with rights and with civil administration, although this changed over time. Neither are manorial boundaries co-incident with other boundaries, particularly ecclesiastical, meaning there may be several manors in an ecclesiastical parish, or indeed the other way around. A manorial holding may also be fragmented, not comprising a single block of land.[8] Other jurisdictions, for example Forest Law, could overlie the manor and hence the English Common Law.[9] 'Manorialisation' was in an advanced state in Wessex before 1066. The manor as an institution should not be confused with 'feudalism',[10] one reason being that the manorial

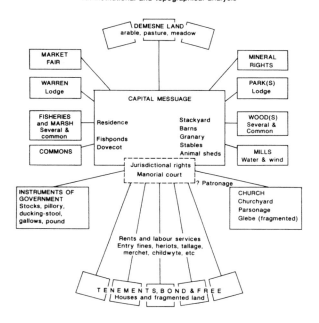

Figure 3.1. Institutional and topographical analysis of the medieval manor (in M. Bailey, 2002. *The English Manor, c.1200-c.1500*, Manchester University Press, Manchester, p.xiv)

---

[7]   In modern politics, governance is often associated with changes in state-society relations and the shift from centralised government to hierarchies, networks, markets and communities. See J. Pierre and B.G. Peters (2000) *Governance, Politics and the State*. Macmillan, Basingstoke, Chapter 1.
[8]   Victoria County History (2018) The Manor <https://www.victoriacountyhistory.ac.uk/explore/themes/government-and-administration/manor> Accessed November 2018.
[9]   H. Cook (2018) *New Forest: the forging of a landscape*. Windgather Press, Oxford, 77.
[10]   H. Cook (2018) *New Forest: the forging of a landscape*. Windgather Press, Oxford, Chapter 3.

system outlived any form of feudal organisation or perceived practice by many centuries in England. The manor as an organisation was essentially and directly economic.[11]

An institutional and topographical analysis of the medieval manor is displayed in Figure 3.1. The 'capital messuage' was an enclosed area containing the seigneurial (lord of the manor's) residential complex, including mostly functional elements. Both 'demesne' (the lord's own arable land) and that of the peasants were typically distributed in unenclosed parcels of land throughout an open field system, generally as strips organised in large fields. Other points of interest in Figure 3.1 include the 'glebe' land which supported the parish priest, the institutions of government administered locally via the manorial courts (see below), the conditions of the tenants (bonded and free) and the 'capital messuage' (main dwelling with its outbuildings).

The customs of the manor were the set of rules by which each manor was governed and administered, for example the regulations governing the inheritance of land holdings. Land in a typical manor was of three kinds:

- The lord's 'demesne' land that was directly under his control for the benefit of his household
- That of dependent villeins or serfs who paid for it through labour obligations or cash rent
- That leased to free tenants who did not usually owe labour obligations and were free to inherit property, but who nonetheless were still subject to the customs of the manor and hence fell under its court's jurisdiction[12]

The manorial system ruled the land until the beginning of the 17th century. Although its origins were feudal, it outlived that system. It may also be viewed as a kind of early 'environmental agency' because it regulated tenurial conditions, marshalled natural resources and organised communal agriculture (especially in manors comprised of nucleated villages with open field agriculture-based manors). Indeed, the English manor has been described as a model for sustainable agriculture, trading high productivity for sustainability, social stability, and a diversity of livelihood strategies.[13]

The manor has been likened to, or even proposed to be derived from, the Roman villa system. Villas represented a clear central place for the administration of agriculture and other economic activities under the control of a wealthy individual, generally a Romanised Briton. One view of manorialisation certainly suggests that it developed from the late Roman period when small farmers and landless labourers sought protection at a time of instability caused by invaders.[14] However, in England there is no evidence for Roman villas as centres for the administration of justice and no clear link through to the early medieval period. The villa is generally presumed to be the focus of a country estate 'geared largely to the production of

---

[11] A. Jones (1972) The rise and fall of the nmanorial system, a critical comment. *Journal of Economic History* 32 (4), 938-944.

[12] M. Bailey (2002). *The English Manor, c.1250-c.1500*. Manchester University Press, Manchester, 26.

[13] J.N. Pretty (1990) Sustainable development in the middle ages: the English Manor *Agricultural History Review* 30, 1-19.

[14] P. Sarris (2004) The Origins of the Manorial Economy: New Insights from Late Antiquity, *English Historical Review* 119, 279–311.

an agricultural surplus'.[15] Roman villas tended to be built within 10 miles (16km) of a major urban centre.[16] Like major urban centres of the time, their known distribution is dense across 'lowland' southern England.[17] Here, we may presume, lay the markets for produce, suggesting villa development was an economic response to mature local markets. By contrast the distribution of Roman villas was sparse across modern Devon and Cornwall.

In England specifically, the manorial system seems to have emerged in an identifiable form during the 8th and 9th centuries, somewhat later than the Roman period. It likely arose out of chiefdoms held by tribal leaders, as large land areas were handed over to larger churches or to nobles, giving rise to military and ecclesiastical aristocracies.[18] Continuity from the Romano-British period has never been proved, but the notion of a (re-) introduction of continental ideas during the Anglo-Saxon period (for example under Caroliginian influence), including feudalism is popular among historians. Unfortunately, the term 'feudalism' remains contested as it includes both elements of military obligation (ultimately to the king), production of agricultural goods through labour services from the peasantry and the successful operation of the 'manorial system'.[19]

Manorial organisation and function were suited to the related feudalism (ideally when no cash transactions regarding land occurred) but like Open Field Agriculture, it was not of itself feudal, for both survived intact into the age of capitalist farming. The manor had no connection with the military and political concept of the fief, an estate of land, especially one held on condition of feudal service that came to be termed a 'fee'.

In the feudalism, the king 'owned' the entire kingdom. Below him land was 'held', as ownership of land in the modern sense was unknown. Those who held land directly from the king were termed 'tenants-in-chief'. The three offices permitted to do so under the feudal conventions were barons, bishops and abbots (or priors) making the church integral to feudal governance. Barons could be lords of the manor, but frequently beneath them were the lower order lords of the manors (the 'gentry') and small towns who could be knights, but generally not aristocrats.

Royal manors were held directly by the King; Lyndhurst in the New Forest, Britford in Wiltshire, Cheddar and Crewkerne in Somerset, were all Royal manors before 1066. Overall, the system worked in terms of obligations, service or labour and in its purest form. 'feudalism' operated in England between approximately AD900 and AD1200. Holdings down to the manorial system worked essentially in terms of military service owed upwards to the tenants-in-chief, and the king. Another, and interrelated, aspect of 'feudalism' relates to labour obligations on the peasantry.

Labour service was owed further down the ladder by the peasantry who were obliged to offer work in kind or produce to the lord of the manor. Cash rents become more common at the manorial level, although development was never entirely linear and other obligations

---

[15] B. Cunliffe (1993) *Wessex to AD 1000*. Longman, New York, 256.
[16] D. Ross (n..d.) Roman Villas in England. <https://www.britainexpress.com/architecture/roman-villas.htm> Accessed April 2020.
[17] M. Allen *et al.* (2016) The Rural Settlement of Roman Britain: an online resource. <http://archaeologydataservice.ac.uk/archives/view/romangl/map> Accessed April 2020.
[18] M. Bailey (2002) *The English Manor, c.1250-c.1500*. Manchester University Press, Manchester, 11.
[19] P. Halsall (2001) <https://sourcebooks.fordham.edu/sbook1i.asp#Feudalism> Accessed June 2021.

remained in many cases. A significant development was that of 'Bastard Feudalism', which evolved from about AD1200. Here, a lord would reward his retainers among the gentry for their service with a money fief (payment in cash) instead of land, that is regular money payment in return for service, this time excluding any aspects of direct land holding. In essence, this change did not function at the level of tenants-in-chief who were not retained by the king – their loyalty was expected to be a given – but at the next rung down, that is magnates who retained the lesser gentry. The term 'bastard feudalism' therefore means that land is no longer a part of feudal transaction associated with military obligations.[20]

At the bottom of the Anglo-Saxon social hierarchy were the peasants, the thane was lord of the manor and between him and the King was an ealderman (in effect a sheriff). Ealdermen were appointees of the King, and were originally mostly from powerful families. The ealderman was in effect the 'viceroy' within a shire with both legal and administrative powers, as well as major military responsibilities to the King[21]. Feudalism in England likely followed the influence of Merovingian and Carolingian rule on the Continent. Progressive influence is evident within the economy, society and landscape of Anglo-Saxon England. Only after the Norman Conquest, were noblemen holding land directly from the king termed 'tenants-in-chief'.

The Continental model, one of fiefdoms created to ensure military support for the royal houses from barons, suited the need for military protection of the day and was to prove effective against invasion from such as (ironically) the Normans and also the Magyars in central Europe.[22] During the reign of Offa, King of Mercia (reigned 757-796), the influence of Carolingian political thought was being felt in England,[23] so much so that it has been stated 'Forest law was probably of Carolingian origin'.[24] Later, Alfred the Great of Wessex (reigned 871-899), through his struggle against the Danes, achieved a strong centralised authority. He reformed the currency, legal and military structures in Wessex and created the fortified 'burhs' as well as a new fleet of warships.

Alfred's organisational system (continued by his daughter Æthelflæd c. 870–918) would have made the imposition of feudalism easier.

It is also likely that the Normans presided over a more rigorous classification of landholding, including establishing that most land was held heritably in return for military service; that is land 'held *in fee*'. Indeed, late Anglo-Saxon England was an efficient nation state, no doubt ripe for conquest and something the succeeding Normans were to build on, for they retained many state functions, including the well-established legal system.[25]

The 'Domesday' Survey of 1086 purported to list every manor including reference to lordship, freeholders, smallholders, and tenants. Entries included areas of pasture, meadow ploughed

[20] Oxford Reference (2020) Bastard feudalism <https://www.oxfordreference.com/view/10.1093/oi/authority.20110803095450812> Accessed June 2021.
[21] Britannia online (2018) Ealderman <https://www.britannica.com/topic/ealderman> Accessed April 2020.
[22] M.M. Postan (1972) *The Medieval Economy and Society*. Weidenfeld and Nicholson, London, 85, 90.
[23] J. Story (2003) *Carolingian connections: Anglo-saxon England and Caroligian Francia 750-850*. Ashgate Publishing, Farnham, Chapter 5.
[24] C.A. Baker (2001) *The Compendium to British History*, Routledge, Abingdon, 529.
[25] J. Campbell (2000) *The Anglo-Saxon State*. Hambledon, London, 10.

land, woodland, livestock (pigs, sheep. cattle), fishing rights, eels, and mills. The Domesday Commissioners asked and recorded answers to the following ten questions:

- What is the manor called?
- Who held it in the time of King Edward (in 1066)?
- Who holds it now (in 1086)?
- How many hides[26] are there in the manor?
- How many plough teams on the demesne (that is the local lord's own land) and among the men in the rest of the village?
- How many freeman, sokemen, villans, cottagers and slaves?
- How much woodland, meadow, pasture, mills and fisheries?
- How much has been added to or taken away from the manor?
- How much was the whole worth (1066) and how much now (1086)?
- How much had or has each freeman and each sokeman?[27]

At Domesday, Wessex has been described as 'an area of large consolidated manors with many villages belonging to a single lord, and with a peasantry bound to labour service' One example being the Bishop of Winchester who, together with the monks of Winchester, held an enormous estate representing the ancient patrimony of 'the most important church in the capital city of Wessex'[28]. There were fewer free peasants here than in the east of England and more serfs, maybe reflecting a downgrading of the peasantry due to the Norman conquest.[29]

The medieval manor was essentially an administrative unit of a landed estate in the days before there was any notion of local government. The lord of the manor had customary and legal rights over the manorial land and its resources. These rights included markets and fairs, warrens, fisheries, commons, mineral rights, parks, woodlands, water mills and windmills (Figure 3.1). Windmills were introduced in England around the 12th century; water power was much older.[30] These rights were exercised through a manorial court, although during the Tudor period, many of the civil functions were transferred to the parish.

By contrast, ecclesiastical institutions held some land '*in alms*' that is for religious service. Much later continued manorial strength would be demonstrated by the important role played by their lords in the introduction of water meadows in the early decades of the 17th century.[31]

From about 1600, the lords of the manor lost political power, but they retained land (the 'demesne'), sporting and mineral rights, and the revenue from copyhold properties, particularly quitrents[32], entry fines (payments to the lord of the manor when receiving a copyhold tenancy) and court revenues. The main function of the manor *court* after this

---

[26]   A hide is the standard unit of assessment used for tax purposes. It was meant to represent the amount of land that could support a household, c. 120 acres [c.50 ha]. There were four virgates to every hide.

[27]   National Archives (n.d.) Domesday: Britain's finest treasure <https://www.nationalarchives.gov.uk/domesday/> Accessed April 2020.

[28]   J.H. Bettey (1986) *Wessex from AD 1000*. Longman, New York, 19.

[29]   J.H. Bettey (1986) *Wessex from AD 1000*. Longman, New York, 20.

[30]   M. Berry (2018) History of Windmills <http://www.windmillworld.com/windmills/history.htm> Accessed November 2018.

[31]   J.H. Bettey (1986) *Wessex from AD 1000*. Longman, New York, Chapter 4.

[32]   A tax imposed on occupants of freehold or leased land in lieu of services to a higher landowning authority

time was to register title to copyhold land[33]. The decline of the manorial system forced the Elizabethan central government to give civil responsibilities to the townships and parishes (specifically care for the poor, highways, law and order). [34]

The lord of the manor had the right to hold courts, presided over by their steward if not by the lord in person; the records were kept in manorial court rolls by the parish clerk. There were two kinds of manorial court, the *Court Baron* and the *Court Leet*. The Court Baron enforced payment and services due to the lord, while the Court Leet dealt with the administration of the communal agriculture, keeping law and order, and the customs of the manor.[35]

Although a 'one size fits all' model for the English manor existed nowhere, its governance structure was powerful in the lives of ordinary people and the lesser gentry. The 'manor' was, in theory, a kind of legal, administrative and economic expression (the estate). On the other hand, the medieval 'vill', as an administrative unit under feudalism, that governed by a leet court, while usually synonymous with the manor, might in practice include multiple manors. In the middle ages, the 'village' was a term for a settlement larger than a mere 'hamlet' but was not essentially an administrative unit as was the 'vill';[36] in the later Middle Ages 'manor' had a looser meaning, that was any economic unit of land which could consist of all demesne land without tenants. The existence of the manor did not exclude other, more informal channels of expression and communication among the people, including guilds: work and religious-based associations which helped to regulate village life.[37] These guilds were distinct from the urban guilds that represented networks around crafts and merchants controlled key factors affecting markets, apprenticeships, training and prices.[38]

Relationships between the manor and what occurred on the ground were complicated and changed over time. For example, earlier fields, some prehistoric, could be more irregular in shape having been won as 'assarts' (newly enclosed arable land) from woodland, waste (otherwise uncultivated land) and from scrub, a process commonplace into the medieval period. While, on archaeological evidence, large prehistoric holdings (sometimes even described as 'estates') are believed to have been present in the Bronze Age, nothing is known about prehistoric governance. Where in the middle ages there were open fields, OFA is seen, at least originally, as a product of feudal relations, although it outlived feudalism by centuries. Actual direct evidence for OFA in Domesday Book is not forthcoming, although some earlier 10th century charters may contain clues.[39]

The classic OFA system comprised two or three large fields. The crop rotation in three fields was typically a) pease (pea), lentils or beans, (i.e. a legume that could fix nitrogen), b) a cereal

---

[33] Copyhold tenure was a form of customary tenure of land from the Middle Ages when the land was held according to the custom of the manor, and the "title deed" kept by the tenant was a copy of the relevant entry in the manorial court roll.

[34] Victoria County History (2018) The Manor <https://www.victoriacountyhistory.ac.uk/explore/themes/government-and-administration/manor> Accessed November 2018.

[35] J. Langton and G. Jones (n.d.) *Forests and Chases in England and Wales*, c. 1000 to c. 1850, St John's College, Oxford <http://info.sjc.ox.ac.uk/forests/glossary.htm> Accessed April 2020.

[36] M. Bailey (2002) *The English Manor, c.1250-c.1500*. Manchester University Press, Manchester, 7.

[37] M.M. Postan (1972) *The Medieval Economy and Society*. Weidenfeld and Nicholson, London, 124

[38] Encyclopaedia Britannica (n.d.) Guild <https://www.britannica.com/topic/guild-trade-association> Accessed April 2020.

[39] O. Rackham (1987) *The history of the Countryside*. J.M. Dent, London, 172-176

(typically wheat, barley or oats although oats, barley and beans could be planted on the same strip), and c) fallow. Fallow was introduced in order to rest the land, while animals grazed on the weeds and added dung to build up soil organic matter. In the simpler two field system, the legume planting was absent, hence the system would be less productive long-term. In Somerset, an evolution between two and four field rotations has been described from the post-medieval period before enclosure.[40] At Fifield Bavant in Wiltshire, late 18th century five-course rotation was associated with enclosure.[41]

The OFA pattern of fields dominated the midlands; in Wessex it predominated in the central area running from the Dorset coast approximately north-eastwards. In Dorset, surviving evidence of OFA systems is commonplace, especially on the Isle of Portland.[42] Across the chalklands of Wessex, the manorial system remained strong with open fields and nucleated villages that served to organise and control the agrarian population. Wheat and barley production dominated into the post-medieval period and fertiliser was generally provided by manure from sheep close-folded overnight on the arable land.[43]

The creation of new agricultural land, or more intensively managed agricultural land from waste, would also be under the auspices of the manor, although 'squatter' (illegal) encroachments were known from time to time and especially in the 17th century, a time of gradually increasing population. In Wiltshire and Dorset, manorial records tell of encroachment on road verges, wastes and commons by flimsy dwellings. Petitions to quarter sessions and other accounts pertain to individuals wishing to build a simple dwelling but lacking a 'statutory' four acres of land. There was a real housing crisis affecting the poor.[44] Indeed, what might have been quietly ignored or even condoned by the manor (who after all required labour) was turned into a conscious radical movement by the 'Diggers' who, in 1649-50, started to cultivate certain commons in England, inspired by their leader Gerrard Winstanley. Their communities were broken up and legal action taken.[45]

The manorial system continued operationally into the 17th century in Wessex, as elsewhere. There were also technical (rather than tenurial) developments in agriculture including the introduction of irrigated ('floated') water meadows, enclosures, root crops and new varieties of sheep and cereals. These developments increased into the 18th century, with sainfoin (both fodder crop and legume), clover ryegrass and the introduction of industrial crops (woad, teazels, flax and hemp) as well as experiments in soil improvement. Daniel Defoe, writing in the early18th century, reported increasing areas of arable on former sheep-walk downland between Winchester and Salisbury. While some enclosure was by mutual agreement between tenants, later on an Act of Parliament was passed. The process was later spurred by high corn process during the Napoleonic war period (1793-1815) and the outcome was an expansion

---

[40]   Encyclopaedia Britannia (2020) Three-field system <https://www.britannica.com/topic/three-field-system> Accessed April 2020; M. Aston (1988) Land Use and Field Systems in M. Aston (ed.) *Aspects of the Medieval Landscape of Somerset*. Somerset County Council, Taunton, Chapter 5.

[41]   J. Freeman and J. H. Stevenson (1987) Parishes: Fifield Bavant, in D.A. Crowley (ed.) *A History of the County of Wiltshire Volume 13, South-West Wiltshire: Chalke and Dunworth Hundreds*. Victoria County History, London, 60-66.

[42]   J. H. Bettey (1986) *Wessex from AD 1000*. Longman, New York, 35-36. Portland is joined to the mainland by the shingle bar of Chesil Beach. Such a feature is technically a 'tombolo'. See also ch 9, this volume.

[43]   J. H. Bettey (1986) *Wessex from AD 1000*. Longman, New York, 121-122

[44]   J. H. Bettey (1982) Seventeenth Century Squatters› Dwellings: Some Documentary Evidence, *Vernacular Architecture*, 13 (1), 28-30, DOI: 10.1179/vea.1982.13.1.28

[45]   E. Vallance (2009) *A Radical History of Britain*. Abacus books, London, Chapter 7.

of arable acreage supported by the folding of sheep on the corn fields. Flock size was in turn increased through use of water meadows and feeding the sheep on new fodder crops.[46]

By the early 20th century, the manorial system was in terminal decline. There was a clear movement to establishing freehold in its place. Certain manorial rights, however, were retained by lords of the manor in England and Wales when land became freehold in the early 20th century.[47] This remained (and remains) a bone of contention. For example, upon transfer of tenancy 1892, the new tenant discovered (to his irritation):

> The landlord excepts and reserves to himself all trees, saplings, spires, pollards, bushes, gorse wood and under wood (not hedge wood) whatsoever and all minerals, coprolites, stone, flints, clay, marl, chalk, gravel and sand and all manorial rights and royalties whatsoever with power for himself and all persons authorised by him with or without horses and carriages to enter the premises at all time to cut down, dig for, remove and carry away the same and also enter for the purposes of viewing, and of executing the repairs and ascertaining the state and condition of the premises.

Additional to the privileges noted above, the landlord also had claim to game and other livestock and access for himself, the gamekeeper and others concerned with sporting.[48]

3.3 Royal forests

Kings and nobles certainly enjoyed hunting in Wessex before the arrival of the Normans in 1066. The creation of royal forests (sometimes 'royal hunting forests') in Wessex was extensive (Figure 3.2), providing recreation and venison (a high-status meat) for aristocrats. It also affected the lives of many people of lower status. The Latin word *foris* essentially means (in this context) 'outside', that is outside the English Common Law, so providing an alternative legal system. Apart from hunting, there was a real economy. Management of trees, including pollarding and coppicing, provided for building and fencing materials. Depending on the geology beneath, 'afforested' areas could also include fishing rights, quarrying, clay extraction, pottery making, and milling. The local population held rights as approved by the Crown such as grazing, pigs, poultry, rights to 'wood pasture' and fuel gathering. It was not until the Charter of the Forests (1217) that commoners' rights were more clearly defined.[49]

Domesday Book for Hampshire (1086) includes a listing of manors which, after an assessment of their estates, comments upon their relationship with the forests. Lyndhurst in Hampshire, for example, was a royal manor in the tenth century and remined that way in 1086.[50] Other areas were taken into the forest, although parts of holdings could remain without, and exceptions were also recorded.[51]

---

[46] J.H. Bettey (1986) *Wessex from AD 1000*. Longman, New York, Chapter 6.

[47] House of Commons Justice Committee (2015) Manorial Rights <https://publications.parliament.uk/pa/cm201415/cmselect/cmjust/657/657.pdf> Accessed April 2020

[48] E. Rice (2008) *Roses and Shamrocks*. Five Crowns publishing, Gillingham, 4-5.

[49] H. Cook (2018) *New Forest: the forging of a landscape*. Windgather Press, Oxford, Chapter 5; K. Barker (2020) pers. comm.

[50] The Real New Forest guide (2020) <http://www.thenewforestguide.co.uk/forest-villages/lyndhurst-royal-links/> Accessed April 2020.

[51] J. Morris ed (1982). *Domesday Book: Hampshire*. Phillimore, London, 54.

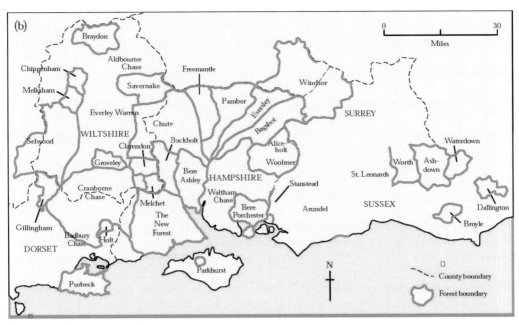

Figure 3.2. Forests and chases in central southern England, late 16th Century (J. Langton and G. Jones, Forests and Chases in England and Wales, c. 1000 to c. 1850, St John's College, Oxford http://info.sjc.ox.ac.uk/forests/glossary.htm)

In Wiltshire alone, for example, by the 13th century there were forests along the southeast corner of the county including Savernake, Chute, Clarendon and Melchet on the Hampshire border, while the New Forest is largely all within that county. Large areas of Dorset were also under 'forest law' and by the 12th century there were forests at Powerstock, Blackmore Vale, Gillingham, Cranborne Chase and on the Isle of Purbeck. Somerset had forests at Neroche (Blackdown hills), Quantock, Mendip, Selwood and Exmoor in the west of the county.[52] The list goes on!

Norman kings had the right to hunt in forests, while nobles generally had the right to hunt in 'chases' where there were similar rights over a tract of land.

Parks and free warrens,[53] functioned similarly, generally held by many lords of the manor outside the bounds, or with permission of the Crown. 'Forest' was not a land-use term: it was legal. Such areas exhibited a range of land covers including woods, scrub, heath, grassland and wetland. There are visible remains of deer park pales, coppice banks, lodges, settlement pattern and buildings. However, Domesday Book supplies the first consistent documentary evidence that points to ownership and management.[54]

---

[52] J.H. Bettey (1986) *Wessex from AD 1000*. Longman, New York, 15.
[53] 'Free warren' was a royal franchise granted to a manorial lord allowing the holder to hunt small game (rabbit, hare, pheasant and partridge) within a designated vill <https://www.manorialcounselltd.co.uk/tag/free-warren/>
[54] S. Davies, L. Walker and L. Coleman (1998) *The New Forest Historical Landscape*. Wessex Archaeology, Salisbury, 20.

The historic use of the term 'forest' was therefore different from the modern understanding which refers to a close stand of trees. To the Normans it was a legal concept applied to an area where specific laws pertained, and these laws reflected the pre-occupations of the Crown. A Royal forest was:

> the hunting preserve of the king or lord-marcher, subject to forest law but not necessarily woodland; originally an area of land in which only the owner had the right to hunt deer and boar. Special laws were applied in this area which was outside the jurisdiction of common law.[55]

Crimes under Forest Law were essentially of two kinds: those of 'vert', infringements against the vegetation, and those of venison, affecting or poaching or otherwise adversely affecting the beasts. The ordinary manorial court process continued within the New Forest, with forest laws supplementing the common law, rather than replacing it.[56] The first surviving legal texts relating to the forest laws that date from 1100, at the very start of the reign of Henry I (reigned 1100-1135).[57]

Offences included clearing of land (illegal assarts), cutting of wood, burning, hunting, carrying of bows and spears in the forest, and loosing of livestock; there were also rules relating to de-clawing of dogs and the discovery of hide or flesh in the forest. Subsequent monarchs, especially Henry II (reigned 1154– 1189) were most enthusiastic about their implementation. Henry in part codified the existing laws under the Assize of the Forest (1184) and he introduced a system of courts, the highest being called the Forest Eyre.[58]

Forest laws were restricted to specific areas and were fundamentally about securing preservation of game for the Crown.[59] No original draft of the laws has survived from Norman times,[60] and, although scholars argue about dating and the provenance of material included within, the earliest known version of the Forest Laws probably relates to the Assize of the Forest (or Woodstock) from 1184, and is thus thought to be from the reign on Henry II.[61] The 1166 Assize of Clarendon was concerned with producing an evidentiary legal system, and hence is significant in the development of English Law. The Assize of the Forest appears to summarise previous laws. Punishments included blinding and mutilation, but it is debatable as to whether these were ever followed through, the main objective now being to raise revenue for the Crown.

The first article states[62]: 'First he forbids that that anyone shall transgress against him in regard to his hunting rights or his forests in any respect; and he wills that no trust shall be put in the fact that hitherto he has had mercy for the sake of their chattels upon those who have

[55] S. Davies, L. Walker and L. Coleman (1998) *The New Forest Historical Landscape*. Wessex Archaeology, Salisbury, 20.

[56] P.D.A Harvey (1979) in D.J. Stagg (ed.) *A Calendar of New Forest Documents 1244-1334*. Hampshire Record Series III, Hampshire County Council, Winchester, ix.

[57] Mercian Archaeological services CIS Forest Laws (2003) <http://www.mercian-as.co.uk/forest_law.html> Accessed April 2020.

[58] H. Cook (2018) *New Forest: the forging of a landscape*. Windgather Press, Oxford, Chapter 5.

[59] E.M. Yates (1985). *The Landscape of the New Forest in Medieval Times*. Dept. of Geography, Kings College, London, 6.

[60] P.J. Neville Havins (1976). *The Forests of England*. Robert Hale, London, 24-26.

[61] D.J. Stagg (1979) *A Calendar of New Forest Documents 1244-1334*. Hampshire Record Series III, Hampshire County Council, Winchester, 7.

[62] D.C. Douglas and G.W. Greenaway (eds) (1981) *English Historical Documents* (2nd ed.). David Charles, London, 451-3.

offended against him in regard to his hunting rights and his forests. For if anyone shall offend against him hereafter and be convicted thereof, he wills that full justice be exacted from the offender as was done in the time of King Henry, his grandfather'

Crown interests are tied up with interventions in management, governance, and economy. In the New Forest, for the period 1105 to 1254 there was great need to supply the Royal Court.[63] Although documentary information is limited, venison was regularly sent to London and elsewhere for Christmas feasts. Locally, Clarendon Palace could supply its needs from its own forest.[64]

There were to be no bows, arrows, hounds or harriers in his forests, except by license from the king or other duly authorised person. No owner of a wood within the forest could sell or give away anything out of the wood to its wasting or destruction, beyond his own needs. Demesne woods were protected by the penalty of an offender 'answering with his own body' and not payment of a fine.[65]

The New Forest provided four specific functions in the middle ages: hunting for the King free from interference, a supply of meat, revenues for the exchequer and, from the 14th century at least, income from the sale of timber and underwood products from 'coppices' within the perambulation.[66] However, it should be noted that 'hunting lodges' were not all 'royal' in purpose, being used by the lower orders![67]

Exploitation of the underwood was principally for domestic use and supplying small-scale wood crafts. Following enclosure, wooded areas were leased to an individual. After 7-9 years they could again be thrown open for grazing. While wood was cut and removed, the action of creating compartments (*coupes*) and thereby allowing cyclical cutting followed by re-growth from existing coppice stools is unknown[68].

Commoners with grazing rights in and around forests were potentially in conflict with both Crown interests protecting the vert, and later with exchequer/state interests where broadleaved planting became important, generally from the 17th century. Rights to graze animals in areas that would become royal forest originated in pre-Conquest times and were affirmed in the Charter of the Forest (1217).[69]

In 1609, John Norden undertook a survey of New Forest coppices. This new and innovative level of auditing and presumed intervention would support the claim that the forest law was de facto in decline by the 16th century.[70] Coppices were let to woodcutters who cut the underwood and pollards but not saplings, for Norden expressed his concern that these be left to grow to tall trees. However, underwood management and coppicing were not a success and

---

[63] R.P. Reeves (2010) The administration of large holdings in the New Forest 1130-c1430. Unpublished MA thesis, University of Winchester, 61.
[64] R.P. Reeves (2013) pers. comm.
[65] P.J. Neville Havins (1976) *The Forests of England*. Robert Hale, London, 24-26.
[66] P. Roberts (2002) *Minstead: Life in a New Forest Community*. Nova Foresta Publishing, Ashurst.
[67] P. Everill and D. Ashby (2019) Excavation and Survey at Church Place, Denny Wait, in the New Forest in 2016/17, *Hampshire Studies* 74, 115-136.
[68] R. Reeves (2013) pers. comm.
[69] H. Cook (2018) *New Forest: the forging of a landscape*. Windgather Press, Oxford, Chapter 5.
[70] C.R. Tubbs (2001) *New Forest, the history, ecology and conservation*. New Forest Ninth Centenary Trust, Lyndhurst, 81.

there was a later attempt to make new pollards illegal. The New Forest and other royal forests were to become a source of timber, largely for the navy and especially from the late 17th century. This would be driven by the Exchequer rather than the Crown nationally; only the New Forest and Forest of Deane remain royal forests technically under forest law, and the New Forest still has a system of administration, albeit reformed, that owes its origins to medieval times. The 'Verderer's Court' still meets regularly at Lyndhurst.[71]

The New Forest authorities were always concerned with protection of the 'vert' because it provided cover for venison. However, the interest in coppicing from the time of Elizabeth I was clearly an attempt to generate wood products. The switch of function was slowly realised. For example, an offence against the venison is recorded from 1488[72] and timber for the navy had been gathered long before, in 1416/17[73] for Henry V's flagship *Grace Dieu*. Tudor monarchs were, like Henry II, concerned with managing resources in order to increase revenue for the Exchequer and in this context we see the rise of notions of environmental auditing. The Office of General Surveyor was established in 1512, early in the reign of Henry VIII .[74] Subsequently there were concerns over the sale of wood, with the establishment of the Court of General Surveyors in 1547. These developments, almost exactly spanning Henry's reign, were to pave the way for a *divisum imperium* where the Exchequer would muscle in on the ancient system of governance inherited from Norman times.

Problems inevitably emerged from this divided jurisdiction and the post was created of Surveyor General of Woods, Forests, Parks and Chases, along with Deputy Surveyors, until the abolition of the post of Lord Warden (who was in charge of Crown interests in the forests) in the 19th century. The inevitable clash in governance between anciently defined officers and emerging state interest was to handicap the running of the New Forest for centuries; the new officials appointed by the Exchequer were much resented. There follows an account of the historic Forest administration that came under the control of the Lord Warden, appointed by the Crown.[75]

Naturally, the role of Crown officials varied from time to time and from royal forest to royal forest. However, an early account of officers of a royal forest (from the Staffordshire Forest Pleas, 1262) concerning Cannock Chase (actually a royal forest) provides an indicative list.[76] The complexity of management of the land covers and enforcement of laws within a forest may be illustrated by the range and roles of officials. All the officers of the forest listed below had the power of arresting malefactors and they comprised:

1.   The Steward or Chief Forester was of great importance and was ultimately in charge.

[71]   H. Cook (2018) *New Forest: the forging of a landscape.* Windgather Press, Oxford, Chapter 6.

[72]   R. Reeves (2008) *To Enquire and Conspire, New Forest Documents 1533-1615.* New Forest Record Series volume V. New Forest Ninth Century Trust, Lyndhurst, xxxiii.

[73]   R. Reeves (2008) *To Enquire and Conspire, New Forest Documents 1533-1615.* New Forest Record Series volume V. New Forest Ninth Century Trust, Lyndhurst, xxiv.

[74]   R. Reeves, (2008) *To Enquire and Conspire, New Forest Documents 1533-1615.* New Forest Record Series volume V. New Forest Ninth Century Trust, Lyndhurst, xviii.

[75]   R. Reeves R. (2008). *To Enquire and Conspire, New Forest Documents 1533-1615.* New Forest Record Series volume V. New Forest Ninth Century Trust, Lyndhurst, xxxviii.

[76]   G. Wrottesley (1884) Staffordshire Forest Pleas: Introduction, in G. Wrottesley (ed.) *Staffordshire Historical Collections, volume 5, pt. 1.* Staffordshire Record Society, Stafford, 123-135.

2. The foresters were actually a diverse group: some held the office by heredity, others were probably appointed by the king. These were also called 'ryders' or 'rangers.' The 'riding forester' liaised between the more senior foresters-in-fee and foot foresters who oversaw prevention of timber thefts.

3. The office of Verderer was a judicial officer of the forest chosen by the Crown, later by the freeholders of the county. Verderers enforced forest law to protect the vert and venison among other duties.

4. The reguarders (or regarders) were officers of the forest whose duties were to keep a roll on which was written all the ancient assarts, wastes, and purprestures[77]. Regarder surveys were typically undertaken every three years.

5. The agisters, in effect regulated grazing, preventing animal trespass and receiving and accounting for the money paid for herbage and pannage on the King's demesne lands and woods. This was paid for grazing of the herbage of the woods and pastures, and the mast (acorns) of trees, the latter was called pannage.

6. The Preservators were a post-medieval addition. Around the turn of the 16th century there was evident concern for the maintenance of forests, due to both their physical state and bad practice that was attributed to corruption. These are precursors to John Evelyn's post-Restoration interest in 'silviculture'.[78]

7. The woodward (wodewardus) looked after the woods and vert, although the origins of the post are unclear. In the medieval period he was a private forester looking after the King's rights in *private woodlands*, however the relationship with other officers was complicated and changed with time.

John Manwood (d. 1610), a contemporary of John Norden (c.1547–1625), was aptly named, for his book 'A Treatise on the Law of the Forests' was published in 1598 and ran to several editions.[79] A barrister by profession, Manwood had been Justice in Eyre of the New Forest under Elizabeth I, and he provides valuable information on medieval forest laws, offices, beasts and customs.[80] The function of the Chief Justice in Eyre was basically to fix the fines and punishments of those who had been previously convicted at the swainmotes. The justices in eyre, or their deputies, continued down to 1635; they were virtually ended by the Act for the Limitation of Forests (1640), although Charles II attempted to revive them.[81]

In an Act of 1542, management of Crown woods for timber had been formalised: the Court of Surveyors and the post of Surveyor General of the Crown Woods were assigned to manage 'forest districts'. These precursors of the 'Office of Woods' (ultimately this became the modern Forestry Commission) were created to increase the commercial function of Crown woodlands under the direction of the Exchequer. Under the new arrangements they had responsibilities for creating 'profit of the king' through sale of timber.[82]

---

[77] A 'purpestre' or 'perpresture' was an unlicensed inclosure or building within the forest or an encroachment upon land. <http://info.sjc.ox.ac.uk/forests/glossary.htm#F> Accessed June 2021.
[78] J. Evelyn (1664) *Sylva, or a Discourse of Forest Trees*. John Martyn, London.
[79] J. Manwood (1598). *A Treatise and Discourse of the Laws of the Forest*, 143 (modernised spelling).
[80] J. Langton and G. Jones (n.d.) Seeing the Forest for the trees, St John's College, Oxford. <http://info.sjc.ox.ac.uk/forests/glossary.htm> Accessed April 2020.
[81] H. Cook (2018) *New Forest: the forging of a landscape*. Windgather Press, Oxford, Chapter 6.
[82] Stagg, D.J. (1989) Silvicultural Inclosures in the New Forest to 1780. *Proceedings of the Hampshire Field Club and Archaeological Society* 45, 135-145.

The Surveyor General was answerable to the Exchequer and hence often in conflict with the Lord Warden who continued to administer the forest law, which was geared towards the preservation of the game and thus to conservative actions in respect of forest management. The creation of the post of Surveyor General in turn gave rise to a second hierarchy of officers, known as deputy surveyors. The ensuing Statute of Woods in 1544[83] extended the period of enclosure to seven years in order to exclude grazing animals from woods over a 24-year growth period, and from woods of lesser age for shorter periods. This effectively recognised the value of woods in an economic sense - distinct from hunting.

The Office of the Preservators has its origin in the Exchequer's drive towards control of wood sales on Crown estates; the commission to set up the office was published in 1567. Indeed, their presence in the New Forest was not realised until recent scholarship,[84] although they played a significant role, often overlapping with that of regarders, and surviving until about the start of the 17th century. Preservators therefore represented a significant manifestation of the *divsum imperium* between the older governance under the Lord Warden (including the regarders) and the Deputy Surveyor, dating from Tudor times. Up to 18 Preservators were appointed by Verderers, it would seem, from among the regarders, or the yeomanry.

The impact of John Evelyn's book 'Sylvia' (1664) was profound.[85] It had been printed and promoted by the Royal Society at a time when timber was being replaced by new fuels such as coal. Evelyn was very much a 'Restoration man'. He expressed concerned about neglect, and hence lost opportunity for profit from more marginal lands;[86] his aim was to maximise timber production in the national interest and advocated the widespread establishment of plantations.[87]

Evelyn's book on forest trees and the propagation of timber was originally delivered as a paper to the Royal Society. Probably the first treatise on timber production, it reflects the concerns of the day. The author castigated the decline of England's woods during the Commonwealth and was clearly responding to a need to produce more timber. Evelyn was furthermore a friend of that other diarist, Samuel Pepys, who held the offices of clerk of the office of Privy Seal and clerk of the Acts of the Navy from 1660.

Probably the most embarrassing event in British Naval History, and worse than the loss of the Mary Rose (see chapter 9), occurred in June 1667, when the Dutch fleet crippled the Royal Navy in a raid on the Chatham Naval Dockyard when it was located on the Medway in Kent.[88] The requirements of the navy for timber superseded the forest law and led to its effective breakdown. The commoners had claimed the right to graze stock on the New Forest throughout the year and were not apparently constrained by the 'fence month'. Daniel Defoe passed through the Forest around 1720, remarking that there was plenty of timber on the shore of one river that was bound for the Portsmouth shipyards; prior to that there had been

[83] H. Cook (2018) *New Forest: the forging of a landscape*. Windgather Press, Oxford, Chapter 6.
[84] P. Roberts (2002) *Minstead: Life in a New Forest Community*. Nova Foresta Publishing, Ashurst.
[85] J. Evelyn (1664) *Sylva, or a Discourse of Forest Trees*. John Martyn, London.
[86] J. Evelyn (1664) *Sylva, or a Discourse of Forest Trees*. John Martyn, London, Chapter 1.
[87] O. Rackham (2006) *Woodlands*. William Collins, London, [449].
[88] J. Brain (n.d.) Raid on Medway <https://www.historic-uk.com/HistoryUK/HistoryofEngland/Raid-On-Medway/> Accessed April 2020.

much felling during the reign of William III.[89] Defoe could find 'no signs of decay of our woods, or of the danger of our wanting timber in England', something that conflicted with the official (government) view of resource scarcity.

There remained problems arguably greater than those that beset food production. At best, the forest system was an unreformed medieval legal system, at worst there was evident corruption exacted by forest officers. Yet, ironically, a Commission of Enquiry advised against disafforestation of the New Forest, probably because the Crown award (compensation) would have been small, compared with that which might have been received by the commoners. However, the Crown was in favour of powers to increase silviculture. The Commissioners of Woods, Forests, Land Revenues, Works, and Buildings or in short, the Office of Woods, was established in 1832, and over the next 70 years, was to strengthen its authority by re-enacting lapsed powers and reducing commoners' rights. The Office of Woods would become the Forestry Commission in 1919.[90] The matter of modern forestry will be returned to in Chapter 10.

3.4 The Church in the Middle Ages

From material culture in Wessex, it is evident that Christianity was present by the end of the Roman period.[91] The ruling dynasty of Wessex probably had some British ancestry (see Introduction), and although the early kings were not themselves Christian, it is plausible there was some Christian continuity among the Britons. This has been cautiously suggested for Sherborne, which may have had its origins in a community called *Lanprobi*, developing as a minster church and a Benedictine Abbey.[92] It is otherwise generally assumed that paganism dominated the Anglo-Saxon Kingdoms before St Augustine.

The Augustinian mission to convert the English occurred between 596-7 and was successful in Kent. Mercia probably became Christian probably during the later 7th century, Wessex was around the same time. Alfred the Great was King of Wessex from 871 and King of the Anglo-Saxons from *c.* 886 until AD899. His role in establishing Wessex in history is well-known, as is his identity as a Christian monarch set against the heathen Danes (see the Introduction). Monastic institutions in Anglo-Saxon England, like their Norman successors, were endowed with considerable estates in order to support the communities.[93]

By the 9th century, England was clearly a Christian country and under continental influence. However, the extent of Christian continuity from the late Roman period is unclear. Nonetheless, once Christianity became established in Wessex, not only did it have strong influence over ordinary people, but the church would soon became a major landowner, and by the ninth century held considerable estates.[94]

[89] B. Vesey-Fitzgerald (1966) *Portrait of the New Forest*. Robert Hale, London, 93.
[90] H. Cook (2018) *New Forest: the forging of a landscape*. Windgather Press, Oxford, Chapter 6.
[91] B. Yorke (1995) *Wessex in the Early Middle Ages*. Leicester University Press, London, 16.
[92] M. Grimmer (2005) British Christian continuity in Anglo-Saxon England: the case of Sherborne/Lanprobi, *Journal of the Australian Early Medieval Association* 1, 51-64.
[93] M. Aston (2000) Monasteries in Somerset, in C.J. Webster (ed.) *Somerset Archaeology: Papers to Mark 150 Years of the Somerset Archaeological and Natural History Society*. Somerset Archaeological and Natural History Society, Taunton, 99-104.
[94] B. Yorke (1995) *Wessex in the Early Middle Ages*. Leicester University Press, London, 56, 174, 194; K. Barker (2020) pers. comm.

The Anglo-Saxon kings evidently hunted in their Wessex demesnes, and there were officers appointed for the royal sport.[95] It is furthermore likely that demesne woods were attached to royal estates, such as that of Cheddar, which was to become a part of the Mendip (or Cheddar) Forest post-1066.

In the *Life of St Dunstan* (909–988) a miraculous event prevents King Eadmund's horse from following stag and hounds blindly over the cliffs at Cheddar Gorge.[96] The new king had not treated Dunstan well because other courtiers were jealous of his favoured position and contrived to have him expelled from the court. Eadmund could only prevent a serious accident by promising to make amends to the Saint. His horse then stopped at the very edge of the cliff and consequently Eadmund appointed Dunstan Abbot of Glastonbury. The story tells us immediately of the link between senior clerics and emergent feudalism.

Both Christianity and the coming of the Normans had the potential to alter demand for beverages:

> Vines have been cultivated in England, possibly by the Belgae, but certainly since Roman times. They were neglected by the Anglo-Saxons, who preferred ale, but not surprisingly, interest recovered again with the coming of the French at the Norman Conquest.[97]

Quelle surprise! Consumption by the church apparently did not stimulate much demand, although vineyards are mentioned in Domesday and were associated with the head manors of great lords. A 'good vineyard' was to be found at Wilcot, belonging to the Sheriff of Wiltshire. Domestic production has also to be balanced against the fluctuations of trade with France.

In a largely agrarian society, self-sufficiency is the starting point. While the rural economy produced surplus goods for market, importation was expensive and limited to luxury products.[98] Such goods included herbs, spices and rice that were not readily produced in England. Wine, although much imported, was produced in medieval England and this may have been linked to the 'Medieval Warm Period' which preceded the 'Little Ice Age'. Wine production is recorded in Domesday, although the reasons for a decline in wine production in England during the middle ages are complex.[99]

In pre-Reformation times (at least in theory), The Church enjoyed domestically produced ingredients for religious observance (bread and wine) as well as fishponds throughout the land providing for abstinence from meat on Fridays. The ordinary parish priest, meanwhile, was supported at the manorial level.

Bishops and the heads of the great monasteries held land in the same way as did other feudal lords. The vast holdings of the Bishops of Winchester are an example, and there are good

---

[95] H. Cook (2018) *New Forest: the forging of a landscape*. Windgather Press, Oxford, 62.
[96] <https://www.newadvent.org/cathen/05199a.htm> accessed June 2021.
[97] J. Thirsk (1997) *Alternative Agriculture*. Oxford University Press, Oxford, 135.
[98] Castle and Manor House Resources (2010-2014) Medieval Food and Cooking. <http://www.castlesandmanorhouses.com/life_04_food.htm> Accessed April 2020.
[99] Real Climate (2006) Medieval warmth and English wine. <http://www.realclimate.org/index.php/archives/2006/07/medieval-warmth-and-english-wine/> Accessed April 2020.

records from their demesne land; there are likewise records for the Benedictine Priory of St Swithun at Winchester. These estates show similarities with the great estates of southern England, particularly the great chalkland manors, being characterised by mixed farming, with large sheep flocks, and late leasing of the demesne.[100]

Rather than merely acting as landlords, there are good examples of some monasteries acting directly as agents of landscape change. The larger monastic houses in Wessex certainly possessed wealth and power. Founded before the Norman Conquest, the Benedictine Abbey of Glastonbury was extraordinarily wealthy. Together with Muchelney Abbey, Athelney Abbey and the secular canons of the Dean and Chapter of Wells (all in Somerset), these institutions were responsible for an early phase of draining of the Somerset Levels (see Chapter 4).

Near to Meare in Somerset is the 'Abbot's Fish House' (Figure 3.3). This building is on the site of a drained lake ('mere') on the Levels. Built for Glastonbury Abbey in the 1330s, it was likely an occasional residence for the abbey's officials concerned with the lake and its fishery, rather than being built for processing or storing fish. After 1539 the Fish House was largely put to agricultural use until gutted by fire in the 19th century. We are reminded by this of the importance of fish in the diet of religious communities inclined to refrain from meat consumption.[101]

The Cistercian reform, an attempt to return to the simplicity of the Rule of St Benedict, was based in hard work, simplicity, and prayer. Cistercians were introduced into England when William Gifford was Bishop of Winchester (1100-29). While the most famous early Cistercian was St Bernard of Clairvaux, it is said that another co-founder of the Order, Stephen Harding, was born in Somerset.[102]

Farming by the Cistercians may be described as entrepreneurial and their economic activities in many spheres are well-known. Organised internationally, they typically occupied remote locations in areas of poor and economically marginal land. The first Cistercian Abbey in England was at Waverley in Surrey (1128) near to the Hampshire border. Others in Wessex include Quarr in the Isle of Wight (1132), Forde Abbey in Dorset (1141), Cleeve (1198) and Beaulieu in Hampshire (1203).The Cistercian model of working the land involved a religious group called *conversi* (lay brothers) who worked on outlying farmsteads, called 'granges', concentrating on sheep husbandry and hence wool production, so vital in the economy of medieval England.[103] The remains of a grange complex of Beaulieu Abbey may be seen at St. Leonard's some 6km from the main Abbey. Here the remains of a barn, grange and chapel for the lay brothers can be seen. The Abbey pond was at Sowley.[104]

The *conversi* system was not hidebound by tradition and this 'direct labour force' were in a better position than ordinary peasants to innovate in agricultural practice. One unfortunate

---

[100] J. Hare (2006) The Bishop and the Prior: demesne agriculture in medieval Hampshire. *The Agricultural History Review* 54 (2), 187-212.

[101] English Heritage (n.d.) History of Mere Fish House < https://www.english-heritage.org.uk/visit/places/meare-fish-house/history/> Accessed April 2020.

[102] J.H. Bettey (1986) *Wessex from AD 1000*. Longman, New York, 69

[103] J.H. Bettey (1986) *Wessex from AD 1000*. Longman, New York.

[104] New Forest Explorer's Guide (n.d.) Beaulieu history - an introduction. <http://www.newforestexplorersguide.co.uk/heritage/beaulieu/introduction.html> Accessed April 2020.

Figure 3.3. The 'Abbot's Fish House' at Mere, Somerset (Wikimedia Commons)

result of Cistercian settlement was actually the removal of established settlements in order to create a wilderness within which to work.[105] The unpalatable message is that monastic houses in England were capable of unchristian acts in seeking communion with God. The most impressive and well-known Cistercian remains are in Yorkshire in isolated locations such as Fountains and Rievaulx.

3.5 Enclosure

Making enclosures (for present purposes the modern term is 'fields') is ancient and is one product of assarting since prehistory. While Mesolithic people probably made clearings in the wildwood, creating fields in a systematic fashion is strongly associated with agriculture. Making enclosures has been enthusiastically undertaken since at least the Bronze Age in Britain, and is proven from the Neolithic in Ireland. The way 'Enclosure' (sometimes 'Inclosure') is defined in the present, generally refers to agrarian change since Tudor times. There were classically three major changes:

- The selective breeding of livestock
- New systems of cropping (typically involving turnips and clover)
- The removal of common property rights to land

---

[105] T. Williamson and L. Bellamy (1987) *Property and Landscape: A Social History of Land Ownership and the English Countryside.* George Philip, London, 53.

All are interrelated but the third aspect affected the rural economy in a revolutionary way, effectively removing all vestiges of feudal agrarian practice in England, and led to resentment and to social change. The poem at the top of this chapter demonstrates this point.

The Crown, Exchequer, manors, the Church, colleges, monasteries and secular landowners extended power over the lives of ordinary folk. Karl Marx saw feudalism as a transitional form of society between tribal organisation (as we imagine was the case in Iron Age Wessex) and capitalism. The enclosure of land was a pre-eminent part of this process of change which hit English agrarian society as hard as it changed the physical rural landscape.

Feudal societies operated, for better or worse, largely because of the power and interests of those in charge. True, the interests of villeins and commoners were affirmed from time to time. Peasants after all, required organisation. And they could be difficult as witnessed by the 'Peasants' Revolt of 1381. Largely caused by the raising of a poll tax in order to pay for the Hundred Years' War with France, those in rebellion were in reality a cross-section of English society, although the demands of the peasantry are especially recalled.[106] Their demands included an effective end to feudal bondage.[107]

A further impact on the landscape was the great population loss caused by the Black Death 1348-9 (there were later visitations). As well as abandonment of some villages, there was an extensive switch to pastoral farming as arable was no longer required on such a large scale, and by the middle of the following century the profit obtained from sheep farming was very much evident, although the impact in Wessex was likely less than elsewhere in England.[108] A further incentive to peasants inclined to break with feudal bonds was to move to towns; although not a peasant, the historical character Dick Whittington serves to illustrate the wealth which could be gained through migration.[109]

The fecundity of the core of Wessex was recorded by John Coker of Mappowder in Dorset, writing in the first half of the 17th century. Coker could comment on the Dorset countryside as being well-watered, the rivers full of fish and its produce of cattle, sheep, and corn supplying the London Market and clothiers exporting cloth to France.[110]

Changes throughout the 17th century were all part of the broader national picture, and for rural people (the bulk of the population) it would be enclosure of both open field arable and common land that would prove problematic. Abolition of feudal tenure and the lack of security for copyhold tenure, would be a part of dramatic changes that would be realised into the 18th century. Ultimately enclosure benefited larger tenants at the expense of smaller.[111] Private landlords would lease out their estates rather than cultivate themselves. Agriculture

---

[106] M. Schlauch (1940) The Revolt of 1381 in England. *Science and Society* 4(4), 414-432.
[107] B. Hunter (1981) In defence of the Peasants' Revolt <http://www.billhunterweb.org.uk/articles/In_defence_of_peasants_revolt.htm> Accessed April 2020.
[108] Taylor C. (1988) Commentary to W.G. Hoskins, *The Making of the English Landscape*. Hodder and Stoughton, London, 96-7.
[109] Museum of London (2007) Dick Whittington: the true story <http://www.bbc.co.uk/gloucestershire/content/articles/2005/06/16/about_dick_whittington_feature.shtml> Accessed April 2020.
[110] L.E. Tavener (1940) Dorset, in L.D. Stamp (ed.) *The Land Utilisation of Britain*. Geographical publications, London, 247.
[111] C. Hill (1980) *The Century of Revolution 1603-1714* (2nd ed.). van Nostrand Reinhold, Ne York, 175, 230.

had transitioned to a truly capitalist basis with the rise of private property as understood in a modern sense.

The 'Second Agricultural Revolution' (of the 17th century) is itself a difficult concept for those who believe in steady progress (in this case for food production). It certainly had several components and the changes that really started with Tudor enclosures continued into the 18th century. After all, there are no records for the Neolithic agricultural revolution but from the archaeological record we can infer that it took some time to transform both economy and landscape.

A key source from the mid-17th century is Walter Blith, a former Parliamentarian army officer who had been concerned with sequestering Royalist estates. Blith managed to capture the key processes in 'Six Peeces of Improvement', outlined as follows in his 1653 edition of the *English Improver Improved*:[112]

> All clearly demonstrated from Principles of reason, Integrity, and the late but most Real Experiences; and held forth at an Inconsiderable charge to the Profits accrewing thereby, under 'Six Peeces of Improvement.
>
> 1. By floating and Watering such Land as lieth capable thereof
> 2. By Drayning Fen, reducing Bog and regaining Sea-lands
> 3. By such enclosures as prevents Depopulation, & advanceth all Interests
> 4. By tillage of some Land lost for want of, and pasturing others destroyed with Plowing
> 5. By a Discovery of all Soyles and composts with their nature and use
> 6. By doubling the growth of Wood by a new Plantation'

Blith encapsulates the Second Agricultural Revolution in full, during a period of catastrophic social and political change. 'By such enclosures as prevents Depopulation, & advanceth all Interests' is interesting, for it promotes land tenurial change while acknowledging that there should be no losers. This may reflect Blith's more egalitarian leanings than his adversaries on the Royalist side. Recognition of problems is one thing, but sadly there were many losers.

The lords and gentry were incentivised to cause open and communally managed areas to be broken up and re-distributed into parcels of land, generally separated by hedges (sometimes fences) and on reclaimed wetlands by freshly dug ditches. The enclosure process would be long-term in England. An estimate for proportions of the countryside enclosed over time is:[113]

- Already enclosed in 1500 c.45.0%
- Enclosed 1500-1599 c. 2.0%
- Enclosed 1600-1699 c.24.0%
- Enclosed 1700-1799 c.13.0%
- Enclosed 1800-1914 c.11.4%
- Commons remaining in 1914 c. 4.6%

---

[112] W. Blith (1653). *The English improver improved*. John Wright, London, preface.
[113] J.R. Wordie (1983) The Chronology of English Enclosure, 1500-1914 *The Economic History Review* 36 (4), 483-505.

The figure for 1600-1699 may be exaggerated because 17th century enclosure involved local agreements not always so easy to identify from documentary evidence; such uncertainty has been discussed by Turner.[114] Nonetheless this reflected a very high rate of enclosure. Between 18th and 19th centuries, Enclosure Acts were passed for specific parishes, Parliamentary Enclosure finished the process, mostly between 1750 and 1830. In woodland areas, generally open fields had vanished by the end of the 17th century[115].

To summarise the legal mechanisms involved:[116]

Private enclosure agreements, which enabled early enclosure of many common lands, pastures and manorial wastes, whether by popular agreement or compulsion, left no formal record although there may be references in private estate or manorial records. Where contested, there may be legal records at courts such as the Court of Requests or Star Chamber.

Piecemeal enclosure is the term used when a landlord gradually bought out small farmers or exchanged arable strips. Alternatively, **general enclosure** is a general term used when a whole Parish of open fields and commons made a decision to enclose, but at the request of more powerful members of the community. Landholding became re-organised and re-allocated. Very regular field pattern emerged because surveyors became involved.[117]

Enclosures by enrolled decree or agreement Enrolment means to officially record a judgment at court. Enclosure awards[118] were sometimes enrolled in the law courts of record. From the mid-16th century, enclosures were commonly enrolled by decree of one of the equity courts, especially Chancery and Exchequer. Sometimes fictitious quarrels were created in order to bring the case before the court, although most were by agreement.[119] In Hampshire it appears this was still a significant process of enclosure into the 18th century.[120]

Enclosure by private Act of Parliament There were a few private enclosure acts made in the 16th century, largely concerned with the drainage and enclosure of marshes, although acts confirming enclosures by decree are sometimes found from the following century. Private enclosure acts for waste, common land and open fields were more frequent after 1750.

Enclosure by public Act of Parliament From 1801, public general enclosure acts were passed. These normally specified where awards were to be deposited or enrolled, either by one of the courts of record or with the local clerk of the peace (clerk of the quarter sessions). The General Enclosure Act of 1845 appointed permanent enclosure commissioners who were authorised to

---

[114] M. Turner (1986) Parliamentary Enclosures: Gains and Costs. *ReFRESH* 3 <http://www.ehs.org.uk/dotAsset/d62ccd6c-115b-4b21-8802-15b02f32cb5f.pdf> Accessed April 2020.

[115] O. Rackham (1987) *The history of the Countryside*. J.M. Dent, London, 5.

[116] National Archives (n.d.) Enclosure Awards < https://www.nationalarchives.gov.uk/help-with-your-research/research-guides/enclosure-awards/ > Accessed April 2020.

[117] T. Williamson (2000) Understanding Enclosure, *Landscapes* 1 (1), 56-79. DOI: 10.1179/lan.2000.1.1.56

[118] Enclosure awards are legal documents recording the ownership and distribution of land. They may detail land owned by churches, by schools and charities as well as roads, rights of way, drainage, land boundaries, different types of land tenure and liability to tithe. National Archives (n.d.) Enclosure Awards <https://www.nationalarchives.gov.uk/help-with-your-research/research-guides/enclosure-awards/> Accessed April 2020.

[119] National Archives (n.d.) Enclosure Awards <https://www.nationalarchives.gov.uk/help-with-your-research/research-guides/enclosure-awards/> Accessed April 2020.

[120] J. Chapman and S. Seelinger (1995). Formal Agreements and the Enclosure Process: The Evidence from Hampshire. *Agricultural History Review* 43(1), 35-46.

issue enclosure awards without submitting them to Parliament for approval. Manorial wastes and lands subject to indefinite rights of common were excluded but were covered by later legislation.

Parliamentary enclosure was enabling legislation for particular parishes to enclose, typically during the period approximately 1750 to 1850. The term **enclosure award** refers to this legislation and not to any financial arrangement. Where capital was required, this could be raised by land sales, sometimes the levying of a rate,[121] or capital raised through the banks.[122]

Enclosure could occur in several stages at different times, and land holding varies. For example, in the Woodford valley north of Salisbury in 1839, Upper Woodford had four open fields. In 1955 it was still not subdivided into smaller parcels and the landscape still has an open aspect today. By contrast, the former manor of Heale (between Middle and Upper Woodford) contained fields with pasture on the steep sides of Heale Hill. The fields of Heale were still divided into small allotments at the beginning of the 17th century and there was one freehold tenant. There was further enclosure of meadows by the river, and in the 1660s water meadows were constructed. By the beginning of the 19th century they had been drawn together into one compact farm, although there was further subdivision and re-unification up to 1920.[123]

At West Milton, Dorset, there are strip lynchets on the slope of Pitcher's Hill, immediately south of West Milton village, that extend for about 410m, with terraces which are about 9 to 12 metres wide. Here, straight enclosure hedges run, topographically blind, across the lynchets on the grassed hillside that were medieval ploughland, probably dating from before the Black Death and representing a former extension to arable land from the top of the hill (Figure 3.4). These hedges also constrain the former ploughland top and bottom.

Close scrutiny shows a holloway track that once provided access to the strips from the right of the photograph, and a bank across the valley bottom suggesting subsequent removal of a field boundary.[124] An enclosure award for Powerstock is dated 1861 but this is of limited area.[125] In comparison with the English Midlands, Parliamentary enclosure in Wessex was of lesser importance.[126]

From the Middle Ages to the 20th century, 'sheep and corn' husbandry predominated throughout the parish of Amesbury. The settlements of Amesbury and West Amesbury each had open fields and common pastures, and this was probably the case for Countess Court Manor and Ratfyn. Arable farming was undertaken on the chalkland nearest to the settlement, with extensive downland pasture further east or west and meadows beside the Avon. Between

---

[121] B.J. Buchanan (1982). The Financing of Parliamentary Waste Land Enclosure: Some Evidence from North Somerset, 1770–1830 *The Agricultural History Review* 30 (2), 112-126.

[122] M.E. Turner (1973) The Cost of Parliamentary Enclosure in Buckinghamshire. *Agricultural History Review* 21(1), 35-46.

[123] E. Crittall (1962) Woodford in E. Crittall (ed.) *A History of the County of Wiltshire, Volume 6*. Victoria County History, London, 221-227.

[124] Anon. (1952) Powerstock, in *An Inventory of the Historical Monuments in Dorset, Volume 1, West*. The Stationery Office, London, 181-186.

[125] National Archives (n.d.). Enclosure award and map of Powerstock (parish), Dorset. <http://discovery.nationalarchives.gov.uk/details/r/C3829120> Accessed April 2020.

[126] M. Turner (1986) Parliamentary Enclosures: Gains and Costs. *ReFRESH* 3 <http://www.ehs.org.uk/dotAsset/d62ccd6c-115b-4b21-8802-15b02f32cb5f.pdf> Accessed April 2020.

Figure 3.4. Enclosure fields and strip lynchets, West Milton, Dorset (the author)

1635 and 1725 the division of Amesbury's lands between several and commonable, the arrangement of open fields and common downs and the division of the land among the farms changed little. The watering of meadows may have begun around 1658 when hatches on the Amesbury—Normanton boundary were licensed by the Earldom manor court. However, large areas of the downland pasture had been ploughed by the early 18th C and this process continued to the mid-20th century.

It appears that after 1725 Charles, Duke of Queensberry, adopted the policy of merging all the open-field land which he owned into a single farm (Figure 3.5). A farmhouse (the Red House) had evidently been built by 1726. The land was in copyhold. To accelerate the process, Queensberry leased land from some copyholders and leaseholders, bought out others, and became tenant of another landowner. In the 1750s, nearly a third of the parcels in the open fields were in Red House farm, and by 1760 Queensberry owned nearly all Amesbury's land and proceeded to bring it in hand, putting an end to open fields as well as common husbandry. Most of the former open fields and common downs were laid out as two several farms: Red House and Southam.[127]

To summarise, 18th century enclosure around Amesbury and West Amesbury in Wiltshire was on the estate of the Duke of Queensberry and hence strong seigneurial control enabled systematic enclosure.

---

[127] A.P. Baggs, J. Freeman and J.H. Stevenson (1995) Parishes: Amesbury, in D.A. Crowley (ed.) *A History of the County of Wiltshire: Volume 15, Amesbury Hundred, Branch and Dole Hundred.* Victoria County History, London, 13-55.

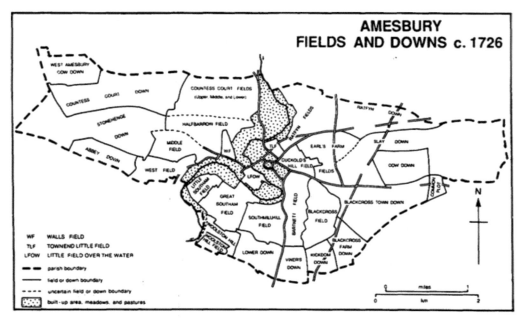

Figure 3.5. Amesbury in the 1720s from Victoria County History (in A P Baggs, J. Freeman and J.H. Stevenson, 1995, 'Parishes: Amesbury', in *A History of the County of Wiltshire: Volume 15, Amesbury Hundred, Branch and Dole Hundred*, ed. D.A. Crowley, London, pp. 13-55. British History Online http://www.british-history.ac.uk/vch/wilts/vol15/pp13-55)

Figure 3.6. Parliamentary enclosure fields near to Fovant in Wiltshire (the author)

Figure 3.6 shows parliamentary enclosure fields near to Fovant taken from Fovant Down, Wiltshire, looking northwest. Note the removal of former hedges, especially on the right of the picture, to create a large, modern arable field. Enclosure awards were granted in 1787 following the work of surveyors appointed by the Parliamentary Enclosure Commissioners; the area photographed was until that time in open fields.[128]

In Wiltshire, there were parliamentary enclosures in all of Fovant, Alvediston, Broad Chalke, and Bower Chalke in 1792. However in 1567–8 about 11% of the land of Chalke was already in severalty, and 13% by 1631–2. At nearby Alvediston, about two-thirds was already in severalty, while there were further enclosures of field and downland there in the early 18th century. At Fovant specifically, about 59% of the land was already in severalty in 1632, and here, as at Alvediston, much of the severally held areas were hedged. Enclosure was well advanced before the 18th century, so to presume it all occurred at that time is an exaggeration.[129]

While the exact impacts of the enclosure movement have been debated, the generally accepted outcome is that it deprived the poor of access to land (both open field and common) as a means of sustenance, providing a pool of surplus rural labour who were furthermore available to tramp to towns in search of new livelihoods. By contrast, the winners were able to accumulate wealth through agricultural improvement.[130]

3.6 Internal Drainage Boards

Land drainage and flood protection have an ancient origin, although most of today's Internal Drainage Boards (IDBs) were established following the passing of the influential Land Drainage Act (1930).[131] The forerunners of IDBs date back to Henry III, who established a commission for the drainage of Romney Marsh in Kent and East Sussex in 1252.[132]

On Romney Marsh the 'jurats', an organised body of men able to repair sea walls and control ditches, are known to have been effective in 1252. They raised money for their operations by charging a tax called the 'scot'. Similar institutions for flood defence existed elsewhere, such as on the Thames Marshes, where in 1390 a commission was established to inspect and repair flood defences and ditches. The first commission of sewers was established in Lincolnshire in 1258 following experience on Romney Marsh. The ensuing courts ensured that landowners should maintain flood defences by maintaining dykes, walls and bridges in the Fenland. The role of central government in these matters is reflected in the formal establishment of the commissions in 1427. In 1532, a Statute of Sewers was passed, reinforcing legislation at a time when monastic lords were responsible for much drainage and reclamation. Monastic power was soon to be broken by the Reformation, with resulting increased flood hazards in some areas, yet the courts survived into the 20th century.[133] Over time, a balance had to be struck

---

[128] Fovant History Interest Group (2017) <http://www.fovanthistory.org/entry_point.html> Accessed April 2020.
[129] E. Crittall (1959) Agriculture 1500-1793 in E. Crittall (ed.) *A History of the County of Wiltshire: Volume 4*, Victoria County History, London, 43-64.
[130] M. Turner (1986) Parliamentary Enclosures: Gains and Costs. *ReFRESH* 3 <http://www.ehs.org.uk/dotAsset/d62ccd6c-115b-4b21-8802-15b02f32cb5f.pdf> Accessed April 2020.
[131] J. Sheail (1999). Water management systems, drainage and conservation, in H. Cook and T. Williamson (eds) *Water Management in the English Landscape*. Edinburgh University Press, Edinburgh, Chapter 15.
[132] H.F. Cook (2017) *The protection and conservation of water resources* (2nd ed.) Wiley-Blackwell, Chichester, Chapter 3.
[133] H.F. Cook (2017) *The protection and conservation of water resources* (2nd ed.) Wiley-Blackwell, Chichester, Chapter 3.

between the protection of the individual and the furtherance of the common good, and no balance could be struck without the cooperation of all the interests involved. In the Romney Marsh area, as in so many similar reclaimed wetlands, the imperative of protecting from flooding enabled agricultural development to the present.[134]

A modern IDB is a type of local public authority responsible for managing water levels in England where there is a special need for drainage: generally land reclaimed from the sea or within alluvial river valleys. The IDBs have permissive powers to manage water levels within their 'drainage district', by maintaining rivers, drainage channels, culverts, sluices, weirs, embankments and pumping stations. They also have an important role keeping watercourses clear of obstructions with the aim of reducing flood risk. In Wessex they have a regional responsibility for the Somerset Levels and Moors.[135] Their activities and responsibilities are principally governed by the Land Drainage Act 1991 as amended by subsequent legislation.[136]

3.7 Governmental involvement and institutions

Organisations concerned with governance have been outlined, these being in royal forests the Crown, the Exchequer and Office of Woods, later the Forestry Commission and manors. As has been demonstrated, manors may be seen as environmental agency-type organisations that controlled natural resources, labour, enacting laws and regulating the lives of many ordinary people. In governance terms, manors represent hierarchical systems. Over time, economic monetarisation seen particularly as cash rents replacing 'payment in kind' gave rise (eventually) to capitalism yet the manor showed itself surprisingly enduring. The king distributed land to his 'tenants-in-chief' who in turn permitted lesser gentry, and ultimately villeins or other un-free men to hold land, although the latter generally paid cash in rent.[137]

While ownership, in the modern sense of proprietarily commanding one's freehold land, dates largely from the mid-17th century[138], in the agrarian society of the Middle Ages, land was more accurately said to be 'held' in exchange for military obligation, or indeed labour service or payment of rent further down the social ladder.

Copyhold tenancies evolved from customary tenure; freehold would eventually become the right to hold land without feudal obligations, although freehold tenants still paid some rent to the lord of the manor. The church was likewise a major landowner that also collected tithes and land that was set-aside specifically for the support of a parish priest was termed the 'glebe'.[139]

Feudal relations are complicated by another factor: alongside the obligation-led feudalism, a market economy operated, largely in the towns and cities. The latter would emerge strongly

---

[134] H.F. Cook (2010) Boom, slump and intervention: changing agricultural landscapes on Romney Marsh, 1790 to 1990, in M.P. Waller, E. Edwards and L.L. Barber (eds) *Romney Marsh: Persistence and Change in a Coastal Lowland*. Romney Marsh Research Trust, Sevenoaks, 155-183.
[135] See Chapter 4 (this volume).
[136] Association of Drainage Authorities (2017) Introduction to Internal Drainage Boards <https://www.ada.org.uk/wp-content/uploads/2017/12/IDBs_An_Introduction_A5_2017_web.pdf > Accessed April 2020.
[137] M.M. Postan (1972) *The Medieval Economy and Society*. Weidenfeld and Nicholson, London, 106-7.
[138] C. Hill (1972) The World Turned Upside Down Penguin Books, Harmondsworth, 15.
[139] Law Insider (n.d.) <https://www.lawinsider.com/dictionary/glebe-land> Accessed June 2021.

in the later medieval period as a factor in economic development in both England and Europe. Market economic elements were not along the lines of *laissez faire* neo-liberalism as understood today, for they required Royal Charters to operate. There were also monopolistic and restrictive practices operated by guilds that controlled labour markets.[140]

Towns would affect the governance of England. Established largely from the Roman period onwards, they were a focus for economic – largely mercantile – activity and required strong governmental arrangements. In the middle ages, 'ancient boroughs' were important towns that were established by royal charters,[141] and which were to a large degree self-governing.[142] Urban governance arose from the 'shire moot' (regular county-based meetings) of Anglo-Saxon times where legal cases were heard and local matters discussed. They were also attended by the local lords and bishops, the sheriff, and four representatives of each village. After the Norman Conquest, this meeting became known as the County Court. Moots introduced the idea of representative government at the local level.[143]

Evolution of urban governance into the Norman and later medieval period owed much to Alfred and his burhs. A borough provided economic safety as well as a central place for trade and manufacturing including minting money.[144] Location was important. Self-evidently, ports in Wessex such as Bridgwater, its more successful neighbour Bristol, Poole and Southampton were all at coastal locations where draught was sufficient for shipping. Medieval Old Sarum was located for defensive reasons (it once included a mint!) within an Iron Age hillfort some 3km from New Sarum (modern Salisbury). And there was a Norman cathedral within the ramparts.

With hindsight, the decision in 1075, to develop Old Sarum within an Iron Age hill fort on a hilltop may have suited Norman military objectives including a will to dominate the landscape. However, it has been described as 'folly' in the longer-term, particularly in respect of other considerations, be they about space, ecclesiastical or commercial interests, climatic or water supply.[145] Or some combination. There are many accounts of the reasons to move the Norman settlement, many not so plausible. Non-mythical reasons for moving the hilltop city to a new site in the river valley are dominated by the decision to move the cathedral, and not unrelated is the need to better locate the market within the existing road network. Co-incidentally the relocation was to a place where the Bishop held land, and where a case has been made for 'proto-urban' activity prior to 1220.[146]

The success of 'new towns' recognised at the time, was in contrast to a decline in the fortunes of Old Sarum that was evident in 1200. Moving the city was enabled by a Papal Bull dated 1218,

---

[140] M.M. Postan (1972) *The Medieval Economy and Society*. Weidenfeld and Nicholson, London, Chapter 12.

[141] Encyclopaedia Britannica (n.d.) 'Charter' <https://www.britannica.com/topic/charter-document> Accessed April 2020.

[142] A 'charter' is a document granting certain specified rights, powers, privileges, or functions from the sovereign power of a state to an individual, corporation, city, guild, university or other unit of local organisation. Typically, these included therefore the right to hold a market of self-government (as for example, a 'mayor and corporation').

[143] UK Parliament (n.d.) Birth of the English Parliament: Anglo Saxon Origins <www.parliament.uk/about/living-heritage/evolutionofparliament/originsofparliament/birthofparliament/overview/origins/> Accessed April 2020.

[144] D. M. Stenton (1965) *English society in the early middle ages*. 4[th]ed Penguin London ch 4

[145] J. Chandler (2001) *Endless Street*. Hobnob Press, Salisbury, Chapter 1.

[146] A. Langlands (2014) Placing the burh in Searobyrg: rethinking the urban topography of early medieval Salisbury. *Wiltshire Archaeological & Natural History Magazine* 107, 91-105.

construction of the new cathedral commenced a year later, and in 1227 a Royal charter was granted, confirming the right to hold both fairs and markets and the Bishop stood to benefit financially from both.[147] Today markets are held in New Sarum (Salisbury) twice weekly.

The decline of Old Sarum and its development, including its status as a 'rotten borough' (abolished under the Reform Act 1832), to a settlement today blending the suburb of Stratford-sub-Castle with a surrounding rurality is a matter of current research by archaeologists and local historians. There was certainly a long-term decline after 1220, but not abandonment of the site. There were attempts to keep the borough going from 1229, and by 1275 'both Old Salisbury and Wilton were complaining that the bishop was allowing too many markets in New Salisbury, and so monopolising trade'.[148] Stone from the former cathedral was re-used, and the castle demolished in 1322, while in 1377 there were only 10 recorded poll tax payers.[149] Current geophysical investigation is revealing the ground-plan of the medieval city.[150]

In the Middle Ages, landlords' rights were exercised through a manorial court; during the Tudor period, many of the civil functions were transferred to townships and the parish as responsibilities for the poor, highways and law and order. They became the responsibility of the new civil parish councils upon the separation of ecclesiastical and civil parishes in the 19th century, similarly town councils and county councils were established. Separation of church and state in local government was thereby achieved.

The Local Government Act 1888 established county councils and county borough councils in England and Wales.[151] Subsequently, under another Local Government Act 1894,[152] civil parish councils were established, separating them from what would become the parochial church councils responsible for the running of Anglican ecclesiastical parishes. The Act established urban and rural districts and provided for the election of parish councillors in rural areas including rationalising their borders. Both kinds of new districts were based on earlier sanitary authorities that were responsible for public health matters.

Governance at local, town or country level is one thing, but most significant is the role of central government in determining the form of national policy through statute. This can be viewed in many ways, but essentially the 19th century saw the rise first of pollution legislation, then legislation to support the voluntary sector, typified by the National Trust (created 1895), with the first Act of Parliament supporting the new organisation in 1907, and early legislation to permit the formation of river catchment organisations, the Thames Conservancy Act 1857 and Lee Conservancy Act 1868.[153] Other examples of influential environmental legislation that would contribute to development of the statutory planning process included the Restriction

---

[147] J. Chandler (2001) *Endless Street*. Hobnob Press, Salisbury, 95. S. Hannath (2010) The Cathedral Rocks, self-published.
[148] E. Critall (1962) Old Salisbury: The borough, in E. Crittall (ed.) *A History of the County of Wiltshire: Volume 6*, Victoria County History, London, 62-63.
[149] E. Crittall (1959) Poll-Tax payers of 1377, in E. Crittall (ed.) *A History of the County of Wiltshire: Volume 4*, Victoria County History, London, 304-313.
[150] K. Strutt, D. Barker, A. Langlands and T. Sly (2018) The Old Sarum Landscapes Project Research Report No. 3 < https://www.researchgate.net/publication/330106143> accessed June 2021.
[151] Legislation (n.d.) Local Government Act 1888 <http://www.legislation.gov.uk/ukpga/Vict/51-52/41> Accessed April 2020.
[152] Legislation (n.d.) Local Government Act 1894 (n.d.) < http://www.legislation.gov.uk/ukpga/Vict/56-57/73 > Accessed April 2020
[153] H.F. Cook (2017) *The Protection and Conservation of Water Resources*. (2nd ed.) Wiley-Blackwell, Chichester.

of Ribbon Development Act 1935, Town and Country Planning Act 1947, National Parks and Access to the Countryside act 1949, the Wildlife and Countryside Act 1981 and the Countryside and Rights of Way Act 2000.

The single most important manifestation of 'statism' is the rise of regulatory and 'competent authorities' to enact legislation, implement policies, monitor and otherwise regulate. Their histories are most complicated, however, in England there exists the Environment Agency (*inter alia* pollution control and flood defence with responsibilities relating to the protection and enhancement of the environment in England), Natural England (nature conservation and advice to central government) and Historic England (governmental adviser on the historic environment and its interpretation), while the English Heritage Trust manages historic buildings, monuments and sites.

Voluntary (third) sector engagement grew out of Victorian concern for the natural environment although mainly became realised in the 20th century. In the present day it includes country wildlife trusts, rivers trusts, and nation-wide organisations such as the Royal Society for the Protection of Birds, Wetlands and Wildfowl Trust and Woodlands Trust. Other organisations concerned with voluntarism or campaigning include The Conservation Volunteers (formerly British Trust for Conservation Volunteers) and the Campaign to Protect Rural England. Small charities also exist, for example the Harnham Water meadows Trust.[154] These matters will be returned to in chapter 10.

The policy environment largely generated by such organisations is inevitably complicated. However, one example relevant here is the National Character Area (NCA) profiles in England that comprise 159 distinct 'natural areas' (Chapter 6). Each is defined by a unique combination of landscape, biodiversity, geodiversity, history, and cultural and economic activity and their boundaries follow natural lines in the landscape rather than administrative boundaries. While essentially a planning tool, NCA is a ready-made multifaceted physical, land utilisation and ecological classification scheme that will be referred to in this book.[155]

3.8 Science, technology and landscape-scale change 1700-1940

The above summary from Walter Blith encapsulates both the technological optimism of the mid-17th century (Chapter 3.5) and the associated agrarian change. In 'Six Peeces of Improvement' we can imagine a massive shift in landscape management, towards a fully productive landscape as understood in a modern sense. Revolutions that appear to take place over decades, even centuries are difficult to understand as such. The Author prefers rather a model of steady progress that is punctuated by economic spurs to incentivise production of food, extract minerals or produce timber etc. The driver is generally population increase and/or industrialisation. Whiggish progress is probably a better description. Walter Blith, a revolutionary army officer, had regard for the poor.

---

[154] H. Cook and A. Inman (2012) The voluntary sector and conservation for England: Achievements, expanding roles and uncertain future. *Journal of Environmental Management* 112, 170-177. http://dx.doi.org/10.1016/j.jenvman.2012.07.013

[155] Natural England (2014) National Character Area profiles: data for local decision making. < https://www.gov.uk/government/publications/national-character-area-profiles-data-for-local-decision-making/national-character-area-profiles> Accessed April 2020.

If the improving action occurred in the turbulent seventeenth century, then the profit was realised in the subsequent centuries. An institutional driver for scientific endeavour was the establishment of the Royal Society at the time of the Restoration. The first meeting was held on 28th November 1660 followed a lecture at Gresham College by Christopher Wren. Joined by other leading polymaths (in natural philosophy and medicine) including Robert Boyle and John Wilkins, the group received royal approval becoming, from 1663, 'The Royal Society of London for Improving Natural Knowledge'.[156] Notwithstanding such developments, prominent scientists with agricultural interests have been elected Fellows of the Royal Society. These included Arthur Young (1741-1820), John Dalton (1766-1844) and much later (Ralph) Louis Wain (1911-2000) of Wye College.

The mid-18th century Dorset antiquary Revd. John Hutchins (1698–1773) could comment that the 'down, vale and heath will always be distinct' and he talks of the productive sheep and corn husbandry: in winter sheep are fed on grass fodder, arable is widespread including barley for malt liquors that cannot be equalled elsewhere. The enclosed vale of Blackmoor had less arable but was fertile and 'profitable for the breeding and fattening black cattle'. Cider orchards too were important, and 'it is true of late years that many spots have been greatly improved'; Hutchins lists three improvers and commends their skill including the plantation of fir trees, and the construction of new roads.[157] Dorset was benefitting from the improvements of the age.

Upward social mobility was evident from the late medieval period, and for some it continued. From the latter part of the reign of King Henry VIII monastic estates had fallen into the hands of lay owners. This momentous change in the countryside gave opportunity for a nouveau riche class to start large-scale farming. In the succeeding century there was a strong tendency for local gentry to give way to large landowners, who were able to expand by acquiring land held by less successful smallholders; a major factor was taxation on gentry and small freeholders notably in the arable sector.

The 'Glorious Revolution' of 1688 which placed William of Orange on the throne, offered the prospect of improved economic stability. This paved the way for the 18th century to see the rise of great estates. Successful landowners often occupied offices of state, providing what were in effect subsidies for already lucrative agricultural operations. In England, a social elite arose that would express itself in new designed landscapes on the back of agricultural development, particularly enclosure:

> This is seen most clearly in the evolution of the landscape park, but it also applied to the wider rural landscape. The design of schools, churches, cottages and barns all began to reflect the personal tastes of individual landowners.

In the manorial hall of the medieval period during feasts, the lord of the manor and his family would have dined in the company of yeomen and certain tenants. Yet by Georgian times,

---

[156] The Royal Society (2019) History of the Royal Society <https://royalsociety.org/about-us/history/> Accessed April 2020.
[157] L.E. Tavener (1940) Dorset, in L.D. Stamp (ed.) The Land Utilisation of Britain. Geographical publications, London, 248.

the country house was a quiet place with discrete servants waiting upon elite guests. The landowners were remote from their tenants and others and often set in parkland.[158]

Wessex was no different, for:

> The seventeenth and eighteenth centuries saw the gentry and aristocracy of the region greatly increase their power, influence and prestige. As owners of great estates, landlords, employers of labour and dispensers of patronage and as members of Parliament, Justices of the peace, royal agents, lords lieutenants, sheriffs, commissioners for Musters, coastal defences, sewers, and countless other governmental duties, even the lesser gentry dominated the political and social life of their neighbourhoods. Above all through the Commission of the Peace, the fulcrum of local administration, the gentry exercised complete power over local government.[159]

Houses displaying characteristics of the period include Kingston Lacey, Wimborne St Giles, Hinton St George, The Vyne, Stourhead, Fonthill, Longleat and Mottisfont (both the latter were former Augustinian priories). The parklands of Longleat were designed by Lancelot 'Capability' Brown. Other, non-agricultural sources of income are represented within this list: Stourhead was built in 1741 by the banker Henry Hoare, while the Beckfords of Fonthill made an absolute fortune from slave plantations in the West Indies.

By the 18th century, there were certainly rewards for some, and in order to improve agricultural practice in the region, the Bath and West of England Agricultural Society 'For the encouragement of Agriculture, Arts, Manufactures and Commerce' was founded in 1777. It had considerable influence promoting agricultural shows, meetings, lectures, practical demonstrations and publications.[160] The Bath and West and Southern Counties Society letters and papers on Agriculture, planting &c., from 1780 and later the *Journal of the Bath and West of England Society for the encouragement of agriculture, arts, manufactures and commerce*.

Nationally, the Board of Agriculture was formed by Royal Charter in 1793. This was a consequence of lobbying by Sir John Sinclair. Seen as a forerunner to the Department for Food and Rural Affairs, its composition was dominated by aristocratic and landowning interests. Lord Lonsdale, the Duke of Bedford and Coke of Holkham were all members as well as the great agricultural improver Arthur Young FRS. Sinclair 'believed that a public society would have more weight than a private institution for the promotion of agriculture',[161] yet the Board was still seen as a government department. In its role to give impetus to the New Agriculture, although it had no bureaucratic function, public money was involved as was private subscription: it received a grant of £3000 per annum from Parliament. The Board commissioned agricultural surveys on a county basis to ascertain the full state of agriculture,

---

[158] T. Williamson and L. Bellemy (1987) *Property and Landscape : A Social History of Land Ownership and the English Countryside*. George Philip, London, Chapter 6.
[159] J.H. Bettey (1986) *Wessex from AD 1000*. Longman, New York , 209-214.
[160] J.H. Bettey (1986) *Wessex from AD 1000*. Longman, New York, 198.
[161] H.F. Cook (2010) Boom, slump and intervention: changing agricultural landscapes on Romney Marsh, 1790 to 1990, in M.P. Waller, E. Edwards and L.L. Barber (eds) *Romney Marsh: Persistence and Change in a Coastal Lowland*. Romney Marsh Research Trust, Sevenoaks, 155-183.

promote the most advantageous methods of agriculture, act as a general magazine for agricultural knowledge and inspire local societies.[162]

By the 1790s, the Board of Agriculture could still report extensive unenclosed downland in Dorset, covered by numerous sheep, although the prevalence of not only corn, but also beans, pease, vetches, turnips, clover ley, flax and hemp and the practice of fallow point to a progressive agriculture with rotation practiced in enclosed areas including the Blackmoor Vale.[163] The local agricultural economy, especially near to the ports, supported the manufacture of twine, string, pack thread, cordage, ropes and sails, hammocks, bags and tarpaulins in the neighbourhood of Bridport and Beaminster. Such products supplied the fishing and maritime industries. Land values and rental on land were highest near to towns, indicating a local stimulus to the agricultural economy.[164]

Government spending during the Napoleonic wars both increased the size of the army and navy, and strengthened defenses, particularly along the south coast, including the construction of Martello Towers, most notably along the coast of Kent and Sussex between 1805 and 1812.[165] Increased taxation caused desperate misery and food prices were inflated. While industries associated with the war effort benefitted, there was unemployment caused by wartime trade restrictions, and the increased mechanisation.[166] Discontent would come to a head in the actions of the Luddites in the industrializing north of England (1811-16) but not yet in the south.

By 1815, Enclosure Acts had efficiently transformed many areas but there remained open areas, or large enclosures, especially on the downs. Heathland largely also remained unenclosed, unless planted with firs where enclosures were made from sod banks. During the period of the Napoleonic Wars, land rental greatly increased, as did the practice of crop rotation, including the Norfolk Rotation developed in the 17th century (wheat, turnips, barley, and clover or under-grass).[167]

By the 19th century Wessex had become famed for its great estates. Large landowners could play major roles in the region often on the back of agricultural profits, although gained at the expense of the rural population.

Demand in the economy fell after 1815 as the disturbance of the Napoleonic Period came to an end with peace in continental Europe. However, the agricultural distress caused by the Enclosure Acts caused social and economic problems by 1830.[168] Although the war had begun

---

[162] H.F. Cook (2010) Boom, slump and intervention: changing agricultural landscapes on Romney Marsh, 1790 to 1990, in M.P. Waller, E. Edwards and L.L. Barber (eds) *Romney Marsh: Persistence and Change in a Coastal Lowland*. Romney Marsh Research Trust, Sevenoaks, 155-183..

[163] L.E. Tavener (1940) Dorset, in L.D. Stamp (ed.) *The Land Utilisation of Britain*. Geographical publications, London, 249.

[164] L.E. Tavener (1940) Dorset, in L.D. Stamp (ed.) *The Land Utilisation of Britain*. Geographical publications, London, 250.

[165] J.G. Load (1990) Dymchurch Martello Tower No 24 English Heritage <https://theromneymarsh.net/assets/fileman/Uploads/Documents/Martello_Tower_No.24%20eh%20booklet.pdf> Accessed June 2021.

[166] R. Mather (2014) The impact of the Napoleonic Wars in Britain. <https://www.bl.uk/romantics-and-victorians/articles/the-impact-of-the-napoleonic-wars-in-britain> Accessed June 2021.

[167] L.E. Tavener (1940) Dorset, in L.D. Stamp (ed.) *The Land Utilisation of Britain*. Geographical publications, London, 251.

[168] L.E. Tavener (1940) Dorset, in L.D. Stamp (ed.) *The Land Utilisation of Britain*. Geographical publications, London, 252.

with the ostensible aim of protecting the liberties of the British people, towards the end, many queried whether the corrupt aristocracy were the only ones to benefit.[169]

Poor rural conditions caused the Swing Riots of 1830 which affected Hampshire, Wiltshire and Dorset and can be compared with the actions of the earlier Luddites and the Tolpuddle Martyrs, convicted for forming a union and transported in 1834. The population was rising throughout the 19th century and so the application of science in agriculture was seen to be desirable. In 1838 the Royal Agricultural Society of England was formed and this joined the Board of Agriculture in promoting scientific agricultural investigations.[170] It was evidently important once more to create a body to promote the scientific development of Agriculture, and that it should be independent of government.

Between the 1830s and 1850 agriculture prospered. An economic depression between 1850-1851,[171] although serious, was short-lived in many parts of England; it apparently had little effect in Dorset. The recession occurred in the aftermath of the Repeal of the Corn Laws by Robert Peel's government (1846) and was a subject of discussion among the farming community of the day with many blaming the removal of protectionism from home producers.[172]

The period 1850 to 1875 would see a decisive switch back towards arable farming in Dorset much as it did in Wiltshire (see Chapter 5).[173] The county surveyors of agriculture during Napoleonic times, a one-off survey of 1850-1851 by James Caird and the systematic collection of Agricultural area statistics on a parish basis from 1866 (although there had been earlier censuses),[174] all not only demonstrate State intervention and interest in food security but also would provide a basis for direction of agricultural policy.

A Parliamentary enquiry of 1874 showed (for the first time since Domesday) just how much land was concentrated into the hands of the few. For example, the Maquis of Bath who lived at Longleat owned 55,000 acres (136,000ha). The Earl of Pembroke who lived at Wilton was not far behind. The counties of Wiltshire, Dorset and Somerset were particularly affected by concentration of ownership into single estates. Joseph Bettey states:

'The country house with its well-wooded parkland remained one of the most widespread and characteristic features of the whole region, and great houses like Ashdown Park, Stratfield Saye, Beaulieu, Dunster Castle, Hinton St George, Montacute, Melbury, Wimborne St Giles, Longleat, Wilton and scores of others sustained by rents from their farms and manors continued to emphasise the power, wealth and influence of their owners.'[175]

In Dorset, the area of down pasture varied from around one third to one half of a farm area and was generally promoted because sheep, be they fed on water meadows or on downland,

---

[169] R. Mather (2014) The impact of the Napoleonic Wars in Britain. <https://www.bl.uk/romantics-and-victorians/articles/the-impact-of-the-napoleonic-wars-in-britain> Accessed June 2021.
[170] Royal Agricultural Society of England (n.d.) 'History' <https://www.rase.org.uk/history/> Accessed April 2020.
[171] J. Caird (1852) *English Agriculture 1850-51*. Longmans, London, 58-64.
[172] J. Caird (1852) *English Agriculture 1850-51*. Longmans, London, 28.
[173] L.E. Tavener (1940) Dorset, in L.D. Stamp (ed.) *The Land Utilisation of Britain*. Geographical publications, London, 253.
[174] National Archives (n.d.) Agricultural statistics of England and Wales <http://www.nationalarchives.gov.uk/help-with-your-research/research-guides/agricultural-statistics-england-wales/> Accessed April 2020.
[175] J.H. Bettey (1986) *Wessex from AD 1000*. Longman, New York, 280.

were folded on wheat and barley. In reality the period 1853 to 1875 was profitable, because the interiors of Australia, New Zealand and north America were not yet developed so as to export meat and grain. There were no imports from Russia and the American Civil War impacted or delayed exports. Subsequently, the balance between pasture and arable changed, with pasture gradually increasing after the mid-1870s as the home industry was affected by imports.[176]

The period 1875 to 1937 included the serious agricultural slump of 1879, which followed some poor harvests for Dorset, and, except for the First World War (see Chapter 1), virtually all field crops declined dramatically, with the exception of mangolds (aka mangelwurzels) which are used to feed livestock. Total cattle numbers increased, sheep decreased, and pigs and horses varied but overall numbers remained steady. It was wheat and sheep that seem to have suffered most, and the situation in Wiltshire was comparable.[177]

In Chapter 1, the notion of governmental intervention was introduced. As far as Wessex is concerned, The First Land Utilisation Survey of the 1930s provides a good base-level for agricultural land use in a relative agricultural recession, for comparison with times both before and after. For Wessex, two examples will serve to illustrate this point. Figure 3.7 shows the classification scheme adopted for publication.

While many categories of land use are lumped together and some may seem puzzling, the basic distinction between tilled land, meadowland and pasture, kinds of woodland and heath and moorland is intuitive and clear, although there is an eliding of market gardens and 'arable land'. More conventionally 'arable' would be understood as cereal, legume or root crop monocultures as well as temporary grass. The identification of 'new' plantation and housing land may seem puzzling, as may the confusion between 'common land' (a legal concept) and heath, moorland and rough pasture.

It is interesting to note how the direction of land use change may be inferred by new woodland planting and new housing development. Furthermore, the notion of housing density that might permit food production from gardens is something that would come into its own in the 'Dig for Victory Campaign' of the Second World War, as would the allotments, also recorded here.

Inspection of Figures 3.8 and 3.9 shows surprising similarities over almost 90 years of land use. There has been little urban development, and there remains a characteristic nucleated settlement pattern. Arable fields marked in salmon recur across much of the satellite image of the area (Figure 3.9), appearing as similarly coloured large fields differentiated from areas of woodland. Yellow areas in Figure 3.8 show there was, and remains, largely downland rough pasture.

Areas of woodland east of Cranborne reflect the change of geology to a range of sediments of the lower Tertiary age of the Hampshire Basin, which tend to produce less fertile soils than those underlain by chalk. Many areas of woodland west and north of Cranborne, however,

---

[176] L.E. Tavener (1940) Dorset, in L.D. Stamp (ed.) *The Land Utilisation of Britain*. Geographical publications, London, 253.
[177] L.E. Tavener (1940) Dorset, in L.D. Stamp (ed.) *The Land Utilisation of Britain*. Geographical publications, London, 253-255; A. H. Fry (1940) Wiltshire in L.D. Stamp (ed.) *The Land Utilisation of Britain*. Geographical publications, London, Chapter III; Chapter 5 (this volume).

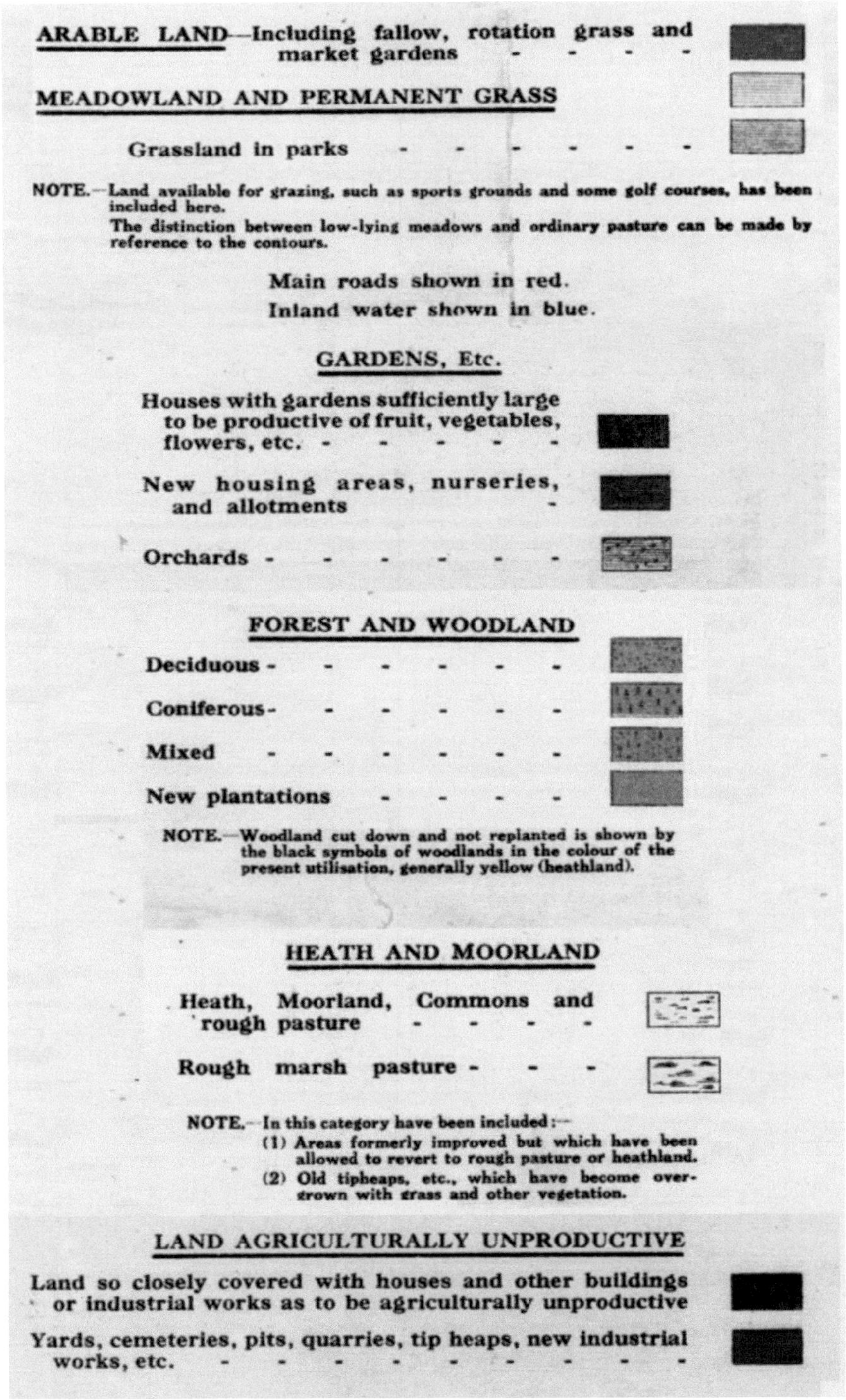

ARABLE LAND—Including fallow, rotation grass and market gardens - - - -

MEADOWLAND AND PERMANENT GRASS

Grassland in parks - - - - - -

NOTE.—Land available for grazing, such as sports grounds and some golf courses, has been included here.

The distinction between low-lying meadows and ordinary pasture can be made by reference to the contours.

Main roads shown in red.

Inland water shown in blue.

GARDENS, Etc.

Houses with gardens sufficiently large to be productive of fruit, vegetables, flowers, etc. - - - - -

New housing areas, nurseries, and allotments -

Orchards - - - - - -

FOREST AND WOODLAND

Deciduous - - - - - -

Coniferous - - - - - -

Mixed - - - - -

New plantations - - - -

NOTE.—Woodland cut down and not replanted is shown by the black symbols of woodlands in the colour of the present utilisation, generally yellow (heathland).

HEATH AND MOORLAND

Heath, Moorland, Commons and rough pasture - - - -

Rough marsh pasture - - -

NOTE.—In this category have been included:—
(1) Areas formerly improved but which have been allowed to revert to rough pasture or heathland.
(2) Old tipheaps, etc., which have become overgrown with grass and other vegetation.

LAND AGRICULTURALLY UNPRODUCTIVE

Land so closely covered with houses and other buildings or industrial works as to be agriculturally unproductive

Yards, cemeteries, pits, quarries, tip heaps, new industrial works, etc. - - - - - - - -

Figure 3.7. Legend to the First Land Utilisation Survey of the 1930s (public domain)

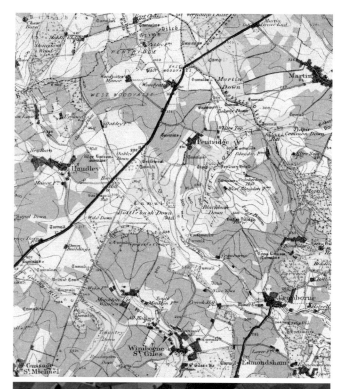

Figure 3.8. Land use in the early 1930s between Wimborne St Giles and Martin Down. Land Utilisation Survey of Great Britain. Sheet 131, Ringwood (public domain)

Figure 3.9. Modern Satellite image of area displayed in Figure 3.8 (Google Earth)

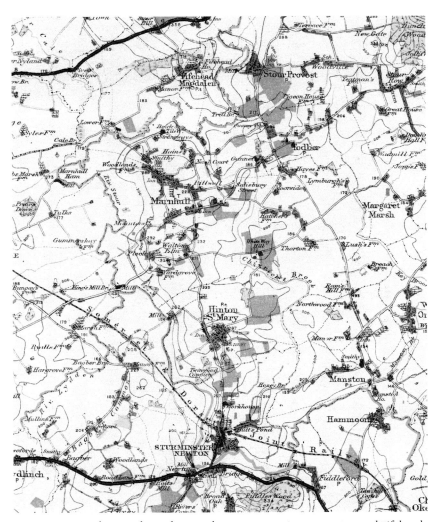

Figure 3.10. Land use in the early 1930s between Sturminster Newton and Fifehead Magdalen, Dorset in the 1930s. Sheet 130, Yeovil and Blandford (public domain)

grow not in chalk soils, but in the overlying Clay-with-Flints. Around the village of Pentridge (as elsewhere on Cranborne Chase) are abundant prehistoric earthworks including tumuli and field systems.[178] These serve to show the antiquity of agriculture in the area. The famous Fir Tree Field Shaft (Chapter 2.3) that possibly shows the transition to agriculture occurs close to the area marked 'Down Buildings' to the south of (Sixpenny) Handley' and west of the Roman Road with a distinctive raised flint, stone and chalk 'agger' marked as the 'Ackling Dyke'.[179] The persistence of arable agriculture in this area since the middle 20th century is a testament

[178] Historic England (2020) Penbury Knoll camp on Pentridge Hill. <https://historicengland.org.uk/listing/the-list/list-entry/1002717> Accessed April 2020.

[179] Historic England (2020) Part of Ackling Dyke (Roman road), including Roman road on Oakley Down. <https://historicengland.org.uk/listing/the-list/list-entry/1003309> Accessed April 2020.

Figure 3.11. Modern Satellite image of area displayed in Figure 3.10 (Google Earth)

to the suitability of chalkland soils, we can also confidently state this is something of great antiquity.

If the Cranborne Chase example shows more continuity than change over the last century, arable expansion on a dramatic scale can be seen in Figures 3.10 and 3.11. This is easily explained in terms of geology and soil. The 'Corallian ridge' runs north-south through the picture and in the 1930s supported a few areas of arable cultivation, notably between Sturminster Newton and Fifehead Magdalen. This feature, cut through by Chivrick's Brook, is a ridge comprising a complex succession of interdigitating limestones, marls, sandstones, sands, siltstones, silts, spiculites (developed from sponges) and mudstones up to 100m in thickness, and also containing a thickness of aquifer.

This lithology means that this ridge rises above both the Dorset Stour floodplain located to the west and underlain by Kellaways Formation and Oxford Clay Formation, mudstone, siltstone and sandstone, but also the land to the east of the Corallian ridge, where the solid geology comprises West Walton Formation, Ampthill Clay Formation and Kimmeridge Clay Formation of similar lithologies.[180] While the soils developed on these are complex and variable they are highly suited to a range of land uses, including cereals.[181] This contrasts with the majority of soils east and west of the ridge, which are most suited to grassland and woodlands.[182] Inspection of Figure 3.11, however, shows a considerably larger area of arable land at present

---

[180] British Geological Survey (2020) Geology of Britain viewer <http://mapapps.bgs.ac.uk/geologyofbritain/home. html > Accessed April 2020.
[181] D.C. Findlay, G.J.N Colborne, D.W. Cope, T.R. Harrod, D.V. Hogan, and S.J. Staines (1984) *Soils and their Use in South West England*. Soil Survey of England and Wales, Harpenden, 1:250,000 map, legend and memoir 162-163.
[182] D.C. Findlay, G.J.N Colborne, D.W. Cope, T.R. Harrod, D.V. Hogan, and S.J. Staines (1984) *Soils and their Use in South West England*. Soil Survey of England and Wales, Harpenden, 143- 144.

when compared with the early 1930s including land east and west of the ridge (Figure 3.10). The expansion of arable area on suitable soils has continued since 1940.

3.9 Some conclusions

Feudalism enabled certain families to become wealthy through military service to the king. This was to change as the economy moved more towards a capitalistic mode of production. From approximately Tudor times onward, the path was open for agricultural development, and this would be realised by the 18th century with its huge estates and luxurious stately homes. While progress towards a 'privatised' landscape was brought about through unequally shared (including total dispossession) profits for the Wessex peasantry, who suffered considerably from enclosure, there were other ways. The 18th century gave ample options for beneficiaries from the developing financial system to 'buy in', as industrialists would later. There were even less wholesome ways and means to fast track into the aristocracy.

The acquisition of wealth through ownership of slave plantations has been mentioned. For example, the Codrington Family owned plantations in Barbados and were able to acquire land and build at Doddington Park in south Gloucestershire.[183] Still more notorious were the Beckfords (see Chapter 5). Wessex was to prove a kind of entrepot for wealth thus acquired, something historians of the landscape need to fully appreciate, although estates which benefitted from the profits of slavery are to be found across Britain. Needless to say, by the 19th century industry was another available option to acquire land, prestigious houses or both, and in recent decades successful actors and rock musicians have followed suit. These include 'Sting', who owns Lake House in Wiltshire.

The manorial system was doomed and with 18th century 'power in the land' came state interest, something that can be traced from the Board of Agriculture, through the nineteenth century Corn Laws to the collection of agricultural statistics and a general desire to acquire mapped and economic information as well as regulation of the environment. Because of scientific advances and the systematic collection of agricultural information into the 20th century, there is good information about land-use change, which has been supplemented by aerial photography and satellite imagery. The collation of agricultural and land use data has directed landscape change since 1940 in a manner that has equalled, or exceeded, those changes wrought by enclosure of common land and OFA. The nineteenth and twentieth centuries experienced a rise in state intervention through of statutory bodies and strong voluntary sector organisations that would impact the countryside of Wessex.

This chapter has introduced some key topics around the historic governance of Wessex. The remainder of this book will systematically describe the historic function and development of the Wessex Landscape based upon the landscape types and agro-ecological descriptions so far outlined.

---

[183] R. Thorne ed. (1986) The History of Parliament: the House of Commons, 1790-1820. <http://www.histparl.ac.uk/volume/1790-1820/member/bethell-codrington-christopher-1764-1843> Accessed April 2020.

Chapter 4

Floodplains, levels and marshes

4.1 Lowlands of the West

River management, economic development on floodplains and managed and reclaimed wetlands all play a vital role in the Wessex landscape, providing for transport, fisheries, watercress production, water supply, waterpower and irrigation. This chapter shows how water made the lowland areas of Wessex, and why the management of water must provide sufficiency of supply while doing its best to control excess. Effective water management should be a major cultural achievement that impacts the modern landscape. More recently, water management imperatives for conservation purposes sit alongside historic practices with direct economic benefits. Historic management systems also provide an impressive array of ecological benefits.

Throughout England and Wales, extensive areas such as Romney Marsh, the Broadlands of Norfolk, the Fens of East Anglia and - in Wessex - the Somerset Levels and Moors, as well as (opposite the intervening Severn Estuary) the Gwent Levels of South Wales, form coastal lowlands prone to flooding from both landward and seaward sides. Here, the interplay of sediment delivered by the rivers and coastal processes cause floodplains to coalesce.[1] Inland marshes and fens, the creation of coastal saltmarsh and land reclamation are all characteristic of lowland landscapes resulting from environmental management.

The practice of protecting areas reclaimed from the sea by the construction of sea walls designed to prevent the ingress of saline water, and also providing some hydrological control of freshwater on the landward side, is ancient and may often be traced back to Romano-British times.[2] From the medieval period land reclamation was progressed, including by monastic institutions, who had the resources to develop areas reclaimed from the sea, while later, lay owners had the resources to undertake regional-scale alteration, generally for agriculture. In any event, the regulating institutions for land drainage and sea defences in England that came to be termed Commissioners of Sewers had their origins in the 13th century.[3]

As noted in the Introduction, mean rainfall across Wessex varies greatly; the possible Bristol Channel tsunami was described in Chapter 1. Some of the more significant flooding events over the last 100 years occurred in 2012, 2000, 1997, 1960 and 1929 which remind us of the origins of wetlands. In early 2014, the area saw widespread flooding, particularly within the Parrett and Tone river catchments. It has been estimated that there were more than 100 million cubic metres of floodwater covering an area of 65 km². Residents of Northmoor (Moorland,

[1]   H.F. Cook and H. Moorby (1993) English marshlands reclaimed for grazing: a review of the physical environment. *Journal of Environmental Management* 38, 55-72.
[2]   S. Rippon (2006) Taming a Wetland Wilderness - Romano-British and Medieval reclamation in the Somerset Levels and Moors, in P. Hill-Cottingham, D. Briggs, R. Brunning, A. King and G. Rix (eds) *The Somerset Wetlands: An Ever Changing Environment*. Somerset Archaeological & Natural History Society, Taunton, 47-56.
[3]   H. Cook and T. Williamson (1999) Introduction: Landscape environment and history in H. Cook and T. Williamson (eds) *Water Management in the English Landscape*. Edinburgh University Press, Edinburgh.

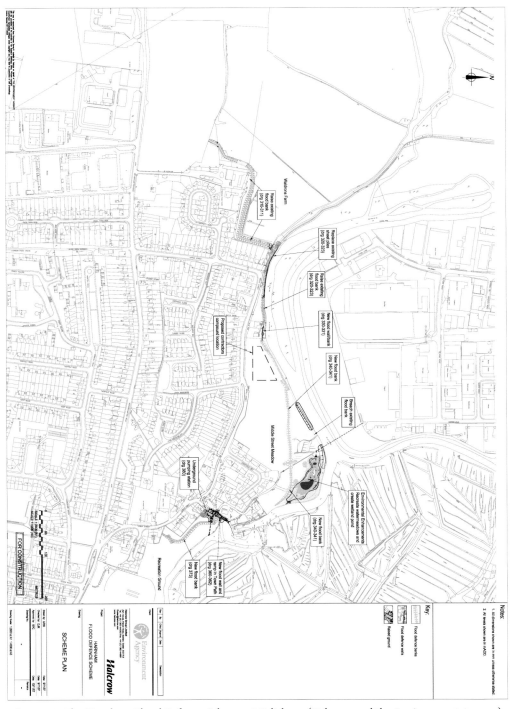

Figure 4.1. The Harnham Flood Defence Scheme at Salisbury (Halcrow and the Environment Agency).

Chadmead and Fordgate) left their homes at the height of the flood and communities were cut off by floodwater.[4]

Further inland, expectations of flooding may be less dramatic. However, in general, public policy in the UK is to defend urban areas but, with careful consideration, permit certain rural areas to flood. This may be seen in microcosm in Figure 4.1. The Harnham Flood Defence Scheme (2008)[5] defends the urban area with bunds, a detention pond and associated channels, while the Harnham Water Meadows (northeast side of the extract) are located on a 40ha alluvial island, bordered by rivers: the north branch of the Nadder, its south branch and the Salisbury Avon.

This part of the Nadder valley is located on the edge of National Character Area (NCA) Profile 132: Salisbury Plain and West Wiltshire Downs, geologically dominated by chalk and drained by the Salisbury Avon river system, including the Nadder at West Harnham.[6] The water meadows will flood during times of high river discharge. While the houses and businesses of West Harnham are protected from a 1:200-year flood event, the Cathedral and its Close (to the Eastnortheast, not shown) should experience reduced flood-risk due to detention by the water meadows.

It is helpful to differentiate four kinds of land use and cover on natural floodplains:

- *Areas of natural floodplain vegetation seemingly unaffected by management.* These are rare, if not absent, in Britain. However, examples of climax vegetation that has been re-established over time may be found behind the artificial coastal dam at Lymington Reed Bed (see Chapter 9), on wet woodland around the Decoy at Itchen Valley Country Park near Eastleigh, some small islands in Wessex rivers which are not grazable and the wet woodland at Mellow Farm, Dockenfield, Surrey.[7]
- *Flood meadows inundated only when the river is in spate.* In their natural form, soil nutrients come from the river water or else are derived via the digestive tracts of grazing animals. An example is the Christchurch meadows on the lower Avon valley.
- *Grazing marshes* are areas where soil water levels are managed through a networked system of ditches that may be augmented by pumping or lifting of water. Common in Wessex, especially across the Somerset Levels, their conservation value is high, including habitat provision for birdlife. Unlike areas further east in England, investment in pumping to increase productivity for arable farming has hardly been adopted.[8]
- *Irrigated ('floated') water meadows* are engineered systems designed to irrigate pasture. Examples include the water meadows at West Harnham, Lower Woodford and Britford near Salisbury and at Twyford Mead, near Winchester, Hampshire.

[4]   Environment Agency (2015) Somerset Levels and Moors: reducing the risk of flooding <https://www.gov.uk/government/publications/somerset-levels-and-moors-reducing-the-risk-of-flooding/somerset-levels-and-moors-reducing-the-risk-of-flooding> Accessed April 2020.
[5]   Environment Agency (n.d) Harnham Flood Defence Scheme <http://teamvanoord.com/downloads/MG024%20Harnham%20FDS.pdf > Accessed April 2020.
[6]   Natural England (2014) National Character Area Profiles: data for local decision making. < https://www.gov.uk/government/publications/national-character-area-profiles-data-for-local-decision-making/national-character-area-profiles >Accessed April 2020.
[7]   K. Stearne (2019) pers. comm.
[8]   J. Purseglove (2015) *Taming the Flood* (2nd ed.) William Collins, London, 300-308.

Areas of purely **natural vegetation** include forms of 'primary wetland' (Table 4.1). In Wessex, most areas will have had human intervention at some time in the past 500 years or more. Vegetation communities that are considered 'natural' are probably more likely an outcome of modern neglect than a continuation of pre-agricultural land cover. To put it another way, there is a tendency towards conditions of a locally environmentally stable 'climax vegetation'. With human intervention, developed plant communities are termed 'plagioclimax' where an unmodified climax (stable) vegetation condition is not reached due to human intervention. Habitats such as reedbeds, alder carr and riverine wetlands may all experience minimal intervention. Interestingly, ideas of 'rewilding' appear to look back to generally unspecified time period.[9] Simply abandoning river and floodplain management that may have been in place for millennia to the vagaries of weather and vegetation succession would prove controversial to many stakeholders in river management.

A balance should be sought between historic (and generally sustainable) management and proven ecosystem services such as nutrient stripping from water, carbon sequestration and habitat provision for biodiversity. Floodplains potentially offer a wide range of habitat restorations based in topography, soils and hydrology.[10] The historical economic value of such landscapes lies in relatively unmanaged climax vegetation. Wood products, reeds, fishing, fouling and withy cutting all provided 'primary' wetland products for human exploitation. In modern times they provide a range of ecosystem services and habitats.

Flood meadows represent a greater degree of human intervention historically, specifically grazing or cutting of hay crops in the summer and the provision of ditches to assist drainage. They are therefore unequivocally created as a part of an agricultural system, providing not only grazing throughout much of the year but also hay and silage for animals. 'Meadow' (as opposed to pasture) is a category recorded for each manor in the Domesday survey (1086) as well as other river-related features such as fisheries and mills,[11] emphasising their economic importance. Whilst it is unlikely there was much cultivation on active floodplains in earlier times, livestock would be a natural choice, because animals are able to move when a river is in spate. Flood meadows provide for a range of ecosystem services including floral diversity, habitats for animals and birds, and floodwater detention silt and nutrient sinks.[12]

Grazing marshes, like riverine flood meadows, are derivatives of primary wetlands. Classic 'grazing marsh' is a hydrologically modified grassland area given over to grazing[13]. They operate through a network of ditches, created so as to rid the subsoil of water at the field-scale and linked, with or without a capacity to pump or lift water, to an arterial watercourse,[14]

9   Wilderness Foundation (2018) <https://www.wildernessfoundation.org.uk/wilderness-action/wild-britain/> Accessed April 2020.
10   H.F. Cook (2010) Floodplain agricultural systems: functionality, heritage and conservation. *Journal of Flood Risk Management* 3, 1-9 DOI:10.1111/j.1753-318X.2010.01069.x
11   National Archives (n.d.) Survey and making of Domesday. <http://www.nationalarchives.gov.uk/domesday/discover-domesday/making-of-domesday.htm> Accessed April 2020.
12   H.F. Cook (2007) Soil Nutrient and Sediment Dynamics on the Kentish Stour. *Water and Environmental Journal* 21, 173-181; Cook, H.F. (2010). Floodplain agricultural systems: functionality, heritage and conservation. *Journal of Flood Risk Management* 3, 1-9. DOI:10.1111/j.1753-318X.2010.01069.x
13   H.F. Cook and H. Moorby (1993) English marshlands reclaimed for grazing: a review of the physical environment. *Journal of Environmental Management* 38, 55-72.
14   H. Cook (1999) Hydrological management in reclaimed wetlands In: Cook, H.F. and Williamson, T. (eds). *Water Management in the English Landscape*. Edinburgh University Press, Edinburgh, 84-100.

while the creation of embankments prevents inundation. It also follows that extensive areas of these kinds of landscape are coastal. In Wessex they are to be found along the south coast (for example in the lower Avon Valley below Ringwood), on the Somerset Levels and Moors described below, in the lower Dorset river Frome around Wareham and in certain small areas adjacent to the Bristol Channel, for example Sparkhayes Marsh at Porlock in Somerset described below. Examples of embanked grazing marsh are to be found at Lymington, Keyhaven and Beaulieu on the North Solent National Nature Reserve.[15]

The term **water meadow** may carelessly be applied to any area of floodplain grassland adjacent to a river that is inundated when the river is in spate, so that that the floodplain meadow is operating as a second (larger) channel. However, in this book the term is used to describe areas that have been engineered so that irrigation is applied by choice in accordance with the demands of the farming system in question. True (sometimes termed 'floated') water meadows were intensive grassland production systems and required considerable labour inputs, which are described below.

4.2 River Valley floodplains

A river that appears to be 'single thread' may thus be regarded as having two channels. One is the channel normally occupied and floored by the riverbed, the other is the high flow channel floored by not only the riverbed but also its banks and floodplain proper. In ecological terms there is an 'ecotone' where the transition between the floodplain and river channel provides for fringing vegetation, including reedbeds. The overall floodplain is an interface between the fluvial environment and surrounding higher land.

Features of a typically developed single channel river valley floodplain are illustrated in Figure 4.2. The processes are familiar: meandering leads to extreme sinuosity whereby meander bends catch up, one with another, in a dynamic that shortens the stretch of a river and locally increases the stream gradient. The 'cut-off' channels remain as oxbow lakes that provide for wetland habitats and will gradually fill as vegetation develops. Unmodified floodplains are a primary wetland, but modification, such as ditch-digging, moves the classification towards secondary wetlands. When managed for grazing, the description 'flood meadows' is used.

The flood meadows at Cricklade North Meadow in Wiltshire lie at the confluence of the rivers Thames and Churn and they enjoy protection as Sites of Special Scientific Interest (SSSI), Special Area of Conservation (SAC) and a National Nature Reserve (NNR).[16] It forms part of the complex Landscape Character Area 108: Upper Thames Clay Vales, characterised by contrasting landscapes, including 'enclosed pastures of the claylands with wet valleys, mixed farming, hedges, hedge trees and field trees and more settled, open, arable lands. Mature field oaks give a parkland feel in many places and there are internationally important wetland meadows.'[17] At North Meadow, agricultural management has probably remained unchanged for centuries; the use of agrochemicals and sward improvement has been minimal or absent.

[15] K. Stearne (2019) pers. comm.

[16] R.S. Wolstenholme (n.d.) The History of North Meadow, Cricklade. <https://anhso.org.uk/wp-content/uploads/Fritillary/frit5-cricklade.pdf> Accessed May 2021.

[17] Natural England (2014) National Character Area Profiles: data for local decision making. < https://www.gov.uk/government/publications/national-character-area-profiles-data-for-local-decision-making/national-character-area-profiles >Accessed April 2020.

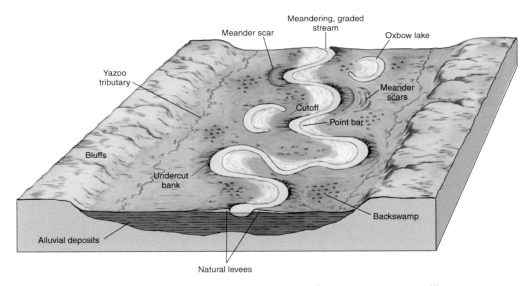

Figure 4.2. The anatomy of a river floodplain (© Pearson Prentice Hall)

The site is today valued for its floral diversity, particularly for supporting a population of snakes head fritillaries (*Fritillaria meleagris*). Other terms applied to the site are 'hay meadows' and 'Lammas meadows'.[18]

The term 'Lammas' (Old English *hlaef-mass* or 'loaf-mass', sometimes 'the feast of the first fruits') is historically moveable but related to harvest festival. It is also remembered from pre-Christian times in Gaelic as *Lughnasadh* or *Lughnasa*. Lammas refers not only to a harvest festival but also describes a particular type of land tenure, whereby the owner (in effect historically the lord of the manor in which the meadow lies) divides the meadow into parcels of land referred to as 'lots' or 'doles', the boundaries of which are demarked by boundary stones or 'meerstones'. The owner then sells the rights to the hay crop to local farmers who are responsible for harvesting the hay in each allotment. The hay is cut after 1st July when the wildflowers have set seed, then the meadow becomes common pasture and the livestock of those with commonable rights are entitled to graze the entire meadow.[19] These commonable rights begin on (Old) Lammas day (August 12th) and end on 12th February, around Candlemas or the pre-Christian festival of *Imbolc* (February 1st), when the meadows are laid up for hay until July. An island in the river Avon just north of Ringwood in Hampshire was a similarly divided tenure until 1960s.[20]

Early evidence that land in North Meadow was divided into allotments comes from glebe 'terriers.' A land terrier is a record system for an institution's land and property holdings and may comprise both written records and maps. The 1588 terrier for Cricklade St Sampson

---

[18] Cricklade Bloomers (n.d) Cricklade North Meadow <http://www.crickladeinbloom.co.uk/north_meadow.html> Accessed April 2020.

[19] Cricklade Town Council (n.d.) Cricklade- the manorial court and North meadow Accessed April 2020 <http://www. crickladetowncouncil.gov.uk/cricklade/north-meadow.html> Accessed May 2021.

[20] K. Stearne (2019) pers. comm.

Figure 4.3. 'Meerstone' on the Cricklade meadows (the author).

incorporated Lammas tithes of cattle grazed in common, in accordance with the ancient custom, and the glebe of 1608 for Cricklade St Mary refers to a share in Northmead of about an acre (c. 0.4ha). There were Inclosure Acts in both 1814 and 1824. The 1814 Act enabled certain persons to exercise either common grazing rights or the right of first vesture. The 1824 Act, however, split the meadow up into allotments, and freehold remained not with the allotment holders but with the Manor. Since then, Manorial rights have vanished, meaning that the freehold rights are now exercised by the allotment holders.[21]

Despite Enclosure Acts affecting the district, the ancient Lammas routine continued, with North Meadow remaining unenclosed and retaining its commonable rights. Today it is managed for the ecology of the sward by Natural England.

A further feature of these meadows was the redistribution of doles so that each owner had the same acreage of mead but in one consolidated unit rather than several small strips. The meadows were united in about 1898 when Cricklade was formed of two parishes. In order that plots could be demarcated without interfering with common grazing, small dole stones were put in place at intervals along the boundaries. Some meerstones remain in position and some are inscribed with initials which allude to names registered in the Enclosure Award of 1824.[22]

[21] R.S. Wolstenhome (n.d.) The History of North Meadow, Cricklade. <https://anhso.org.uk/wp-content/uploads/ Fritillary/frit5-cricklade.pdf> Accessed May 2021.
[22] Historic England (2021) North meadow boundary stones. <https://historicengland.org.uk/listing/the-list/list-entry/1356087> Accessed May 2021.

Regulation remains, unusually, through Cricklade's Court Leet, a rare survival.[23] Figure 4.3 shows the view across the Cricklade meadows with a meerstone in the foreground.

4.3 Multiple channel rivers

One feature of floodplains in Wessex is where the channels apparently naturally break up to produce multiple-channelled irregular networks across the floodplain. These are 'anastomosing' or branching (but then re-joining) channels.[24] Figure 4.4a shows the river channel pattern for the Stour near Charlton Marshall, Dorset, which flows from north to south. It is likely that a conventionally meandering stream, constrained against the eastern valley side by the modern Keynsham Mill, proved inadequate to carry the imposed sediment load caused by accelerated soil erosion, probably during prehistory (Chapters 1 and 2). The result was the formation of secondary channels through a process termed 'avulsion', when during a flood event, a new channel is formed across the floodplain. Channels on the western side were thus formed, capturing streams flowing from the west.[25]

Figure 4.4b shows the Nadder channel network between Wilton and Salisbury in 1838. The large single island (including the Harnham Water Meadows) is downstream at the eastern end and the southern channel of the Nadder is the likely consequence of avulsion from an originally meandering section of river. No scale is provided and the provenance of the map is uncertain, but it may form a part of tithe award mapping. The general channel pattern is much as it appears in the present day, but it lacks at least one cut-off channel, known to have been artificially cut; channel modifications including constructed banks are commonplace.

Before the river emerges from Wilton Park (left side of Figure 4.4b), evidence of channel alterations undertaken for aesthetic purposes are clear. Around the Harnham Water Meadows two mills stood in the 1830s, called Minty's Mill and Bell's Mill (the Harnham Mill or Old Mill on the south side and Fisherton Mill to the north respectively). Upstream of Harnham mills are recorded at Quidhampton ('Bone Mill') and Bemerton Mill.

In both examples (Figure 4.4a and 4.4b), the original channel pattern arose from inadvertent modification. The likely cause of this multiple-channelled network is sediment loading, which aggrades the channel floor with fine sediment, reducing capacity and producing a tendency to overtop the channel, carving a new channel through the existing banks. Extreme events would not permit all the flow to return to the original channel, producing a second channel across the floodplain by avulsion.[26] This process involves a complex interplay of river channels, mills and water meadows as well as providing opportunities to develop navigation. Upstream of Salisbury, there are no known improvements made to facilitate navigation, but a 20th century flood relief channel is located just west of the Harnham Water Meadows.[27]

---

[23] Cricklade Court Leet Charity (2019) <http://www.crickladecourtleet.org.uk/>Accessed April 2020

[24] These are not the same as patterns that result from parallel artificial cuts associated with watermeadow carriers, overflow channels, mill leats or canals for transport.

[25] H. Cook (2018) River channel planforms and floodplains: a study in the Wessex landscape, *Landscape History*, 39 (1), 5-24. DOI: 10.1080/01433768.2018.1466548

[26] H.F. Cook (2008) Evolution of a floodplain landscape: A case study of the Harnham Water Meadows at Salisbury England. *Landscapes* 9 (1), 50-73; H. Cook (2018) River channel planform and floodplains: a study in the Wessex landscape. *Landscape History* 39, 5-25.

[27] H. Cook (2018) River channel planform and floodplains: a study in the Wessex landscape. *Landscape History* 39, 5-25.

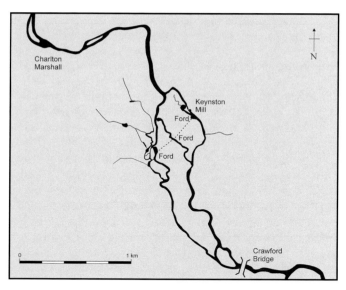

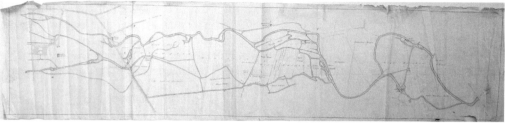

Figure 4.4. Figure 4.4a (top) river channels on the Dorset Stour Near to Charlton Marshall, channels, and islands from the Dorset Stour. This valley forms part of NCA 134: Dorset Downs and Cranborne Chase, which is overall dominated by chalk (author). Figure 4.4b (bottom) the Nadder channel network between Wilton and Salisbury in 1838 (length depicted c. 5km), flow is Westnorthwest-Eastsoutheast or left to right). (Salisbury Museum).

Floodplain environments may have undergone a range of alterations in the past, especially in connection with agriculture. For instance, the SSSI area of the Harnham Water Meadows at Salisbury includes areas of former water meadow that have lost their topography, partly through surface peat shrinkage and disturbance, including 'poaching' by animal hooves and possibly shallow ploughing during the Second World War.[28]

This SSSI provides a floodplain habitat managed for floral diversity instead of historic water meadow. The loss of specific water meadow topography has had the effect of this area effectively reverting to flood meadow. Another example, from Hampshire, is the redundant water meadow system at the Itchen Valley Country Park, which dates from around 1700. Here

---

[28] H.F. Cook, M. Cowan and T. Tatton-Brown (2008) *Harnham Water Meadows: History and Description.* Hobnob Press, Salisbury.

the soil is likely to have experienced problems associated with peat soils (especially poaching) that are located in the topographic lows.[29]

4.4 The Somerset Levels and Moors: Context

The Somerset Levels and Moors[30] (NCA 142)[31] extend across parts of the north and centre of the historical county of Somerset, reaching from Clevedon near Bristol in the north to Glastonbury in the east and Ilchester and Langport in the south. They are dissected by the Carboniferous limestone Mendip Hills (NCA 141), which display classic karst features, and the younger limestone ridges of the Mid-Somerset Hills (NCA 143), notably the Polden Hills. The Somerset Levels and Moors are a flat, alluvial (both marine and river in origin) and peat lowland landscape dominated by 'rhynes' (drainage ditches), canals, reedbeds, willows and meadows, extending across parts of the north and centre of the historical county of Somerset.

The area is approximately 650 km² in area and is punctuated by east-west ridges of hills. The main drainage outlets are the rivers Axe, Brue, the (artificial) Huntspill, Parrett, Tone, Yeo and the King's Sedgemoor Drain (an artificial channel into which the River Cary runs). The Polden Hills separate the area to the south (drained by the River Parrett), and north of these the drainage is to the rivers Axe and Brue, while the Mendip Hills separate the North Somerset Levels (between Weston-super-Mare and Bristol) from the Somerset Levels and Moors south of the Mendip Hills. The general geography of the strongly agricultural Somerset region, including the Central Somerset Levels (and Moors) and the western and North Somerset Level north of Mendip, is shown in Figure 4.5. Significantly, the Levels and Moors occupy a central region in the County, with Exmoor to the west, the Mendips to the north and Jurassic and Cretaceous 'scarplands' to the south and east. Somerset has been likened to a saucer (Figure 1.1)!

The earlier Holocene deposits date from the post-glacial rise in sea level which flooded the area between about 8000 and 4500 BC, depositing silts and clays (with peat forming where there was a temporary lowering of sea level) around upland areas and sand islands.[32] After this time, the sea retreated and reedswamps led to the formation of peat over marine clays. By about 3500 BC there was carr (comprising birch, willow and alder) in addition to reedswamp and pools. This led to the development of raised bogs which became acid due to being rainfed rather than fed by (calcareous) surface or groundwater.

There are three sources of water that may become unwelcome:

---

[29] H. Cook and K. Young (2011) Watermeadows at the Itchen Valley Country Park near Eastleigh, Hampshire *Proceedings of the Hampshire Field Club and Archaeological Society (Hampshire Studies)* 66, 166-186.

[30] The term 'Levels' in this context applies to low, reclaimed areas comprising mineral alluvial sediments that form typically silty soils, generally referred to as marshland (reclaimed or otherwise). 'Moors', on the other hand, a term normally applied to the upland equivalent of heathlands but in this instance, areas that have accumulated peat. Hence the distinction is one of soil type: the marshlands are technically mineral alluvial, the 'moors' lowland peats. Where the chemistry is alkaline, the peats may be referred to as 'fen peats' as in eastern England. Acidic peat soils are referred to as bog.

[31] Natural England (2014) National Character Area Profiles: data for local decision making. < https://www.gov.uk/government/publications/national-character-area-profiles-data-for-local-decision-making/national-character-area-profiles >Accessed April 2020.

[32] B.J. Coles and J.M. Coles (1989) *The Prehistory of the Somerset Levels.* The Somerset Levels Project, Thorverton.

1. The sea itself (see Chapter 1). While primary saltmarshes are a response to the tidal regime at a point, such events as un-seasonally high tides, storm surges (and even maybe *tsunamis*) present a threat to agricultural activities on the newly won land. Damage is both physical from wave attrition and chemical from salinisation of soil

2. Water from landward uplands. Hydrologically speaking, freshwaters from beyond the marsh area itself present both a threat and opportunity for land reclamation. The threat is from natural flooding at times of high discharge. The opportunity is that when draining the adjacent areas, improved water carriage across a reclaimed area can receive drainage from artificial ditches to drain the reclaimed soils by gravity and lifting /or pumping.

3. Water falling within the marsh area of itself, which can, for convenience, be regarded as a water catchment in its own right.

Excess drainage water must be removed to widen options for agricultural production and prevent flooding, unless management as a wetland is desired. That a network of ditches would able to accommodate sufficient drainage through most years, however, proved insufficient in many regions, stimulating a long and complicated history of wetland drainage, going back to Romano-British times.[33] Depending on local geography, drainers might either construct a

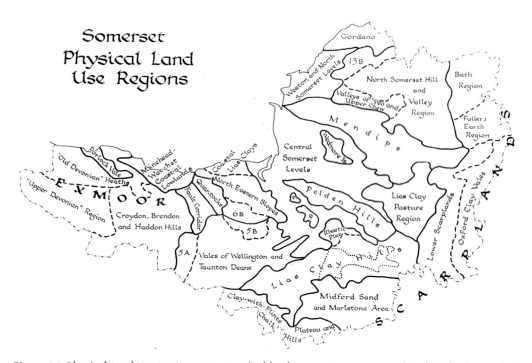

Figure 4.5. Physical Land Use Regions Somerset (public domain, Stuart-MonteathT. (1938). The Land of Britain part 86 Somerset. Geographical Publications. p 12)

---

[33] S. Rippon (1999) Romano-British reclamation of coastal wetlands (Chapter 8) and R. Silvester (1999) Medieval reclamation of marsh and fen (Chapter 9), both in H. Cook and T. Williamson (eds) *Water Management in the English Landscape*. Edinburgh University Press, Edinburgh.

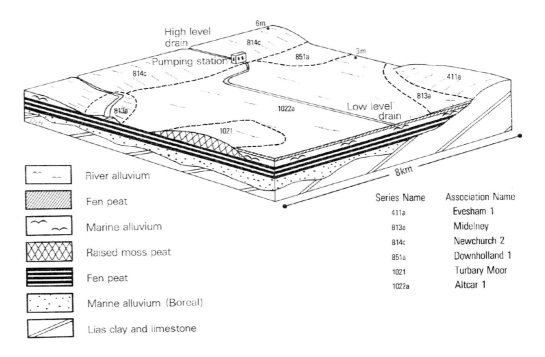

Figure 4.6. Soil Associations on the Somerset Levels and Moors (D.C. Findlay, G.J.N Colborne, D.W. Cope, T.R. Harrod, D.V. Hogan and S.J. Staines 1984. Soils and their Use in South West England. Harpenden: Soil Survey of England and Wales, Bulletin no. 14, 1:250,000 map, legend and memoir, 3)[34]

catchwater[34]channel such as the Royal Military Canal on Romney Marsh,[35] or the (probably) Roman Car Dyke in Fenland.[36] Otherwise rivers crossing the reclaimed wetland would be engineered to accommodate a larger channel capacity; such is the Great Ouse diversion through the Bedford Rivers in Fenland, designed by the Dutch engineer Sir Cornelius Vermuyden, and completed in 1652.[37] Proposals by Vermuyden to drain the Somerset Moor areas were not carried out.

Figure 4.6 shows diagrammatically the main soil associations of the region as well as identifying both coast and surrounding uplands. The centrality of the Somerset Levels is clear. Much enclosure and drainage of the 'moors' (largely reclaimed peat of the Altcarr 1 association, the Turbary Moor association representing former raised bog peat),[38] occurred between about

---

[34] D.C. Findlay, G.J.N Colborne, D.W. Cope, T.R. Harrod, D.V. Hogan and S.J. Staines 1984. *Soils and their Use in South West England*. Soil Survey of England and Wales, 3.

[35] H.F. Cook (2010) Boom, slump and intervention: changing agricultural landscapes on Romney Marsh, 1790 to 1990, in M. Waller (ed.) *Romney Marsh: persistence and change in a coastal lowland*. Sevenoaks, Romney Marsh Research Trust.

[36] P. Daldorph (2019) Fenland Archaeology. <https://www.fensforthefuture.org.uk/the-fens/heritage> Accessed April 2020.

[37] Encyclopaedia Britannica (n.d) Sir Cornelius Vermuyden <https://www.britannica.com/biography/Cornelius-Vermuyden> Accessed April 2020.

[38] H.F. Cook, S.A.F. Bonnett and L.J. Pons (2009) Wetland and floodplain soils: Their characteristics, management and

1750 and 1850. Settlement and drainage of the alluvial 'levels' was generally much older,[39] affecting soils of the Newchurch 2 (marine alluvium), Midlelney (river alluvium over peat) and Downholland 1 (marine alluvium and fen peat) soil associations. Water management has long involved a balancing act between flood management and drainage for agriculture, which today includes wetland conservation and re-instatement.

4.5 Economy and development of the Somerset Levels and Moors

Large-scale drainage of wetlands requires social organisation and capital. The overall motivation would be the control of soil and water management in the interests of agriculture linked to flood alleviation. Originally, prehistoric societies undertook fishing, fowling, animal grazing, salt production and peat cutting in such wetland areas with high ecological productivity. Salt making in Somerset was present during Romano-British times.[40] Dwellers around and on islands within the Somerset levels enjoyed improved access to wetland resources through the construction of wooden trackways between around 3,800 BC and 800 BC, during the Neolithic period and into the later Bronze Age.[41] However, the Nidon's Track, Viper's Track, and Viper's Platforms are possibly Early Iron Age in date.[42] They included links between the Polden Hills in the mid-Levels and dry ground around modern Meare and Burtle. The tracks were largely abandoned by the Middle Iron Age (after about 500 BC) due to flooding caused by climatic change. Peat accumulation then ceased after 400 BC and flooding increased, a situation which has continued. Wetlands are very good at preserving important organic remains, such as the preserved wooden trackways, which may be threatened not only from the repeated operations of arable management, such as ploughing, but also by drying out as a result of deep drainage. Iron Age settlement and activity is important, however. The Glastonbury Lake village and two lake villages at Meare Lake suggest exploitation of wild and domesticated species for food was important and arable foods were likely traded.[43] Access was by trackways and dug-out canoe. Other such settlements are known from elsewhere in Europe in the Bronze and Iron Ages. Otherwise, settlement and reclamation of the Levels is generally much older than that of the peat moors, which may date back to Roman times.[44]

The prehistoric and early medieval economy of the Somerset Levels and Moors was complex. During the winter months there would have been extensive flooding of 'sedge wastelands' and it is possible that the population moved to upland areas during these times. Wetlands have provided for fish, fowl, and mammals such as otter and beaver, and grazing during the drier summer months. On higher land, wheat, barley and beans may have been cultivated. Also

---

future, in E. Maltby and T. Barker (eds), *The Wetlands Handbook*, Blackwell, Oxford, 382-416.
[39] Natural England (2013) NCA Profile:142: Somerset Levels and Moors (NE451). <http://publications.naturalengland.org.uk/publication/12320274> Accessed April 2020.
[40] S. Rippon (2008) Coastal Trade in Roman Britain: the Investigation of Crandon Bridge, Somerset, a Romano-British Trans-shipment Port beside the Severn Estuary Coastal Trade. *Britannia* XXXIX, 85-144.
[41] Avalon Marshes (n.d.) The Sweet Track and other Trackways <http://avalonmarshes.org/the-avalon-marshes/heritage/sweet-track/> Accessed April 2020.
[42] Historic England (1996) Bronze Age and Iron Age timber trackways, 700m north west of Coppice Gate Farm. <https://historicengland.org.uk/listing/the-list/list-entry/1014431> Accessed Aug 2021.
[43] Avalon Marshes (n.d.) Glastonbury Lake village <http://avalonmarshes.org/the-avalon-marshes/heritage/glastonbury-lake-village/> Accessed April 2020.
[44] S. Rippon (1999) Romano-British reclamation of coastal wetlands (Chapter 8) and R. Silvester (1999) Medieval reclamation of marsh and fen (Chapter 9), both in H. Cook and T. Williamson (eds) *Water Management in the English Landscape*. Edinburgh University Press, Edinburgh; footnote 33, above.

available on higher ground was grazing for pigs and timber, including coppice for construction and fuel.[45]

Regional-scale impact on hydrological management probably only began during Romano-British times, with both localised reclamation of creek banks and more large-scale reclamation. Initial seasonal settlement would have been followed by more permanent examples, although there are also limited areas of more regimented 'planned' reclamation evidenced in field boundaries. One example is around a Roman settlement at Puxton on the Somerset levels. After Roman times it is believed grazing occurred during the Anglo-Saxon period.[46] In the medieval period, monasteries had the resources to carry out drainage.

The extent of flooding in the early medieval period may have been considerable, affecting the entire coastal strip and many kilometres inland along the valleys of the Parrett, Brue and Axe.[47] Drainage would inevitably be important to permit effective settlement and economic development. Medieval drainage on the Somerset Levels was probably localised and under the auspices of the monasteries (Chapter 3). For example, in the post-Norman conquest period, wetlands around the northern side of the island area including Meare were drained, probably driven by agricultural expansion into the 12th and 13th centuries. These fields were predominantly meadow and a part of what was, by c. 1300, Glastonbury Abbey's successful pastoral economy.[48]

In Somerset, as elsewhere, monastic landholding ended by 1540, at the Dissolution of the Monasteries. It then fell to secular landowners to fulfil such roles. The 1630s saw the enclosure of part of Alder Moor, near Glastonbury. The risk of freshwater flooding from upstream probably led to a perception of the place as unfavourable marsh and swamp, before the monastic houses commenced drainage improvements.[49]

By the mid-16th century, alder and willow from Aldermoor (or South Moor) was producing timber for hop poles as well as for fuel. One subsequent 21-year lease allowed felling at the end of each seven-year term, with fencing protecting new growth. A lease issued in 1584 required the tenant to plant 600 willows within six years and maintain this number to the end of a 15-year term. In the 1580s Glastonbury farmers had rights to wood products in most of the surrounding moors over a total of about 1,200 ha (3,000 acres) as well as a share of Sedgemoor. South Moor was then largely planted with alder and thorn. By the end of that century the condition of the Moors was deteriorating, partly because felling greatly reduced the number of large oak, ash, and elm on Hulkmoor and near Norwood, leaving few trees in hedgerows. In 1613, a survey concluded that the commons and wastes had been overstocked with cattle, which had contracted murrain (infectious disease affecting cattle).[50] Throughout

[45] P. Rahtz (1993) *Glastonbury*. Batsford/ English Heritage London, 16-17.
[46] S. Rippon (1999) Romano-British reclamation of coastal wetlands (Chapter 8) and R. Silvester (1999) Medieval reclamation of marsh and fen (Chapter 9), both in H. Cook and T. Williamson (eds) *Water Management in the English Landscape*. Edinburgh University Press, Edinburgh; M. Williams (1970) *The Drainage of the Somerset Levels*, Cambridge University Press, Cambridge, 50.
[47] P. Rahtz (1993) *Glastonbury*. Batsford/ English Heritage London, 14.
[48] S. Rippon (2004) Making the Most of a Bad Situation? Glastonbury Abbey, Meare, and the Medieval Exploitation of Wetland Resources in the Somerset Levels. *Medieval Archaeology* 48(1), 91-130. DOI: 10.1179/007660904225022816
[49] M. Williams (1970) *The Drainage of the Somerset Levels*, Cambridge University Press, Cambridge, 46.
[50] M.C. Siraut, A.T. Thacker and E. Williamson (2006) Glastonbury: Parish, in R.W. Dunning (ed.) *A History of the County of Somerset: Volume 9, Glastonbury and Street*, Victoria County History, London, 43-58.

the 17th century, the condition of these moors varied. At this time, they were also used for oak timber and coppice production, charcoal production and hunting purposes. It would be the middle of the 19th century before drainage was considered adequate and there was a boost to dairy and cheese making and summer fattening of animals on the Levels.[51]

Significantly, in 1631 it was agreed that 100 ha (250 acres) out of around 400 ha (c. 1,000 acres) of moor should be allotted to Glastonbury tenants to be enclosed with the purpose of improvement, with similar areas allotted to tenants of Butleigh and Street. The Crown retained the remaining areas and granted a 99-year lease to the courtier James Levington in 1632/3. This share was temporarily enclosed in 1641. In an agreement of 1665, alder and thorns had been removed and grassland was said to be improved. At the same time, access to the South Moor was improved by building a bridge and 'digging boundary ditches', defining the area that was within Glastonbury manor. By 1708 such permanent enclosure was being promoted.[52]

There are two issues: drainage and enclosure. While the first Enclosure Act for Somerset was for the reclamation of Common Moor, near Glastonbury, in 1772, the Commissioners of Sewers were responsible for drainage and flood defence. The commission was founded in the 16th century, and survived in Somerset into the 20th century, but had older antecedents (see Chapter 3).[53] The commonable South Moor, the Common Moor, and Black Acre were enclosed in 1722, and with other areas this added some 360 ha (c. 900 acres) to the enclosed grasslands of the parish. Most of the remaining common land was enclosed in 1783 under an Act of 1778; the final pieces were enclosed by 1800.[54] There has also been a peat cutting industry from early times, which continues in a muted form today.[55]

The years between about 1770 and 1840 were particularly active in terms of drainage development. Extensive works affected the valleys of the Brue and Axe around 1770, followed by a scheme to drain Kings Sedgemoor, including both peat and alluvial sediment soils.[56] Rhynes were dug and embankments raised, but the work was inadequate, with limited results, and there were floods in the 1790s. The agricultural productivity of drained areas in the first half of the 19th century was a matter of comment, and by 1851, the work by the celebrated engineer John Rennie on the Congresbury Yeo 'brought about a great improvement'. Eventually by the 1930s it would be Internal Drainage Boards that largely became responsible for land drainage, in Somerset as elsewhere in reclaimed areas of England.[57] Sea walls may be constructed in connection with drainage and flood protection and must be able to shut

[51] H. Prince (1989) The changing rural landscape 1750-1850 in G.E. Mingay (ed.) *The Agrarian History of England and Wales, Volume VI 1750-1850*. Cambridge University Press, Cambridge, 59-61.
[52] M.C. Siraut, A.T. Thacker and E. Williamson (2006) Glastonbury: Parish, in R.W. Dunning (ed.) *A History of the County of Somerset: Volume 9, Glastonbury and Street*, Victoria County History, London, 43-58.
[53] C. Taylor (1999) Post medieval drainage of marsh and fen, in H. Cook and T. Williamson (eds) *Water Management in the English Landscape*. Edinburgh University Press, Edinburgh, 148.
[54] M.C. Siraut, A.T. Thacker and E. Williamson (2006) Glastonbury: Parish, in R.W. Dunning (ed.) *A History of the County of Somerset: Volume 9, Glastonbury and Street*, Victoria County History, London, 43-58.
[55] H. Cook (2018) *The Protection and Conservation of Water Resources* (2nd ed.) Wiley-Blackwell, Chichester, 251; Avalon Marshes (n.d.) The peat Industry <http://avalonmarshes.org/the-avalon-marshes/heritage/peat/> Accessed April 2020.
[56] M. Williams (1970) *The Drainage of the Somerset Levels*, Cambridge University Press, Cambridge.
[57] ADA (n.d.) Internal Drainage Boards. <http://www.ada.org.uk/member_type/idbs> Accessed February 2018.

at high water to prevent saltwater incursion into the rivers. These constructions should be engineered in terms of height and strength to resist most events encountered.

The practice of 'warping' ('warp' being an old term for silt) has been practiced in Somerset, at least since the late 18th century, as a form or water management. While 'sub-irrigation' was the practice of keeping the water levels high in the ditches during the spring and summer, warping refers to leading turbid freshwater (notably from the rivers Parrett, Tone, Isle, Brue and Yeo) across the surface of the levels. Water was effectively penned in for a period, then released, making this a form of basin irrigation, allowing sediment and nutrients to be trapped in the grass during spring. Warping has been practiced at least since the late 18th century, most notably to improve the peaty moor areas by trapping water, and is closely associated with enclosure.[58]

Figure 4.7 shows the rhynes on the Somerset levels in 1833; there had been considerable drainage prior to 1770 followed by a particularly active period of drainage and enclosure. On the Somerset Levels and Moors, the water catchment area is in the order of four times that of the reclaimed wetlands themselves.[59] Because many of these are essentially extended river floodplains, they are prone to flooding due to hydrological loading from upstream) as well as marine inundation at high tides.

To control excess water, measures include embankment construction, the cutting of new channels (such as the Huntspill river in 1940) and the construction of pumping stations, sluices, flood gates and coastal flood defences. Improvements between 1930 and 1970 and improved pumping efficiency extended the effective drainage season by four months, from between May and September to between March and November. There has been considerable upgrading of such assets since major flooding hit the area in 2013/14.[60] In summer, the drainage system is reversed by causing the rhynes to be augmented through gravity-fed pumping from the main rivers to keep them bank-full and operating as wet fences for livestock and sub-irrigating grassland to increase productivity.[61] Otherwise major periods of drainage occurred during the 17th and 18th centuries, largely across the Moors. Most notably, post-1940, regional drainage improvements would come to bring ecological and landscape considerations to the fore.

The landscape today is dominated by 'secondary' wetlands reclaimed by ditches and embankments (Table 4.1), otherwise there is limited deep drainage for arable cropping[62]. Since the 1980s, with the introduction of Environmentally Sensitive Areas (ESA), water levels have been kept high for ecological gain, in effect restoring the natural wetland functions.[63] Although ESA ended in 2012, replacement agri-environmental schemes are available.[64]

---

[58] M. Williams (1970) *The Draining of the Somerset Levels*, Cambridge University Press, Cambridge, 176-7.
[59] Natural England (2013) NCA Profile:142: Somerset Levels and Moors (NE451). <http://publications.naturalengland.org.uk/publication/12320274> Accessed April 2020.
[60] Environment Agency (2015). Somerset Levels and Moors: reducing the risk of flooding. <https://www.gov.uk/government/publications/somerset-levels-and-moors-reducing-the-risk-of-flooding/somerset-levels-and-moors-reducing-the-risk-of-flooding> Accessed April 2020
[61] J. Purseglove (2015) *Taming the Flood*. William Collins, London, 300.
[62] H.F. Cook, S.A.F. Bonnett and L.J. Pons (2009) Wetland and floodplain soils: Their characteristics, management and future, in E. Maltby and T. Barker (eds), *The Wetlands Handbook*, Blackwell, Oxford, 382-416.
[63] H. Cook (2018) *The Protection and Conservation of Water Resources* (2nd ed.) Wiley-Blackwell, Chichester, Chapter 8.
[64] Natural England (2015) Somerset Levels and Moors (NCA142): Countryside Stewardship statement of priorities. <https://www.gov.uk/government/publications/countryside-stewardship-statement-of-priorities-somerset-levels-

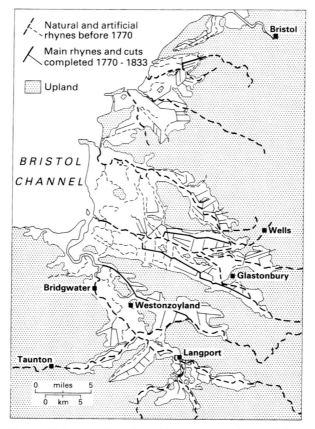

Figure 4.7. Somerset Levels and Moors in 1833 (after Michael Williams, 1970, *The Draining of the Somerset Levels*, Cambridge University Press).

The low-lying Moors provide floodwater storage, and normally protect the towns of Bridgwater and Taunton. Flooding, particularly in winter, is also valuable for from the standpoint of wildfowl. The Wessex Water Authority had major plans to improve drainage for agriculture on the rivers Brue and Parrett between the mid-1970s and early 1980s.[65] This was despite the requirement on the RWAs and IDBs to have regard for the natural beauty and conservation under the Water Act 1973.[66] Pumping is locally employed to lift water from the artificial field ditches (the larger ones are rhynes) up to the embanked arterial watercourses.

Much of the drainage from these channels to the sea is by gravity. Inundation from the sea during the large tidal range experienced in the Bristol Channel is controlled by sluices.

and-moors-nca142> Accessed April 2020.

[65] Wessex Water Authority (1979) Somerset Local Land Drainage District: Land Drainage Survey Report: Bridgwater Somerset.

[66] NRA (1991) *Somerset levels and moors water level management and nature conservation strategy*, NRA, Wessex Region.

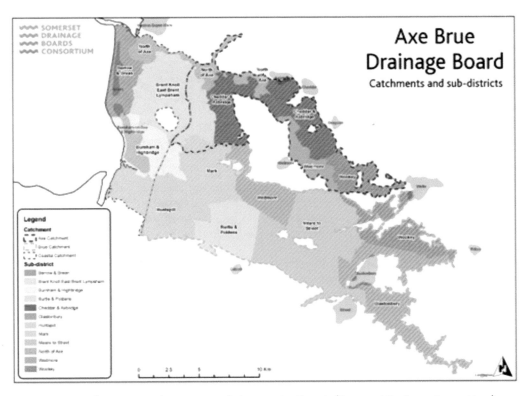

Figure 4.8. The Axe Brue drainage Board since 1st April 2012. (Somerset Drainage Consortium)

Here, unlike the 'Black (peat) Fens' of East Anglia, or similar areas in the Netherlands,[67] peat wastage through oxidation and shrinkage has not led to a significant regional lowering of the land surface. This may be due to the wetter climate, lower arable acreage and maintenance of higher water tables than in the Fenlands,[68] and accounts for the lack of tertiary stage marshland in the area.

Although horticulture and forage maize growing are practical on the Somerset Levels and Moors, widespread arable conversion never occurred, most likely due to a perception of climatic unsuitability; there was also a potential acid sulphate soil problem on West Sedgemoor. Otherwise, eutrophication of watercourses, pesticide contamination and peat wastage might have become serious issues, as in the East Anglian Fens and Norfolk Broadlands.[69] Artificial lifting and pumping were rare on the Somerset levels before the last century, and there were very few windmills engaged in lifting water; only on West Sedgemoor has there been concern for peat shrinkage.

---

[67] H.F. Cook, S.A.F. Bonnett and L.J. Pons (2009) Wetland and floodplain soils: Their characteristics, management and future, in E. Maltby and T. Barker (eds), *The Wetlands Handbook*, Blackwell, Oxford, 382-416.
[68] M. Williams (1970) *The Draining of the Somerset Levels*, Cambridge University Press, Cambridge, 246.
[69] H. Cook and H. Moorby (1993) English marshlands reclaimed for grazing: a review of the physical environment. *Journal of Environmental Management* 38, 55-72; H. Cook (2018) *The Protection and Conservation of Water Resources* (2nd ed.) Wiley-Blackwell, Chichester, 252.

Figure 4.8 shows the Internal Draining Boards for the area of the northern Somerset Levels. For 300 years prior to 1877, the drainage of the area shown in Figure 4.8 had been supervised by the Court of Sewers and their Commissioners. The modern Axe Brue Internal Drainage Board was formed in 1877, as part of the Somersetshire Drainage Act, from the amalgamation of four Drainage Boards: the Lower Axe, Upper Axe, Lower Brue and Upper Brue. The modern IDB covers almost 30,400ha (including over 3,400ha of SSSI), has over 200 water level control structures and manages over 600km of waterways; there are over 11,000 homes and businesses within the area.[70]

Notwithstanding exceptional events, flooding reminds us of the origins of this special lowland region. The persistence of spring and autumn flooding, fear of loss of commoners' rights to cut peat and low agricultural prices may have slowed later major reclamation plans for the Levels and Moors more than elsewhere in England, and the flood problem persists. However, what is bad for property and agriculture is seen as a conservation virtue.

Much of the area lies below the high-water mark and this is not being helped by progressive sea level rise. Artificial pumping was present in the 19th century but was largely installed in the 20th century.[71] Drainage of the Somerset Levels and Moors, unlike the East Anglian Fenlands, has not progressed to the full deep-drained arable 'Tertiary stage' (Table 4.1), so that retaining freshwater on the surface for part of the years is regarded as an ecosystem service, as it supports bird and wildlife conservation.

Figure 4.9 contrasts the regular and later reclamation of the Somerset Moor landscape on peat soils (a) with rectilinear blocks between drainage channels and (b) a meander close to the mouth of the river. There are flood embankments constraining the river on the east side of the figure, and irregular field boundaries (with sinuous boundaries) inside the meander reflect original drainage channels on un-reclaimed marsh.

Willow may be used for securing thatch, basketry, making wicker chairs and even cricket bats, while pollards left in place to secure banks provide stout sticks. The hope was that cultivation would support a home industry. Cultivation seems to have dated largely (but not exclusively) from the early 19th century, and in the 1930s, the area of withy beds was approximately 1600 acres (c.650 ha), with the crop cut during winter. The term 'osier' is preferred for the coarser cut willow used for basketry. Willow was also used for wicker chairs in the 19th century.[72] On the Somerset Levels there was land-allocation for withy beds alongside the river channel embankments within the meander bends.

When viewed from an ecological viewpoint, the extent to which flooding is a 'problem' becomes relative and a matter of user expectation, that is until a serious event occurs. What is appropriate for low-input grazing is not appropriate for intensive grass or arable; intensive

---

[70] Somerset Drainage Boards Consortium (n.d.) Axe Brue Internal Drainage Board <https://somersetdrainageboards. gov.uk/boards-membership/board-areas/axe-brue-internal-drainage-board/> Accessed April 2020.

[71] Natural England (2014). National Character Area Profile 142: The Somerset Levels and Moors <https://www. bleadon.org.uk/media/other/24400/142somersetlmsfinal.pdf> Accessed April 2020

[72] T. Stuart-Menteath (1938) The Land of Britain: part 86, Somerset. Geographical Publications, 65-66; D. Ross (n.d.) Thatching: The traditional British craft < https://www.britainexpress.com/History/thatching.htm > Accessed April 2020.

Figure 4.9. (a) left: Detail of the Westhay Somerset 'Moor' from modern Ordnance survey 1:25,000 mapping. The grid squares are one km in size. (© Crown Copyright Ordnance Survey). All rights reserved. Figure 4.9 (b) right: detail of the 'Levels' Near to the mouth of the river Parrett. (Google Earth).

Table 4.1. An idealised sequence of marshland states caused by systematic reclamation and modification of the hydrological environment

Major wetland category	Wetland form (saline)	Wetland form (freshwater)	Productivity/ land use	Ecological interest	Topsoil properties
PRIMARY	Unenclosed tidal mudflats, creeks, saltmarsh	Riverine/lacustrine mudflat, bog,fen, carr, swamp, uncontrolled flooding	Ecological productivity high: Nutrient enrichment from flood waters and sediment; saltings with grazing	High: Natural habitat for flora and fauna	Unconsolidated sediment. Terrestrial raw soils' or Stage A or B.Completely or practically Unripe' sediment Commonplace. Almost ripe/ half ripe Stage B
SECONDARY	Enclosed non-tidal mudflats with brackish pools and saltmarsh vegetation. Unstable soft mudflats. Firm consolidating salt mudflats (prone to rain flooding). Stabilised, non-saline grassland with network of inter-connected ditches; relatively high watertable prone to winter flooding; reed beds associated with ditches	Flood defences impart control over hydrological regime, Limited flooding managed (i.e. Flood meadows, marshlands). Stabilised, non-saline grassland with network of inter-connected ditches; relatively high watertable prone to winter flooding; reed beds associated with ditches	Economic productivity low without agricultural intensification; grazing marsh/ hay meadow and suitable for reed cultivation under modified watertable regime	Range of semi-natural habitats for flora and fauna, especially birds. Interest high due to relict flora and fauna associated with ditches; important seasonal habitats for birds owing to flooding.	Almost ripe/ half ripe Stage B. Almost ripe/ ripened. Stage B/C soils. Peat associated with valley bottoms, fringing vegetation. Soil structure well developed under permanent pasture. Often in the lee of sandspits and shingle bars.
TERTIARY	Stabilised, non-saline grassland with network of inter-connecting ditches; most suitable for intensive production; lower watertable reduces flood risk; longer growing season than secondary grassland, underdrainage may be present. Deep-drained low water table (deepened ditches) with field underdrainage, allowing for improved arterial drainage	Stabilised, non-saline grassland with network of inter-connecting ditches; most suitable for intensive production; lower watertable reduces flood risk; longer growing season than secondary grassland, underdrainage may be present. Deep-drained low water table (deepened ditches). with field underdrainage, improved arterial drainage	Productivity increased by watertable management, field levelling and fertiliser inputs; intensive grazing and silage cutting. Arable or intensive grassland	Interest low, poor species diversification; eutrophication of watercourses is common	Ripened soils. Soils losing hydromorphic properties including gleying due to reduced watertables

* The terms 'ripening' refers to the degree of consolidation of soft sediment from which a marsh is formed. Stage A is raw, fluid sediment that easily flows between the fingers when squeezed. Soils in this condition are deficient in oxygen. Stage B soils have established vegetation that removes water so that soil structures can develop, for example fissures that permit water infiltration and the introduction of oxygen into a soil. Stage C soils are better oxygenated and typically develop following full drainage and embanking.

farming requires lower water tables and stringent flood controls.[73] Debating land-use issues involves consideration not only of the desired outcome, but of economic inputs. We should ask whether the area is to be managed for semi-natural habitats, floodplain, or intensively managed agriculture, and to what extent these land uses are compatible?

The 'Avalon Marshes' area includes the varied working landscape around Glastonbury and is valued for a range of landscapes, wildlife and heritage features. Located some 4km west of Glastonbury is Ham Wall NNR. Over the past 50 years or so, large scale extraction of peat for horticulture has left 'a scarred industrial landscape'.[74] Since the 1990s, most of this has been transformed into wetlands for wildlife. Interestingly the landscape, one of lakes, reedbeds and mires, would be close to its natural state, pre-extraction and drainage.[75] Managed by the Royal Society for the Protection of Birds (RSPB), birdlife includes marsh harriers, bittern and starlings.

Table 4.1 shows an idealised sequence of marshland states, caused by systematic reclamation and modification of the hydrological environment .[76] It shows how generic 'reclaimed wetlands' can be managed in at least three stages, depending on their intended economic purpose. Reasons for wetland reclamation include coastal defence, control over waterways, flood risk management, improved general transport and communications and options for settlement.

4.6 The Levels and Moors today

The economy of the Somerset Levels and Moors came to be based not so much on fowling, fishing, timber growing and rough grazing, but rather on dairying, cheese making and the fattening of cattle, all of which provided high returns per hectare.[77] For example, parts of the Brue valley were good for grazing and fattening, while the Southern Levels and King's Sedgemoor were summer fattening grounds, although hay was also important.[78]

Because of the higher rainfall, the drainage density of modern ditches is typically $15km.km^2$, or around twice as high as in the drier east of England.[79] The result is a pattern of small fields, small holdings, and fragmented ownership.[80] Relaxation of flood measures to promote conservation during the 20th century have entailed lower engineering costs and gains in ecological 'goods', to be offset by a loss in agricultural production. Flood events across Europe, notably those of the winter of 1994/95 and again in 2013/14, have prompted reconsideration of the natural function of floodplains and a political battle over flood defence priorities after February 2014.

---

[73] H. Cook and T. Williamson (eds) (1999) *Water Management in the English Landscape*. Edinburgh University Press, Edinburgh, footnote 14.

[74] Avalon Marshes (n.d.) <http://avalonmarshes.org/the-avalon-marshes/landscape/> Accessed May 2021.

[75] Avalon Marches (n.d.). Ham Wall National Nature Reserve. <http://avalonmarshes.org/explore/nature-reserves/ham-wall/> Accessed April 2020.

[76] H.F. Cook, S.A.F. Bonnett and L.J. Pons (2009) Wetland and floodplain soils: Their characteristics, management and future, in E. Maltby and T. Barker (eds), *The Wetlands Handbook*, Blackwell, Oxford, 390-3.

[77] H. Prince (1989) The changing rural landscape 1750-1850 in G.E. Mingay (ed.) *The Agrarian History of England and Wales, Volume VI 1750-1850*. Cambridge University Press, Cambridge, 59-61.

[78] M. Williams (1970) *The Draining of the Somerset Levels*, Cambridge University Press, Cambridge, 185.

[79] H. Cook and H. Moorby (1993) English marshlands reclaimed for grazing: a review of the physical environment. *Journal of Environmental Management* 38, 55-72.

[80] J. Purseglove (2015) *Taming the Flood*. William Collins, London, 298

The Somerset Levels and adjoining Moors form the largest area of lowland wet grassland and associated wetland habitat remaining in Britain. They are of outstanding importance for their rich patchwork of reedbeds, mires and fen meadows and network of 8,000km of rivers and ditches. Covering about 35,000 hectares in the floodplains of the Rivers Axe, Brue, Parrett and Tone, the majority of the area is only a few metres above sea level. It drains via a large network of ditches, rhynes (drainage ditches) and rivers. The Levels are particularly important for the aquatic stonewort (*Characea*), and a variety of wetland wildflowers can also be found in the area, including carnivorous sundews, bright yellow irises and vivid pink tufts of meadow thistle.[81]

The Levels also support extremely valuable habitats for migratory birds, wildfowl, otters, fish (especially eels), insect life, and aquatic and meadow flora. Many of the species protected using SSSIs and the Endangered Species Act have expanded. Parts of the Moors are Special Protection Areas for Birds (SPAs) under EU legislation, and the Moors as a whole are designated a Ramsar Site.[82] Coarse fishing is the most important water-based recreation within the area, and locally elver was once caught although this is now banned. Protection of fisheries from agrochemicals is not a problem, although silage effluent, slurries and yard washings are potential pollutants. Care must be taken to avoid inappropriate ditch clearance (particularly disturbing the bottom of ditches) as slow-flowing rivers present problems for aeration and sediment disposal.

Proposed flood risk reduction actions are dredging and river management, alongside land management which recognises that what happens in the upper and mid catchment has an impact on the lowlands and management of urban runoff. Resilience will be built into both infrastructure and communities. However, the most cost-effective mix of management, infrastructure, and resilience actions, is unclear. Resilience needs to cover a range of factors, from road access to Water Level Management Plans and sluice installation and operation; construction materials, especially for roads and protection of power infrastructure, create detention areas and wetland restorations. The flooding in winter 2013/14, which cut off some communities and lasted several weeks, resulted from unseasonably high winter rainfall.[83] Economic damage was considerable. Ultimately, the point is not so much about statistics as about expectations.

Somerset County Council is opposed to peat digging on grounds of wildlife and habitat conservation. It must never be forgotten that tourism is an important industry in Somerset. Key issues for the Somerset Levels and Moors, which bring with them attendant problems for the long-term, are not unlike those for the Norfolk Broads and increasingly the East Anglian Fenland. Modern challenges include:[84]

- Increased risk of flooding (55-60 significant events in 2012-2014) and associated balancing needs of management to protect people, property, agriculture and conservation

---

[81] Plantlife (2020) Somerset levels IPA <https://www.plantlife.org.uk/uk/nature-reserves-important-plant-areas/important-plant-areas/somerset-levels> Accessed April 2020.
[82] Natural England (n.d) <https://designatedsites.naturalengland.org.uk/SiteGeneralDetail.aspx?SiteCode=UK11064&SiteName=king> Accessed April 2020.
[83] BBC News (2014) <https://www.bbc.co.uk/news/uk-25944823> Accessed April 2020.
[84] H. Cook (2018) *The Protection and Conservation of Water Resources* (2nd ed.) Wiley-Blackwell, Chichester, Chapter 8.

- Securing investment to maintain complex infrastructure including pump-drained systems. This is shifting away from protecting purely agricultural land towards people and property. Maintaining main rivers, especially through dredging, particularly the Rivers Parrett and Tone, including establishing 'local ownership' of problems
- Establishing working partnerships with local authorities, the Internal Drainage Board and others to raise funding for dredging
- Planning for climate change: predictions over the next 100 years indicate a significant increase in the risk of serious flooding in Somerset from both increased rainfall and rising sea levels
- Ensuring the future of agriculture on the Levels and Moors (including supporting a local community of predominantly livestock farmers)
- Supporting local industry ('withies' or coppiced willow growing) and managing a reduced peat digging industry
- Managing and encouraging tourism and recreation, especially fishing
- Nature and landscape conservation including restoration of species-rich fen meadow or flood pasture and bird habitats through extensive farming practices

In terms of ecological services, 'live' accumulation peatlands have a great capacity to sequester carbon. However, climate change recently caught up with the Somerset Levels and Moors in a direct and cruel way. For while a modicum of inundation is to be expected on what is basically one coalesced floodplain (Somerset Levels) with drained peat bogs (Somerset Moors), the flooding of February 2014 was unprecedented in modern times. Several interesting points emerged from the angry discourse that ensued, following the damage to livelihoods and to homes. Dredging was undertaken on the Parrett and Tone and a twenty-year 'Flood Action Plan' prepared, under-written by central government, for:

> We see the Somerset Levels and Moors in 2030 as a thriving, nature-rich wetland landscape, with grassland farming taking place on the majority of the land. The impact of extreme weather events is being reduced by land and water management in both upper catchments and the flood plain and by greater community resilience.[85]

The report was updated in 2002-3. There are some 11 Sites of special Scientific Interest within the Somerset Levels and Moors Special Protection Area.[86] Over about 500 years the diversity of habitat and potential economic activity in the wetlands of the Somerset Levels and Moors, through planting and harvesting of food, enclosure and drainage and regional drainage improvement, has given way to a cattle-based economy and more recently to areas that provide 'environmental services' of conservation and (perhaps reluctantly) flood relief.

4.7 Sparkhayes Marsh at Porlock, west Somerset.

If the Somerset Levels and Moors display a regional landscape character, the small coastal wetland area adjacent to Porlock in Somerset is a localised example, responsive to its immediate location. It is set within NCA 145: Exmoor, predominantly a landscape of upland plateaux of Devonian sandstones and slates, terminating in the north at the Bristol Channel,

---

[85] The Somerset Levels and Moors Flood Action Plan (2014) 2nd ed <https://somersetnewsroom.files.wordpress.com/2014/03/20yearactionplanfull3.pdf > Accessed April 2020.
[86] H. Cook (2018) *The Protection and Conservation of Water Resources* (2nd ed.) Wiley-Blackwell, Chichester, 250.

Figure 4.10. Shingle deposition close to the site of the breach in April 2016 (the author).

with a spectacular cliff coastline (Chapter 1). This small coastal wetland is an interesting part of this spectacular -and complex- coastline.

The story of Porlock and Sparkhayes Marshes is conditioned by its position between the coast and the extensive Exmoor region to the south. On the west side is the harbour at Porlock Weir, which may date from the 15th century.[87] The marsh is located behind a shingle bar that no longer offers total protection from the sea due to sea level rise. The Marsh area (less than 1km[2] in area) has for some time been a candidate for 'managed retreat'. Managed retreat is the application of coastal zone management, designed to move existing and planned development out of the path of coastal hazards (e.g., high tides, erosion, flooding, storm surges, sea level rise etc.). The philosophy is one of avoidance or moving out of harm's way, while recognizing coastal zone dynamics should dictate the type of management employed.

At Porlock and elsewhere, this strategy enables the recovery of saltmarsh and coastal ecosystems, as the bar was breached in 1996[88] and has not been artificially repaired, allowing a tidal inlet crossing through the shingle ridge. Such changes long-term are driven by variations

---

[87] D. Corner (2009) *Porlock in Those Days*. Rare Books and Berry, Minehead, 58.
[88] J. Orford (2003) Porlock Gravel barrier, in V.J. May and J.D. Hansom (eds), *Coastal Geomorphology of Great Britain*. Joint Nature Conservation Committee, Peterborough, 266-271.

Figure 4.11. Former drainage ditch emerging from below the
migratory shingle ridge, April 2016 (the author).

in the rate of relative sea level rise which have influenced sediment supply and otherwise affected the dynamics of the shingle bar.[89]

Figure 4.10 shows recent shingle ridges that have developed close to the point of the original breach covering the land surface of the Marsh. The original migratory shingle barrier crossed a landscape dating from approximately 4,000 BC to the 19th century.[90]

Between about 4,500 and 2000 BC, a shingle bar developed with alder carr on its landward side.[91] In prehistoric times there would have been aurochs, elk, red deer and row deer in an

---

[89] S. Jennings, J.D. Orford, M. Canti, R.J.N. Devoy and V. Straker (1998). The role of relative sea level rise and changing sediment supply on Holocene gravel barrier development: the example of Porlock, Somerset, UK. *The Holocene* 8(2), 165-181.

[90] Exmoor National Park (n.d.). A vision for Porlock Marsh. <http://www.exmoor-nationalpark.gov.uk/__data/assets/pdf_file/0011/760565/Vision-Document.pdf> Accessed February 2018.

[91] S. Jennings, J.D. Orford, M. Canti, R.J.N. Devoy and V. Straker (1998). The role of relative sea level rise and changing

Figure 4.12. Remains of lime kiln in at Bossington beach in 2016 (the author).

oak dominated landscape. An offshore submerged forest dates to 3000-4000 BC. Archaeological evidence includes Mesolithic artefacts found sealed beneath a peat layer. Late Mesolithic people living in hunting camps had a choice of food from the uplands and would have made tools from pebbles found on the beach. They likely exploited the marsh for crabs, limpets, and wildfowl as well as catching fish from the sea. At the back of the marsh, Pleistocene (Devensian) solifluction-generated fan material may be observed that could be re-worked in Holocene marsh deposits as sea level rose.[92] Since about 4000 BC deposition has been largely of inorganic sediments, which may have been supplied up-catchment by early farming activities.[93]

This small area had been reclaimed for grazing (secondary marsh) and included a wildfowl decoy (a pond where wildfowl may be trapped and killed), but, during the last 20 years, has been reverting to primary saltmarsh, with pioneering immature saltmarsh vegetation communities readily observed. Figure 4.11 shows a straight drainage ditch, infilled by shingle, which have been revealed as migratory shingle moved inland.

sediment supply on Holocene gravel barrier development: the example of Porlock, Somerset, UK. *The Holocene* 8(2), 165-181.
[92] J. Orford (2003) Porlock Gravel barrier, in V.J. May and J.D. Hansom (eds), *Coastal Geomorphology of Great Britain.* Joint Nature Conservation Committee, Peterborough, 271.
[93] J. Orford (2003) Porlock Gravel barrier, in V.J. May and J.D. Hansom (eds), *Coastal Geomorphology of Great Britain.* Joint Nature Conservation Committee, Peterborough, 266-271.

The wildfowl decoy (there is a plan to restore this) is recorded on older maps of the area. The remains of lime kilns may be found at Bossington beach, part of the shingle bar (Figure 4.12). They were used to produce agriculture lime from limestone and 'culm', in this case the term refers to fuel from anthracite fragments and dust unfit for other purposes, but which, once fired, kept the furnace going for several days. Both geological materials were imported from South Wales, prior to being taken inland to improve the soils of Exmoor, or to produce mortar for construction into the 19th century.[94]

There is a paucity of information about the area in the medieval period. The earliest known map of the area dates from 1710-20,[95] and describes a 'pill' or fishpond, suggesting that later drainage left a smaller area as the wildfowl decoy. This was likely constructed in the 18th century and has since been inundated by the salt-water lagoon. There was originally a near-rectangular lagoon measuring some 90m x 60m with 10 projecting tapering 'arms' 35m- 60m long, which were ditches which could be netted, so that at unfortunate ducks were ensnared at their ends.[96]

Significant drainage works on the marshland date from the nineteenth and early twentieth centuries, including the modern straight ditches, but such works are prone to storm damage. Twentieth century changes include efforts to create a golf course in 1910, which failed, and during the Second World War pillboxes were constructed in the area.[97] It took some time before efforts towards agricultural improvement were abandoned!

Recent work by Exmoor National Park demonstrates that, at the beginning of the 19th century, drainage channels began to be constructed across the west of the marsh, and led the main streams to a single outlet to the sea through a sluice to control the ingress of salt water and permit low intensity livestock grazing (predominantly cattle with some horses and sheep), with limited areas planted with barley and wheat as late as the 1950s. Freshwater reed cultivation, to be sold for thatching, is recorded from the 1960s. During the post-Second World War era, a constant battle to improve drainage and prevent saline water intrusion is evident. As a result, a small mere and reed fen was able to establish itself behind the ridge and would form the basis for the original SSSI designation. This has now itself changed in line with seawater inundation and a loss of freshwater habitats.[98]

4.8 Water meadows

Wiltshire, Dorset and Hampshire contain extensive areas of water meadows, which form an important part of the historical landscape along with canals, mills, watercress beds and other engineered river-system features. A true water meadow is a pasture irrigation system that is irrigated as a deliberate part of agricultural practice. The specialist who maintains and operates the meadows is called the 'drowner'.

---

[94] D. Corner (2009) *Porlock in Those Days*. Rare Books and Berry, Minehead, 21-22.
[95] D. Corner (2009) *Porlock in Those Days*. Rare Books and Berry, Minehead, 8.
[96] H. Riley and R. Wilson-North (2001) *The Field Archaeology of Exmoor*, English Heritage, Swindon 134-135
[97] (Corner op.cit, 39;Exmoor national Park op.cit)
[98] Exmoor National Park (n.d.) A vision for Porlock Marsh <https://www.exmoor-nationalpark.gov.uk/__data/assets/pdf_file/0011/760565/Vision-Document.pdf > Accessed April 2020.

Early water meadows (*e.g.* Pewsey Vale) were of the 'floating-up' design where meadows were flooded with backed-up water. As the technology developed, better design allowed the water to keep flowing over the grass surface and infiltrating into the topsoil. This prevented stagnation, which would eventually kill the grass. Anoxic conditions, for example, will cause the build-up of the hydrocarbon acetylene (C_2H_2) after about 24hrs of waterlogging, causing serious inhibition of nitrogen fixation in soil.[99]

Water meadows may be engineered features on hillsides (termed 'catchworks' or 'catch meadows'), or in the bottom of river valleys on alluvial floodplains (termed 'bedworks'), by far the most familiar design throughout Wessex (Figure 4.13). In bedwork systems, water is made to flow in channels along the tops of constructed ridges, while the familiar hatches and channels control the flow of water on to and across the meadow, so that water trickles through the grass stems ideally at a depth of no more than 25mm. This returns via a series of drains feeding a tail drain returning the flow to the river. The job of the drowner is to maintain an even flow, with irrigation events typically lasting for time periods between three days and one week.

Floated water meadows were used for irrigation in the winter or early in spring to warm the grass sward and bring nutrients and oxygen into the soil. Above 5.5°C, grasses will start to grow in winter and early spring. Nutrients such as nitrogen and phosphorous, as well as carbon in organic matter, are supplied to soils from the river water and may be in either particulate or dissolved forms. The infrastructure is designed to keep the water flowing, ensuring oxygen supply to the topsoil. For this reason, water meadows would have been of great utility during the Little Ace Age, and especially during the Maunder Minimum' c. AD 1650-1715 (Chapter 2).

Usually, grass starts growing about one month earlier than on un-floated floodplain meadows; animals could thus benefit from the 'early bite' of grass.[100] Later in the season, during the summer when the soil was drying out, water meadows were re-watered so that (typically) two cuts of hay were taken and used to feed other animals, typically cattle and horses.[101]

Figure 4.14 shows haymaking on the Harnham Water Meadows at Salisbury. In general, floated water meadows constructed in the lower reaches of river valleys in Wessex (examples also include the lower Meon and Itchen valleys in Hampshire) may require measures to maintain a head of water sufficient for irrigation. Commonly this is achieved using a canal (which may or may not have been intended for transport) that is taken off at a higher level upstream. Nonetheless, low gradients may historically have proven operationally difficult.[102] Such systems are today commonly abandoned for irrigation but function as a modified floodplain environment.[103]

---

[99] H.F. Cook, S.A.F. Bonnett and L.J. Pons (2009) Wetland and floodplain soils: Their characteristics, management and future, in E. Maltby and T. Barker (eds), *The Wetlands Handbook*, Blackwell, Oxford, 390-3.

[100] H. Cook and T. Williamson (eds) (1999) *Water Management in the English Landscape*. Edinburgh University Press, Chapters by R.L. Cutting and I. Cummings, 70-81, and H. Cook, 94-106.

[101] K. Stearne and H. Cook (2015) Water Meadow Management in Wessex: Dynamics of Change from 1800 to the Present Day, *Landscape Research* 40 (3), 377-395. DOI: 10.1080/01426397.2013.818109

[102] Cook and Young (2011) Watermeadows at the Itchen Valley Country Park near Eastleigh, Hampshire *Proceedings of the Hampshire Field Club and Archaeological Society (Hampshire Studies)* 66, 166-186.

[103] H.F. Cook (2010) Floodplain agricultural systems: functionality, heritage and conservation. *Journal of Flood Risk Management* 3, 1-9. DOI:10.1111/j.1753-318X.2010.01069.x

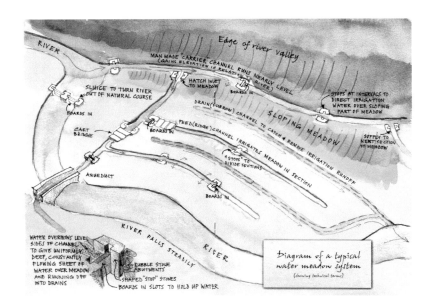

Figure 4.13. Schematic showing catchwork watermeadows (top right of the figure) and valley-bottom, alluvial bedworks between the river and the hillside carrier (Hampshire County Council, artist Mike Clark).

Figure 4.14. Hay making on the Harnham Water meadows c. 1900 (Salisbury Cathedral collection).

Historically, water meadows have proved economically very important. Their construction and operation can be shown to date from the Middle Ages in both England and continental Europe, but their widespread adoption really occurred during the 17th century. Details of their early management is sparse, although a description of an operational system in the Vale of Hereford was published by Rowland Vaughan in 1610.[104]

The spread of the technology in Wessex, especially in Dorset along the river Piddle, from the early 1600s has been well-documented. Actually floating had reached the chalklands by the time Vaughan published his book. The first clear mention of water meadow construction occurs in the court roll of Wylye for 1632, although it is likely there were irrigated meadows in Wiltshire earlier in that century. The antiquarian and farmer John Aubrey apparently introduced meadow irrigation at Broade Chalke in 1635.[105]

The agricultural writer Thomas Davis clearly describes the operation of Wiltshire water meadows at the end of the 18th century.[106] In the 18th and early 19th centuries, ewes and lambs were led away from the valley-bottom meadows in the afternoon to fields of spring wheat or barley on the river terraces and valley sides, so their dung and urine could fertilise the arable land on which they were folded. The system's economic driver was the price of corn rather than sheep products, although these were an important by-product. This is called the 'sheep-corn' system and came to typify descriptions of chalkland agriculture in the 18th and early 19th centuries.[107]

Those who seek to downplay the importance of this model stress that its greatest influence came during the period of the French Revolutionary and Napoleonic Wars, when political, economic, livestock and water management imperatives all conspired to support the operation of such a complex system.[108] Figure 4.14 shows an idealised view of land use in Wiltshire chalkland after the accounts of Davis. The higher downland is marked as 'sheep walk', which provided grazing during the warmer months; the valleys were water meadow, fringed on the river terraces and valley sides by the arable upon which the sheep flocks were folded overnight. The extensive areas of 'pasture or mainly pasture' to the north and north-west represents the 'cheese' country, dominated by dairying, as distinct from the chalk country, dominated by sheep-corn.

The vulnerability of the sheep-corn system to economic downturn and technological change is encapsulated by factors such as rising labour costs, the introduction of imported fertiliser and cheaper production of grain and lamb from north America and Australasia. This largely occurred after the recession in English agriculture from 1879. Seen as a 'fixed asset' on farms, decline in the sheep-corn system presented farmers with an economic problem as to what to do with their water meadows. Some would have simply abandoned watering their meadows, others changed to alternative husbandry, depending on market prices for milk, meat, and hay.

---

[104] J. Bettey (2007) The Floated Water Meadows of Wessex: a Triumph of English Agriculture In: H. Cook and T. Williamson (eds) *Water Meadows: History, Ecology and Conservation*. Windgather Press, Bollington, 8-21; H.F. Cook, K. Stearne and T. Williamson (2003) The Origins of Water Meadows in England. *Agricultural History Review* 51 (2), 155-162.
[105] J. Bettey (2007) The Floated Water Meadows of Wessex: a Triumph of English Agriculture In: H. Cook and T. Williamson (eds) *Water Meadows: History, Ecology and Conservation*. Windgather Press, Bollington, 8-21.
[106] T. Davis (1794) *General View of the Agriculture of Wiltshire*, London.
[107] M. Cowan (2005) *Wiltshire Water Meadows*. Hobnob Press, Salisbury.
[108] G. Bowie (2015) pers. comm.

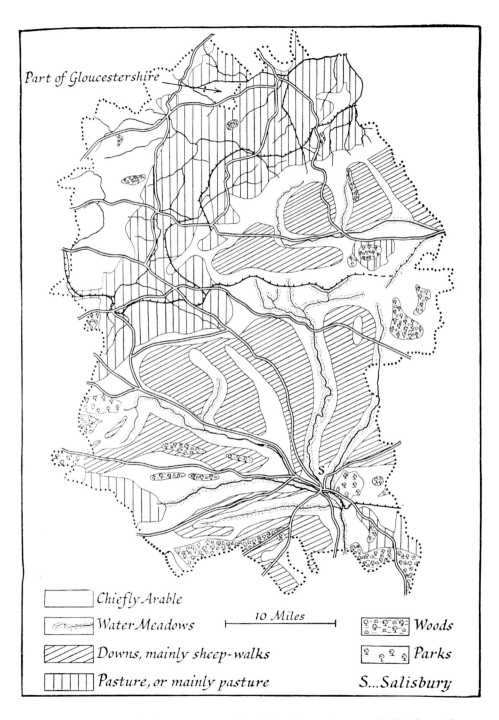

Part of Gloucestershire

Chiefly Arable

Water Meadows

10 Miles

Downs, mainly sheep-walks

Pasture, or mainly pasture

Woods

Parks

S...Salisbury

Figure 4.15. Land utilisation in part of Wiltshire during the 1790s (public domain)

151

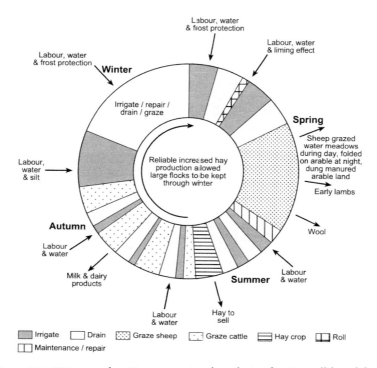

Figure 4.16. Water meadow Management and products after Boswell (1779) (in K. Stearne and H. Cook 2015. Water meadow management in Wessex: dynamics of change from 1800 to the present day. *Landscape Research*, 40(3))

Around Salisbury surviving water meadows largely changed to dairy cattle, something made possible by the opening-up of urban markets following railway construction in the 1840s.[109]

However, the importance of the summer hay crop cannot be underestimated during a time when local transport was largely horse drawn. Elsewhere, on the River Wey near Liphook, Hampshire, the written account makes no mention of sheep but there are references to outputs of hay crops, cattle, and horses, one writer describing two cuts of hay from the period around 1840.[110] The hay was transported by horse along the Portsmouth Road, a different market to that of the chalk country.

Figure 4.15 shows reconstruction of land utilisation in Wiltshire in the 1790s;[111] the southern area is largely NCA 132 (Salisbury Plain and West Wiltshire Downs). In the chalk country the downs are dominated by sheep-walks, the arable located on the valley sides and river terraces, while the valleys floors are mostly water meadow.

---

[109] K. Stearne and H. Cook (2015) Water Meadow Management in Wessex: Dynamics of Change from 1800 to the Present Day, *Landscape Research* 40 (3), 377-395.
[110] N. Bowles (1988) *The Southern Wey: A Guide*. The River Wey Trust, Liphook.
[111] G. East (1937) Land utilisation in England at the end of the eighteenth century. Interpretation of T. Davis, *General View of the Agriculture of Wiltshire*, Richard Phillips, London.

Figure 4.16 shows the annual cycle from Wessex in the late 18th century. It should be noted that the products were not only the early bite grass in early spring, but also a hay crop (sometimes two), aftermath grazing for cattle in late summer. When compared with flood meadows, the productivity and flexibility of irrigated water meadows was considerably improved. A floated meadow could be worth two or more times that of an un-floated meadow in terms of purchase and rental values.[112]

Functioning 'bedwork' systems have been investigated for their impact on water returned to a river in terms of temperature, dissolved oxygen, sediment trapping and the transport of phosphorus. Temperature change in water returned to the river is found to be minimal, and it remains well oxygenated. Water meadow systems reduce the percentage of silt in deposited sediment by an average of 14%, and reduce mobile phosphorus by around one third. Soils show a decline in plant-available phosphorus by one half.[113] We might imagine that the practice of warping on the Somerset Levels has a similar effect on water quality. Water meadows also mitigate local flooding by enabling increased channel capacity, and allowing water to be spread over the surface in more serious flood events.

The outcome of these investigations has been the use of operational and abandoned water meadows to perform a range of functions in the present. Evaluation of the impact on rivers is important because derelict water meadows are typically managed under agri-environmental schemes for flora & fauna, and there is awareness of their historically demonstrated potential for floodwater retention. Clearly the function of water meadows has reflected their economic value over time, providing grazing, hay and meat products from cattle and sheep as well as wool and dairy products. From the 20th century to the present day, water meadow systems have been proven to provide ecosystem services of habitat provision and water quality improvement. The historic water meadows of Wessex are an integral and informative part of a landscape legacy that produces grass in a sustainable fashion and operates to improve the quality of water returned to the river after irrigation.

4.9 Watercress beds and inland fisheries

Watercress (*Nasturtium officinale*) beds were once commonplace throughout the chalklands.[114] The crop became fashionable during the Victorian era, and watercress is still grown in the valleys of the Arle, the Hampshire Avon, Test, Itchen and Ebble valleys and on several Dorset streams. Growing on the Arle was stimulated by the arrival of the railway in 1865,[115] and today a heritage railway connects Alton with Alresford. Around Arlesford, watercress gave an economic stimulus to the area. Groundwater temperatures of 10°C or more are constant and protect the crop from frost in a similar way to watermeadows under irrigation. Although watercress grows naturally in the rivers of Wessex, modern production generally uses water from boreholes and springs to avoid pathogens. Figure 4.17 shows the watercress beds

---

[112] J. Bettey (2007) The Floated Water Meadows of Wessex: a Triumph of English Agriculture In: H. Cook and T. Williamson (eds) *Water Meadows: History, Ecology and Conservation*. Windgather Press, Bollington, 8-21.
[113] H.F. Cook, R.L Cutting and E. Valsami-Jones (2015) Impacts of meadow irrigation on temperature, oxygen, phosphorus and sediment relations for a river *Journal of Flood Risk Management* 10 (4), 463-473. DOI: 10.1111/jfr3.12142
[114] P.D. Smith (1992) Ecological Effects of Discharges from Watercress Farms on the Chalk-Streams of the NRA Wessex Region. <http://www.environmentdata.org/archive/ealit:2808/OBJ/20000103.pdf> Accessed April 2020.
[115] Countryfile (n.d.) <http://www.countryfile.com/days-out/watercress-alresfords-hampshire; http://www.towntrust.org.uk/eel_house.htm> Accessed April 2020.

Figure 4.17. Watercress beds at Broadchalke near Salisbury June 2016 (the author).

at Broadchalke on the river Ebble.[116] The beds, operational since 1880, are today fed from boreholes in the chalk.

Modern fish farms are to be found associated with rivers throughout the region. Along the Hampshire Avon they are associated with former water meadow systems at Britford and north of Downton (Wiltshire) and Bickton mill at Fordingbridge in Hampshire. Eel houses and eel traps are known from the region. At Arlesford, a recently restored eel house built in the 1820s had trapped sea-bound eels and may have still been working in the 1980s. Eels were caught in iron grills beneath the building on dark nights in the water, then loaded into boats which were towed away.[117]

Figure 4.18 shows the 'Eel House' (more correctly the eel trap and sluice house) at Britford on the 'New Cut' of the Avon, southeast of Salisbury. The low, redbrick building, dating from the early part of the 19thC, straddles the river. It was partly rebuilt mid-century.[118] Eels were last caught here in 1973.[119] It also houses the remains of hydro-electric generation equipment

---

[116] BBC News (2014) <https://www.bbc.co.uk/news/uk-25944823> Accessed April 2020.
[117] BBC News (2014) <https://www.bbc.co.uk/news/uk-25944823> Accessed April 2020.
[118] British Listed Buildings (2001) The sluice house and eel trap north of Manor Ditch Britford. <https://www.britishlistedbuildings.co.uk/101334951-sluice-house-and-eel-trap-north-of-manor-ditch-britford> Accessed April 2020.
[119] K. Stearne (2019) pers. comm.

Figure 4.18. The Eel House at Britford in 1999, upstream side (the author).

on the southwest side, used to charge batteries, which were moved to light the 'big house at Britford', and may also have been used to light the stage at night. There was also an eel trap attached to the weir at Middle Woodford mill, north of Salisbury, which was still used in the 1950s.[120]

4.10 Conclusions: landscapes and ecological services of floodplains

The Wessex region is famed for its historic and present ability to manage water. This is the result of a combination of climatic, soil and geological factors, including the many aquifers such as chalk and the Jurassic limestones. The river valleys have a long tradition of water management, associated with floodplain management, meadow irrigation, canal construction, mills, and watercress beds.

The *structures* of the landscapes described above are essentially the result of fluvial or coastal processes that are captured for human purposes. Too much water is problematic, too little fails to perform some functions in agriculture, including warming the soil, supplying nutrients and keeping oxygen fluxes into the soil. This applies to both water meadows and the Somerset Moors. Much of the structures of historic landscapes in river valleys control water for some purpose.

---

[120] E. Crittall (1962) Woodford in E. Crittall (ed.) *A History of the County of Wiltshire, Volume 6.* Victoria County History, London, 221-227.

Functions within the landscape are diverse. Rivers determine transport, water supply and motive power, however the flood meadows, water meadows and grazing marshes described will provide both agricultural products from the livestock sectors, while also recycling nutrients through manure, to include arable land. A modern conception is that of providing ecological services, including habitat provision for birds, animals, and plants. These historic landscapes also improve sustainability through carbon sequestration, water purification and floodwater retention.

Today the *value* lies not only in economic production (meat, milk, grain,) but in the ecological and societal goods provided. These include habitat provision, historic landscapes of water meadows, river valleys and reclaimed marshland as well as non-market values such as ecosystem provision and human well-being. Sparkhayes Marsh near Porlock forms part of a strategy of managed retreat, whereby ecological services are provided in the form of saltmarsh restoration.

The *scale* of description provided varies from small local areas of water meadow or flood meadow to large areas including the extensive floodplain systems around Salisbury or Cricklade. Wetland areas likewise can range from localised wetlands such as those located near Porlock, to the regional and nationally important wetlands of the Somerset Levels and Moors.

Most historic agricultural land associated with floodplains potentially offers ecosystem goods and services, including habitat provision, floodwater detention and a capacity for utilising nutrients and attenuating contaminants and pathogens. The historic floodplain landscapes to be found in Wessex and their main benefits are summarised in Table 4.2:

Table 4.2. Basic ecological services of the major historic floodplain landscapes[121]

System	Hydro-physical properties	Ecological functions and services	Environmental functions and services
Primary marsh	Undrained riverine marsh, unenclosed saltmarsh and mudflats	Tolerant of inundation, variety of habitats: fresh, brackish and saline	Sediment trapping, with complicated nutrient dynamics, refuge for flora & fauna
Secondary marsh	Re-claimed from coastal or riverine marsh, under drainage, minimal floodwater detention	Reduced natural flooding, valued habitats in ditches and/or on grass sward. Coastal grazing marsh tolerant of salt inundation	Overall low nutrient status, brackish ditches near to coast. Capable of trapping sediment, a property utilised in 'warping' and important fringe habitats
Tertiary marsh	Deep-drained landscape, high degree of water table management and flood control	Virtually no ecological or landscape services, intensive horticultural, arable or grass production	Generally a net contributor to eutrophication from intensively managed grass or arable. Fertiliser-free strips effective where no under-drainage outfalls.

[121] H.F. Cook (2010) Floodplain agricultural systems: functionality, heritage and conservation. *Journal of Flood Risk Management* 3, 1-9. DOI:10.1111/j.1753-318X.2010.01069.x.

System	Hydro-physical properties	Ecological functions and services	Environmental functions and services
Undrained bog or fen	Highly retentive of water	High biodiversity value, soils are retentive of water	Overall, low nutrient status. Sequestering carbon through accumulation of organic matter
Drained peatlands	Water retention reduced by drainage and peat shrinkage	Range of habitats in ditches and in grass sward where non-intensive grazing system	Generally low-nutrient, capable of trapping sediment, 'warping'. Sequestering carbon, only where accumulating organic matter
Riverine Flood meadow	Inundated once river overtops its banks	Range of valuable flood-tolerant habitats, grazing systems	Sediment trapping capable of sequestering carbon or phosphorus were arable is reverted to grass
Derelict water meadow	Complicated system of carriers, ridges and drains, no longer operational. Floodwater detention	Range of habitats within complicated man-made topography, soils from mineral alluvium to organic, grazing system	Most properties as flood meadow but with more complicated topography, where this survives. Potentially divers flora and fringe habitat
Active water meadow	Complicated system of carriers, ridges and drains. Diversion of river water through grass of floodplain	As above but sward diversity may be restricted. Water levels controlled, nursery for fish. Grazing system, early-season grass	Diversion an attenuation of peak flows, irrigation water traps sediment (especially silt grade) and reduces phosphorous in water applied. Little effect from an irrigation on temperatures in the river system
Reedbeds	Fringing vegetation, requires waterlogging	Managed for wildlife and reed production and habitat provision	Attenuates a range of potential contaminants, silt nutrients and pathogens
Riparian woodland	Often fragmented woodland, river banks and fringing open water	Historic habitat in many areas as 'fen carr' tolerant of inundation retains water	Attenuates a range of potential contaminants, silt nutrients and pathogens

It is therefore *change* that is perhaps most difficult to characterise in simple terms. However, one can point to the developmental sequence of reclaimed marshlands or peatlands and the development of water management technologies for agriculture that embraces the continuum from simple flood meadows, via the 'warping' on the Somerset Levels, to the full-blown meadow irrigation systems of Wessex. These in turn have variously incorporated the sheep-corn system, developed from turning flocks out early on a floodplain meadow then penning them to arable to provide nutrients and soil conditioner. In addition, Wessex river systems have often been modified for purposes of transport and power, something described elsewhere in this volume.

Chapter 5

The Vales

5.1 Vales in diverse landscapes

The vales are dominated by clays, shales or other soft and erodible rocks. They possess abundant watercourses whose headwaters generally arise from aquifers including Chalk, Upper Greensand and Jurassic limestones (mostly Great and Inferior Oolites and Corallian) or Carboniferous limestone (Figure 6.1).

The words vale, valley, dale and dell are topographic descriptions derived from, or related to, Old English *dæl,* Old Norse *dalr,* Dutch *dal,* German *tal,* French *vallée* and more. Valleys are elongated depressions in the landscape, generally formed by erosion of soil and bedrock as part of the action of draining water. The actual mechanisms of valley formation are complicated, and freeze-thaw action, even ice can play a part, although Wessex was too far south for direct erosion from ice-sheets or valley glaciers.

> In Wiltshire, the geological structure that made for a two-fold division of the county into "Chalk" and "Cheese" forms the basis for a subdivision of its pasture lands. With the spread of agricultural education and the appointment of County Agricultural organisers it is possible that methods of land treatment may be more uniformly practiced, but soil must remain a basic factor. Clays and heavy loams produce rich pastures carrying the most nutritious grasses, and given good drainage are most favourable for dairying.

Thus spake A. H. Fry in 1940, who wrote about land utilisation in Wiltshire at the outbreak of the Second World War (Figure 6.2).[1] He divided pasture in his native county into meadowlands, water meadows (over shallow alluvial soils) and hill pastures. Here we are concerned with meadowlands, mostly developed on heavier soils and away from floodplains.

The term may sound puzzling, for chalk and cheese are quite different. To the unfamiliar, it may even sound like an unconventional culinary experiment. To the initiated however, it is a simple bit of geographical determinism: cheese comes largely from cows, cows graze in valleys, where soils are heavier and include many clays, producing altogether different soils which can support lusher grass. On the other hand, as described in Chapter 6, the 'chalk country' was historically dominated by sheep pasture, with the valleys hosting water meadows. Arable crops, especially wheat and barley, were grown on the better drained soils of the lower part of chalk outcrops at the sides of valleys, and on the near-ubiquitous river terraces, where surface deposits of loess provided water and nutrient retaining soils, commonly with well-draining gravel subsoils.

---

[1] A.H. Fry (1940) Wiltshire, in L.D. Stamp (ed.) *The Report of the Land Utilisation Survey of Britain*, part 87. Geographical Publications. London, 177.

Wiltshire cheese is not as well known today. However, the Oxford and Kimmeridge Clay vale of North Wiltshire (Figure 6.2) is its dairying region *par excellence*.[2] Fry barely mentions cheese production, but we may imagine a proportion of dairy milk was thus used. Between 1867 and 1937, largely a time of prolonged agricultural recession, the number of dairy cattle increased from 50,000 to nearer 100,000, although other cattle remained more constant. During the same period, the sheep population fell dramatically.

Meanwhile, arable agriculture held its own in north-west Wiltshire on the Corallian ridge and Cotswold limestones.[3] However, the arable area overall in the county fell between 1867 and 1937 by more than 50%, with a modest peak around 1917 caused by WW1[4] (see Chapter 6 and Figure 6.2).[5]

Otherwise the vales are in the south Wardour, Wylye, and the north and mid-west grass region (geologically dominated by the Oxford and Kimmeridge Clay), this is largely between the Cotswolds and Chalk, but containing the Corallian ridge. Pewsey Vale (dominated by Upper Greensand) is described as 'mixed farming', as is the Corallian ridge.

Wessex, naturally, is not just about Wiltshire and a comparable map is published for Dorset to which reference should be made (Figure 6.6).[6] In this land use regional distribution, sheep clearly dominate the centre of the county (NCA Profile 134). Dorset's 'cheese country' is found to the north and marked as the Vales of Blackmore and Upper Stour, with dominant dairy farming (NCA 133). The heathlands (Chapter 7) are located to the south of the chalk upland.. The more complicated landscape of West Dorset is to the west. The areas marked 'A' are largely Weymouth Lowlands (NCA 138), west of these are NCA 139: Marshwood and Powerstock. Four areas will be described in terms of the NCA classification areas, being precis of Natural England information.

5.2 National Character Area Profiles and the vales

The Vale of the White Horse is within the NCA 108: Upper Thames Clay Vales, between the town of Swindon and the City of Oxford. To the north is the NCA 109: The Midvale Ridge (comprising limestone), to the south the North Wessex Downs (NCA 130). It is a broad belt of open, gently undulating lowland farmland on predominantly Jurassic and Cretaceous clays, supporting lowland meadow vegetation communities and, at Little Wittenham, one of the most studied great crested newt populations in the UK. The landscapes include enclosed pastures of the claylands, with wet valleys, mixed farming, hedges, hedge trees and field trees as well as more settled, open, arable lands. Mature field oaks give a parkland character in places. There

[2] A.H. Fry (1940) Wiltshire, in L.D. Stamp (ed.) *The Report of the Land Utilisation Survey of Britain*, part 87. Geographical Publications. London, 177-179.

[3] A.H. Fry (1940) Wiltshire, in L.D. Stamp (ed.) *The Report of the Land Utilisation Survey of Britain*, part 87. Geographical Publications. London, 177, 186-7.

[4] A.H. Fry (1940) Wiltshire, in L.D. Stamp (ed.) *The Report of the Land Utilisation Survey of Britain*, part 87. Geographical Publications. London, 177-193.

[5] A.H. Fry (1940) Wiltshire, in L.D. Stamp (ed.) *The Report of the Land Utilisation Survey of Britain*, part 87. Geographical Publications. London, 177-210.

[6] L.E. Tavener (1940) Dorset in L.D. Stamp (ed.) *The Report of the Land Utilisation Survey of Britain*, part 88. Geographical Publications. London. Geographical Publications. London, 276.

are internationally important meadows and wetland sites, which require conservation. Both the wet grassland and wetlands provide opportunities for floodwater detention.[7]

The Vale of Taunton Deane is within NCA 146: Vale of Taunton and Quantock, situated between the Brendon Hills on the edge of Exmoor to the west and the Somerset Levels and Moors to the east. It overlooks the Bristol Channel to the north and the Blackdown Hills to the south and encircles the Quantock Hills.

These uplands contrast with the open character of the clay levels' lush, pastoral character. The area is dominated by mixed farming systems and fragmented semi-natural habitats. It is densely settled, with a largely dispersed pattern of hamlets and scattered farmsteads, linked by sunken winding lanes. Significant towns are Taunton, Wellington, Minehead, Williton and Watchet. There are small patches of limestone and neutral grasslands, small woodlands, hedgerows and rivers, pollarded trees, grazing marshes, coastal saltmarsh and cliff and other maritime habitats.[8]

Marshwood Vale (or Vale of Marshwood, Figure 5.12) is a bowl-shaped valley developed in Lower Lias and, together with Powerstock vale, comprise NCA 139. These are set within a series of Upper Greensand and sand/limestone ridges and hills, which link this area to its neighbours. To the north and west, a broad Greensand ridge separates the area from the Yeovil Scarplands and the Blackdowns NCAs. Eastwards, strata of clays, sands and limestones create terraces, stepping up to the chalk downlands of Dorset and Cranborne Chase, NCA 134. A short but complex coast includes Golden Cap, the highest point on the south coast at 191m, offering long views up and down the coast and inland. West Bay marks the beginning of Chesil Beach, 28km of shingle beach, fronting the geologically complicated Weymouth Lowlands (NCA 138) and joining the Portland limestone dominated Isle of Portland (NCA 137) to the mainland by the shingle of Chesil Beach.[9]

The NCA 139 is within the **Portland Bill to Dawlish Warren coastal cell**, including Chesil Beach (Chapter 1), where littoral drift is generally eastwards and sediment transport along the coast is low and intermittent. The principal transport links in the NCA keep to the higher ground and have avoided the Marshwood Vale, the main axis being along the coast via Bridport and along the Brit Valley to Beaminster. Ham Hill stone (in the Yeovil Scarplands) is a popular building material.

Marshwood Vale has a wooded pasture landscape, with ribbons of woodland and abundant hedgerow oaks. Woodland occurs along the springline, and recent secondary, heathy woodland often caps the Greensand ridges. To the east are more extensive semi-natural woodlands with concentrations around Hooke and Powerstock Common. This area is strongly rural and agricultural. Marshwood Vale supports rich pasture and meadow, with arable graduating to pasture. The eastern side of NCA 139 is dominated by pasture, cattle and sheep being the main grazing stock. Thick and tall hedgerows with regular hedgerow trees dominate. There

[7] Natural England (2014) NCA Profile:108 Upper Thames Clay Vales (NE570) <http://publications.naturalengland.org.uk/publication/5865554770395136> Accessed April 2020.
[8] Natural England (2014) NCA Profile:146. Vale of Taunton and Quantock Fringes (NE550). <http://publications.naturalengland.org.uk/publication/6601735426539520?category=587130> Accessed April 2020.
[9] Natural England (2014) NCA Profile:139 Marshwood and Powerstock Vales (NE517). <http://publications.naturalengland.org.uk/publication/6220303927607296> Accessed April 2020.

is a range of semi-natural habitats, including coastal grassland, scrub and eroding cliff communities; unimproved pasture and meadow on the valley floors; and rough pasture on the hills, ranging from acid to calcareous. There are springline mires, fens, alder woodlands and heathy hill tops.[10]

The Vale of Pewsey, or Pewsey Vale, part of NCA 116: Berkshire and Marlborough Downs (Chapter 6), is an area of Wiltshire centred on the town of Pewsey.[11] The vale separates NCA 132: Salisbury Plain and West Wiltshire Downs from the Marlborough Downs (part of NCA 116: Berkshire and Marlborough Downs). It is about 30km long and around 5km wide. At the western end is the town of Devizes. It is not part of the Downs but is included as part of the North Wessex Downs Area of Outstanding Natural Beauty (AONB). Pewsey Vale includes some fertile soils, supporting arable agriculture and orchards, punctuated by nucleated villages.

Blackmore Vale and Vale of Wardour forms NCA Profile 133. The complex geology of the Vale of Wardour reflects many of the geological exposures to be found on the southern Jurassic Coast, from Mid-Jurassic clays to the Chalk of the Cretaceous. It is a complex mosaic of mixed farming: undulating, lush clay vales dissected by a broken limestone ridge and fringed by Upper Greensand hills and scarps (see also Chapter 6).[12] The geology of Blackmore Vale, which is dominated by the valley of the Dorset Stour, comprises rocks of comparable age (see section 6.2.5 for further information).

5.3 Vales in prehistory and history

The interplay between vale and ridge is complicated. Of nowhere in Wessex can it be said access is impossible, just that areas are more or less accessible. The motivation to create routes varies with economic and settlement requirements. Prehistoric societies hunted across a diverse landscape and migrated and traded according to pragmatically determined available routes; that pragmatism included seasonal considerations, such as the passability of river valleys or wetlands.

Whatever the ecology, cyclicity, density and composition of the 'wildwood' (see Chapters 1 and 2), Upper Palaeolithic and Mesolithic people had to negotiate the underlying land surface in their hunting and migrations. It is generally presumed that communities moved between winter base camps in more wooded areas and summer hunting grounds in upland areas or along the coast where particular resources could be exploited.[13] Such a situation pertained around the present-day Somerset Levels (Chapter 4) and the Solent. Presumed Mesolithic settlement clustered in and around areas of Clay-with-flints (on higher areas) – for example, this is indicated by mapping finds on Cranborne Chase.[14] There have also been finds within valleys, or along the contemporary coastline. It may have been that population increase during the upper Mesolithic led to increased exploitation of upland areas of chalk, while other areas

---

[10] Natural England (2014) NCA Profile:139 Marshwood and Powerstock Vales (NE517). <http://publications.naturalengland.org.uk/publication/6220303927607296> Accessed April 2020.

[11] Natural England (2014) NCA Profile:116 Berkshire and Marlborough Downs (NE482). < http://publications.naturalengland.org.uk/publication/4822422297509888> Accessed April 2020.

[12] Natural England (2014) NCA Profile:133 Blackmore Vale and Vale of Wardour (NE539). <http://publications.naturalengland.org.uk/publication/5858996464386048> Accessed April 2020.

[13] B. Cunliffe (2013) *Britain Begins*. Oxford University Press, 106.

[14] M. Green (2000) *A landscape revealed: 10,000 years on a Chalkland Farm*. Tempus, Stroud, Chapter 1.

including Greensand and coastal sands and brickearth as well as the uplands of Exmoor and its coast were also exploited. Mesolithic peoples were capable of exploiting uplands, coasts, river valleys and estuary environments.[15]

In the succeeding millennia, vales would play an important role in local transport, while those uplands and interfluves that could be were ultimately cleared and settled. The prehistoric tracks that had developed in Wessex by the Iron Age were predominantly on high ground (see Chapter 2.8). The subsequent Roman system of roads commonly crossed the landscape in a 'topographically blind fashion'. A tendency for settlements to be located within river valleys, for reasons of shelter, water supply and water transport, apparently does not signify that trackways automatically followed valley bottoms.

Yet valleys undoubtably had great significance in prehistoric society, and this may have been related more to the significance of water (in rivers or around springs) than to communications. At Langley's Lane, in Bath and Northeast Somerset, a tufa spring has been excavated. Here the ritual deposition of animal bones has been observed dating from the late Mesolithic.[16] A relationship between the river Avon, Stonehenge, 'the Avenue', the large henge monument of Durrington Walls and Woodhenge has been proposed. It is based around symbolic relationships with the ancestors and has currency in contemporary anthropological thought.[17]

Impressive Iron Age hillforts are located near to Warminster: Battlesbury Hillfort, Scratchbury Hillfort and (west of the town) at Cley Hill, while Bratton Camp commands the escarpment on the north side of Salisbury Plain. Here the Wylye Valley provides an obvious corridor through the chalklands. The association of 'hillforts' and valleys in Wessex, as elsewhere, points to boundaries between Iron Age tribal confederations, or at least that valley bottoms formed part of a buffer zone.[18]

Many Wessex valleys seem to be obvious routeways, although this observation requires clarification. Modern valleys, punctuated with settlements joined by modern roads, are the outcome of complicated factors. For one, the floodplains themselves have been severely altered, by drainage, maintenance of riverbanks, or the construction of flood defences, mills and water meadows. For another, there are instances where river terraces provide better drained soils for arable agriculture as well as a basis for tracks and roadways. It is likely the passage through many valleys was only easy during drier months, forcing communications on to higher ground. This point is supported by the extensive networks of drovers' roads across higher ground in our region (Chapter 6).

The Ebble valley in the chalklands located southwest of Salisbury provides a good example of landscape development. Referring to Figure 6.10, it is easily observed that the network of roads is predominantly located away from the valley bottoms. Across much of Wessex the 'uplands' do not preclude access; far from it, they often provide safer and drier passage across the landscape. The antiquity of settled human activity in the area is well-attested; for example

[15] B. Cunliffe (1993) *Wessex to AD 1000*. Longman, London, 26-31.

[16] J. Lewis, C. Rosen, R. Booth, P. Davies, M. Allen and M. Law (2019) Making a significant place: excavations at the Late Mesolithic site of Langley's Lane, Midsomer Norton, Bath and North-East Somerset, *Archaeological Journal* 176 (1), 1-50. DOI: 10.1080/00665983.2019.1551507

[17] M. Pitts (2000) *Hengeworld*. Arrow Books, London, Chapter 27.

[18] H. Cook (2018) *New Forest: The Forging of a Landscape*. Windgather Press, Oxford, Chapter 3.

Ordnance survey mapping reveals a Neolithic long barrow close to the Shaston Drove, 2km due north of Berwick St John and the early Iron Age hillfort of Winklebury lies on a spur southeast of that village, while there are several occurrences of prehistoric field systems and enclosures to be found'.[19]

On Trow Down, in the south side of Alvediston Parish, there are Bronze Age 'bowl' barrows.[20] However, early maps around Alvediston in the west of the Ebble Valley show no roads actually through the valley, although later 18th century mapping shows a network of tracks linking valley bottom roads with access to the combes and the Ox Drove to the south or Shaston / Old Shaftsbury road (sometimes Saxon *herepath* or main road) running east-west to the north, and there is an account of how impassable the valley roads were around 1800.[21] This certainly does not preclude the existence of trackways throughout the valley but points with some certainty to the preference for higher 'ridgeway' routes and drovers' roads for more than very local transit.

It is true that at crossing points, such as fords (which would in time be replaced with bridges), roads cross the valleys linking upland areas on either side. Coombe Bisset is a case in point. The 18th century Old Blandford toll road runs southwest from the bridging point;[22] on Andrews and Dury's 1773 map, a tollgate is marked close to this bridge. Indeed, the modern road (A354) through the dry valley on the south side is 19th century. Furthermore, in 1773, Shaston Drove was part of the London to Exeter road; it was subsequently turnpiked in 1762. By 1788 the trust had lapsed when the Salisbury to Shaftesbury road was turnpiked,[23] with improvements continuing into the succeeding century.[24]

Therefore, north of the Shaston Drove, the route destined to become the modern A30 was a new turnpike road. A segment of this road may be seen in the northwest corner of Figure 6.10, passing Barford Heath. Largely west of the area covered by Figure 6.11, the new turnpike improved access to existing settlements such as Compton Chamberlain, Fovant, Swallowcliff, Sutton Mandeville and Ansty, without actually going near to the centre of such settlements. The concept of a 'bypass' had been born, and progress across the lower ground was more convenient than using the former ridgeway route of Shaston Drove. Otherwise the existing road network (some of which has been shown to be of pre-Norman Conquest date)[25] would largely be inadequate for rapid transport in the developing economy of the late 18th and early 19th centuries.

[19] Discover Chalk valley (2014) <http://www.discoverchalkevalley.org.uk/charming-villages/chicklade/> Accessed April 2020.
[20] J. Freeman and J. H. Stevenson (1987) Alvediston, in D.A. Crowley (ed.) *A History of the County of Wiltshire: Volume 13, South-West Wiltshire: Chalke and Dunworth Hundreds*, Victoria County History, London, 6-16.
[21] B. Trahair (2011) *Alvediston: A History*. Hobnob Press, Salisbury, 16-23.
[22] Wiltshire Community History (2011) Coombe Bisset <https://history.wiltshire.gov.uk/community/getcom.php?id=70> Accessed April 2020.
[23] Wiltshire Community History (2011) Swallowcliffe <https://history.wiltshire.gov.uk/community/getcom.php?id=220> Accessed April 2020.
[24] Anon (n.d.) Our Anstey Landscape <http://www.anstywiltspc.org.uk/wp-content/uploads/Ansty-in-Maps-Pics-and-Diagrams-1.pdf> Accessed April 2020.
[25] J. Freeman and J.H. Stevenson (1987) Ansty, in D.A. Crowley (ed.) *A History of the County of Wiltshire: Volume 13, South-West Wiltshire: Chalke and Dunworth Hundreds*. Victoria County History, London, 93-100.

Figure 5.1. The western end of the Ebble ('Chalke') valley. Width of extract c. 14km.
(1811/1817 Old Series OS map)

Through Broad Chalke and on to Berwick St John in the head of the valley, the Northnorthwest-facing chalk scarp then swings around east of the village to Winkelbury Hill, before swinging westwards once more to form Melbury Down. The chalk outcrop follows this pattern due to gentle east-west flexures affecting the chalk and Upper Greensand south of the Vale of Wardour anticline. The manor of Norrington ('Northampton' on the map) lies immediately west of Alvediston on the south side of White Sheet Hill in Figure 5.1. This lies to the west of the map extract in Figure 6.11.

The Manors of Alvediston and Norrington seem to have been separate entities until the mid-14th century.[26] There was also a Trow Manor within the parish of Alvesdiston in 1086, but Alvediston is not mentioned in Domesday[27] and today Trow Farm lies about 1km Southwest of Alvediston on the road through the village and is marked 'Trough' in Figure 5.1. Wallingford Priory, which held Norrington manor in 1361, also held another estate in Alvediston in 1291, which was probably small, for in 1318 a contribution due from it for the clerical subsidy could not be raised. The Priory still held the estate in 1428.[28]

[26] B. Trahair (2011) *Alvediston: A History*. Hobnob Press, Salisbury, 194.
[27] B. Trahair (2011) *Alvediston: A History*. Hobnob Press, Salisbury, 12.
[28] J. Freeman and J. H. Stevenson (1987) Alvediston, in D.A. Crowley (ed.) *A History of the County of Wiltshire: Volume 13, South-West Wiltshire: Chalke and Dunworth Hundreds*, Victoria County History, London, 6-16.

Figure 5.2. Norrington Manor and farm buildings from the south. To the left of this picture is the deserted medieval village (the author)

Figure 5.3. Holloway heading southwards from Norrington towards the valley road joining Alvediston and Berwick St John (the author).

After the Dissolution of the Monasteries, the Trow estate passed to the Crown, and in 1544 it was granted to George Ludlow. Thomas Gawen may have held it around 1553; it was afterwards merged with his Norrington manor. In the 17th century, Manwood Copse (in the south of Alvesdiston Parish) was said to be part of Staplefoot walk, a deer walk of Cranborne Chase, but the claim was contested by the lord of Norrington manor; a keeper's lodge in or near the copse was perhaps built at that time as part of the manor. A path north of the Ebble river, which today leads westwards from West End, south of Alvediston church and Norrington Farm, marked the route of a valley bottom road which had fallen into disuse by the late 18th century. The modern road runs south of the Ebble and through Alvediston village. Roads also led north from the village and Norrington Farm to the old turnpike road in the late 18th century.[29]

It may be observed that access to Norrington's fine medieval manor house (dating from the 14th century but with many subsequent additions),[30] was both by the Shaston Drove along the chalk ridge to the north and to the south where the east-west road pushes through the valley, just above the floodplain north of the river Ebble. Figure 5.2 shows the manor house and Figure 5.3 shows a holloway leading southwards across a low ridge to the valley bottom road.

Norrington Manor is first recorded in 1198, and it may then have been subsidiary to Alvediston. Like Alvediston, Norrington Manor had open field agriculture with strip cultivation, but the details are unclear, as are the time and reasons for the abandonment of the village, although neither the 14th century Black Death or the 17th century plague can be ruled out.[31] It has the appearance of a deserted medieval village on the ground. There is evidence for a village of medieval origin of 18 to 20 properties with gardens nearby, built along streets, with the lumps and dips in the ground visible in fields south-west of Norrington Manor House.[32] To the north-west of the Manor, 'field system' is marked by 1:25,000 OS mapping.

In this valley to this day, there is no continuous east-west metalled road on the south side of the river between Coombe Bisset and Ebbesborne Wake (Figure 6.11), although there are river crossing points, including a pack-horse bridge at Combe Bisset.[33] The implication of the somewhat complicated road and track network around the valley is that an interplay of valley-side topography and flood risk limited its development in the valley bottom. Until the 18th century, the older network of 'drovers' roads' provided passage, but they did not continue to be fit for purpose.

In Somerset, the Mendip Hills, despite being a prominent topographic feature, are criss-crossed with roads and trackways dating from prehistory (Chapter 6). Access is often via gorges and coombes to the plateau edges. Although no areas of Wessex are inaccessible by ordinary road or track when compared with more mountainous regions, improvement has always been on the agenda, to allow access to rural resources of food, timber, and minerals.

[29] J. Freeman and J. H. Stevenson (1987) Alvediston, in D.A. Crowley (ed.) A History of the County of Wiltshire: Volume 13, South-West Wiltshire: Chalke and Dunworth Hundreds, Victoria County History, London, 6-16.

[30] Historic England (1966) Norrington manor with wall and gate piers. <https://historicengland.org.uk/listing/the-list-entry/1318666> Accessed April 2020.

[31] B. Trahair (2011) Alvediston: A History. Hobnob Press, Salisbury, 197; T. Tatton-Brown (2017) pers. comm.

[32] Wiltshire Community History (2011) Alvediston <https://history.wiltshire.gov.uk/community/getcom2.php?id=6> Accessed April 2020.

[33] E. Crittall (1959) Roads, in E. Crittall (ed.) A History of the County of Wiltshire: Volume 4. Victoria County History, London, 254-271.

Access being possible, if not always easy, would have facilitated settlement and economic development in the region. Over time improvement was achieved by Romans, turnpike trusts, canal constructers, railway builders and motorway networks.

5.4 Regional farming patterns

The mid-twentieth agricultural situation is summarised in Figure 5.4 (key) and extract covering the area of Wessex as considered in this book (Figure 5.5). Compiled in 1939, it summarises agricultural land use at the time of the Second World War and just prior to the modern intensification of British Agriculture.[34] The key is subdivided into 'pasture types' 'arable types' 'intermediate types' (mixed farming) and curiously a 'various' category, of economically marginal land, including marshes, varied farming is located on mixed soils and urban areas. While these divisions clearly reflect the concerns of food security during wartime,

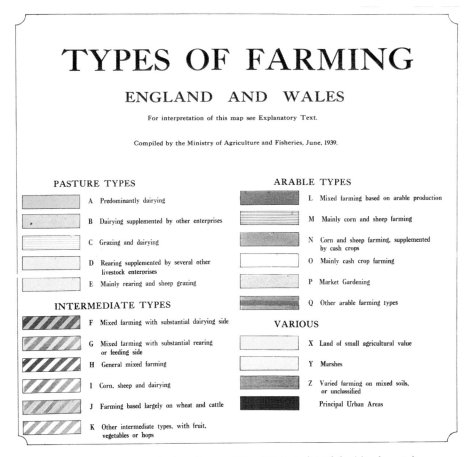

Figure 5.4. Key to Agricultural map of Great Britain (1942) (public domain)

---

[34] Ministry of Agriculture and Fisheries (1942) OS Ten Mile Planning Maps. Sheet 2 - Types of Farming Map <https://commons.wikimedia.org/wiki/File:Ordnance_Survey_Agricultural_Map_of_Great_Britain_Sheet_2,_South,_Published_1942.jpg> Accessed August 2022.

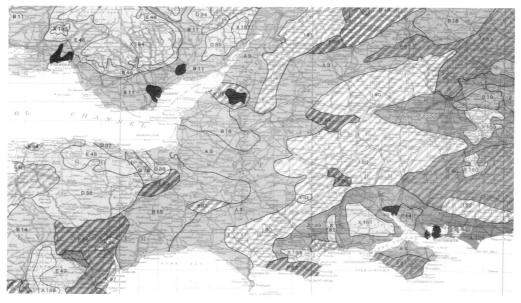

Figure 5.5. Extract of Agricultural map of Great Britain Sheet 2 including Wessex region (1942)
(public domain)

the discernible patterns including sheep and corn (with some dairying) over the chalklands and Cotswolds is clear. The 'Cheese Country' is where dairy production predominates, situated in between and reaching to both coasts. There is upland livestock farming on Exmoor. Finally, the more horticultural or mixed enterprises of Hampshire and south Dorset as well as the approaches to London, are clear.

This pattern had been long established, and for the early modern period regional framework may be discerned from the contributions of J. R. Wordie and G.V. Harrison to Joan Thirsk's edited *magnum opus* on The Agrarian History of England and Wales.[35] Considerations of chalk and cheese country, of vales and plateaus as well as uplands and lowlands were present in the period 1640-1750, between the Second Agricultural Revolution and the dawn of the Industrial Age. Demand was being driven upwards by population increase and urbanisation, while scientific and tenurial revolutions were underway in the Wessex countryside. The early part of the period was dominated by the (English) Civil Wars.

Balances between population size, demand for food and production fluctuated during this period. A times, market demand drove agricultural enterprises, as between 1640 and 1663 (during the Civil Wars) as well as between 1692 and 1713 which experienced relatively high grain prices. On the other hand, between 1725 and 1750 prices could be depressed.[36]

---

[35] J. Thirsk (ed.) (1984) *The Agrarian History of England and Wales* vol. V.I 1640-1750: *Regional Farming systems.* Cambridge University Press, Cambridge.

[36] J. Thirsk (1984) Introduction, in J. Thirsk (ed.) *The Agrarian History of England and Wales* vol. V.I 1640-1750: *Regional Farming systems.* Cambridge University Press, Cambridge.

In the mid-17th century, uncertainty led to innovation in English agriculture, typified by those neatly summarised by Walter Blith in (Chapter 3), and also the introduction of new crops such as clover, sainfoin, lucerne, woad, weld, madder, hops, saffron, liquorice, rape and coleseed, hemp, flax, and orchard and garden fruits. There was also an explosion of literature about agriculture and horticulture.[37] The period 1640-1750 was thus itself one of flux, the prelude to the Industrial Revolution, which would restore demand for meat and grain. The period included the 'Maunder Minimum' cold period within the Little Ice Age, when it is likely that crop yields were greatly reduced and means had to be found to improve the output of English agriculture, including the importation of fertiliser.

Urban growth may have driven demand for vegetables, fruit, and herbs as well as industrial crops including dye plants, hops, coleseed, hemp, flax and timber. All-in-all during this period, livestock farming held up best, as the demand for meat was maintained. Other complexities were introduced through enclosure and the conversion of arable to pasture, or labour shortages sometimes relieved by imported workers.[38]

Changes in this area may have assisted by piecemeal enclosure and there was an increase in leys with clover and ryegrass. Otherwise, areas comprising the clay vales (for example either side of the Corallian ridge and vale of Kennett and north-west Wiltshire) are strongly associated with livestock, including dairying. Osiers were produced and there was also an important coppice industry here. The New Forest is classified in conventional agricultural terms, 'subsistence corn with stock and industries' as is the Reading area where there was some market gardening. The New Forest was mostly a forest-pasture economy.

Other industries in Hampshire and the Isle of Wight included iron smelting, salt production, paper making, quarrying and brickmaking. Cloth production belonged to Wiltshire and other textiles were produced at Wilton. Most clay vales saw a shift towards cattle, especially dairy, and these were most common on heavier soils. Farmers not only took to the innovation of water meadows, but also benefitted from new crops such as clover, ryegrass, and turnips. Fodder and the early bite of grass from water meadows helped close the 'hungry gap' (typically between late March and early May) when historically winter hay stocks were depleted.[39]

The period 1640-1750 in Dorset and Somerset saw specialisation, for example sheep husbandry in Dorset and cattle fattening in the Somerset Levels and Moors. Ley farming and the construction of water meadows allowed an increase in pastoral agriculture in Dorset. In Somerset, innovation included the rise of the cider industry, requiring apple as other orchards; other new crops included hops, teasels, flax, potatoes, and market gardening enterprises. There was also (particularly in the first half of the 17th century) disafforestation and enclosure of 'forest' and moorland areas. These included Gillingham Forest, Neroche Forest, Frome-Selwood Forest and parts of Alder Moor in Somerset. Also, on the Somerset Levels, there was small-scale enclosure of common arable fields and conversion to meadow

---

[37] J. Thirsk (1997) *Alternative Agriculture: A History* Oxford University Press, Oxford, 251-267.
[38] J. Thirsk (1984) Introduction, in J. Thirsk (ed.) *The Agrarian History of England and Wales vol. V.I 1640-1750: Regional Farming systems*. Cambridge University Press, Cambridge.
[39] J.R. Wordie (1984) The South: Oxfordshire, Buckinghamshire, Berkshire, Wiltshire and Hampshire, in J. Thirsk (ed.) *The Agrarian History of England and Wales vol. V.I. 1640 to 1750: Regional Farming Systems*. Cambridge University Press, Cambridge, Chapter 10.

with considerable pasture improvement. Elsewhere, the production of willow, osier and reed were flourishing industries supporting basket-making, faggots, and thatching.[40]

Other improvements to poor quality land included the increased use of manure, burn beating or 'devonshiring' (this was the paring of old turf, burning it and spreading the ashes), marling and liming in preparation for arable planting.[41] In the Vale of Taunton Dean, clover grass was grown. Crop rotation became commonplace in Dorset, where barley came to outstrip wheat production on the chalk downs. Rye was cultivated on restored heathlands. In many parts of the southwest of England, the juxtaposition of cattle and oak woodland led to the development of leather industries and, aside from tanning, certain large estates produced wood and bark. Such was the Dunster Estate in west Somerset.

5.5 A case study: Vale of Taunton Deane

Away from the chalk and limestone country, the Vale of Taunton Deane located south of the Quantocks constitutes a tilted, part wooded plateau which drains eastwards. The Vale is south-east of the Brendon Hills (Exmoor) and (some way) north of the Blackdown Hills. Indeed, the Quantocks plateau of Devonian rocks defines the Vale of Taunton Deane as well as valley of the Doniford Stream to the west (separating the plateau from the Exmoor range) and the lowlands that extend to Bridgwater to the east. The dominant solid geology is of Rhaetic and Keuper marls of late Triassic age developed from the older strata, but also includes some Permian sandstones and Jurassic marls. There is also drift and alluvium derived from the underlying rocks. For present purposes, the Vale will be included with the Vale of Wellington to the south.[42]

The Vale has attracted interest from historians, and the fertility led to relatively high population and settlement density since the Middle Ages, when Taunton Dean was a large manor held by the Bishop of Winchester, which also comprised a fulling mill on the river Tone at Taunton.[43] The climate is also clement and considered mild. At Taunton modern mean annual rainfall is around 875mm.[44] By the 17th century there had been much piecemeal enclosure, producing small family farms, and the settlements were compact and nucleated. The fertile (largely) loamy soils produced wheat, barley, oats, carrots, beans and apples, and cattle were fattened.[45]

To add to this almost Eden-like view of the Vale, meadow irrigation was highly successful; the bottom meadows were excellent, needing little assistance in this respect, but surrounding hillsides evidently benefitted from hillside, or catchwork, meadow irrigation in the 17th century.[46] At the end of the 18th century the Vale was famous for its fertility, including crop

---

[40]   G.V. Harrison (1984) The South-West: Dorset, Somerset, Devon and Cornwall, in J. Thirsk (ed.) *The Agrarian History of England and Wales, vol. V.I 1640-1750: Regional Farming systems.* Cambridge University Press, Cambridge, Chapter 11.
[41]   G.V. Harrison (1984) The South-West: Dorset, Somerset, Devon and Cornwall, in J. Thirsk (ed.) *The Agrarian History of England and Wales, vol. V.I 1640-1750: Regional Farming systems.* Cambridge University Press, Cambridge, Chapter 11.
[42]   T. Stuart-Menteath (1938) Somerset, in L.D. Stamp (ed.) *The Report of the Land Utilisation Survey of Britain*, part 86. Geographical Publications. London, 8-12.
[43]   J.H. Bettey (1986) *Wessex to AD 1000.* Longman, New York, 62.
[44]   Climate-Data.Org (n.d.) Taunton <https://en.climate-data.org/europe/united-kingdom/england/taunton-30365/> Accessed April 2020.
[45]   J.H. Bettey (1986) *Wessex to AD 1000.* Longman, New York, Chapter 4.
[46]   E. Kerridge (1967) *The Agricultural Revolution.* Routledge, Abingdon, Chapter 6.

rotations where wheat was followed by lay grass and turnips, grazed by sheep that supplied a textile industry. The Quantocks produced oak from coppice.[47] Land use in the 1930s was permanent pasture and arable with some leys, capable of producing a range of crops and orchards due to its fertility.[48]

Today the Vale presents a distinct landscape including areas considered under the Agricultural Land Classification (ALC) as 'very good' and 'excellent',[49] and including a very wide range of habitat types. Intensive farming development has meant that few of these occupy a large area; in particular the woodland cover is low and mostly small farm copses. There are, however, many trees and shrubs associated with ancient hedgerows, and semi-natural habitats are generally highly fragmented. The importance of the linear features including the ancient hedgerows and the rivers and streams is critical in maintaining the wildlife interest of the area.[50]

5.6 Case study: Vales within Exmoor

The moors and history of Exmoor are covered in Chapter 7. However, the Moor (which has extensive land considered under ALC as 'poor' and 'very poor') also contains valleys, which could not contrast more with the Vale of Taunton Deane. With mean annual rainfall commonly above 1,000mm, steep slopes and poor soils, agricultural options are limited.

Exmoor is bisected by major valleys.[51] Figure 5.6 shows the (southward flowing) River Quarme Valley to be the most significant, separating Exmoor proper from the Brendon Hills and joining the towns of Tiverton (south of the map) to Porlock on the coast via the modern A396. The village of Dulverton is located near the confluence of the Barle and Quarme. The valleys of the Exe and Barle clearly do not bisect the massif but provide access into the interior of Exmoor. Simonsbath on the Barle is shown in Figure 7.11. Exford (as the name suggests) lies towards the top of the river Exe valley; the river Avill rises on Dunkery Beacon and flows north-eastwards to the coast and a number of other streams flow northwards towards Lynton (in Devon), having risen near to the watershed of the Chains and Central Ridge.

Being an upland area, while the lower reaches of many rivers including the Barle, Quarme and Exe have developed floodplains, their tributaries are frequently above the 'point of alluviation', presenting V-shaped cross-profiles in the upland landscape. This would help to reduce their utility as significant routes along the valley bottoms.

A cursory look at an OS map of Exmoor shows that not only are the modern settlements almost exclusively located in the valleys, but so are the areas of woodland. The upland plateau is largely moorland.[52] It may also be observed that roads and tracks do not automatically

---

[47] J. Billingsley (1794) *General view of the agriculture in the county of Somerset*. W. Smith pub, London, 171.
[48] T. Stuart-Menteath (1938) Somerset, in L.D. Stamp (ed.) *The Report of the Land Utilisation Survey of Britain*, part 86. Geographical Publications, London, 44, 75-76, 88.
[49] Natural England (2010) Agricultural Land Classification Map South West Region (ALC006). <http://publications.naturalengland.org.uk/publication/144017> Accessed April 2020.
[50] DEFRA (2011) 88. Vale of Taunton and Quantock Fringes. <http://adlib.everysite.co.uk/adlib/defra/content.aspx?id=000IL3890W.16NTBZ1SYGQ22M> Accessed April 2020.
[51] H. Riley and R. Wilson-North (2001). *The Field Archaeology of Exmoor*. English Heritage, Swindon.
[52] H. Riley and R. Wilson-North (2001). *The Field Archaeology of Exmoor*. English Heritage, Swindon, 4.

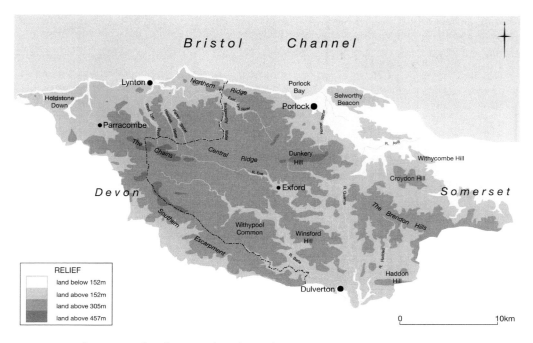

Figure 5.6. The topography of Exmoor (H. Riley and R. Wilson-North, 2001. The field archaeology of Exmoor. English Heritage, Swindon).

follow valleys, as these are often narrow and still in the 'downcutting stage', rather than having developed floodplains. For example, there are only footpaths running south-eastwards from Exford along the valley; the fording point presumably refers to the modern B3224 which crosses the valley running close to the ridge, thence as the B3223 it runs first to Simonsbath before turning northwards to Lynton across the plateau before descending steeply into the valley of the East Lyn River. The topography is suited to roads and tracks across the top of the moors, albeit they are often steep. Road improvement took place in association with the enclosure of the landscape, notably during the time of time of John Knight (Chapter 7).

Figure 5.7 shows the crossing point of the River Barle at the Tarr Steps (a clapper bridge), between Dulverton and Withypool. Again, no major routes follow the valley, however this construction likely dates from the 15th or 16th centuries,[53] and not from prehistory, as was once thought.[54]

The rivers of Exmoor can support fish farms. A trout farm is located in the Haddon Hills, supplied by a mixture of spring water and water from the River Haddeo, which issues

[53] H. Riley (2013) An Historical and Archaeological study of the Tarr Steps, Exmoor National Park <https://archaeologydataservice.ac.uk/archiveDS/archiveDownload?t=arch-1671-1/dissemination/pdf/hazelril1-157448_1.pdf> Accessed April 2020.
[54] Historic England (1959) Tarr Steps <https://historicengland.org.uk/listing/the-list/list-entry/1247822> Accessed April 2020.

Figure 5.7. The Tarr Steps near Dulverton with swimming dog (the author).

from Wimbleball Lake (a dammed water supply reservoir) and natural springs at Hartford, subsequently joining the river Exe.[55]

By way of illustration of the undeveloped nature of certain upland valleys, Figure 5.8 shows the upper part of a valley below Porlock Common, containing only a stream and footpath. The valley runs parallel to the modern A39 as it descends Porlock Hill.

There are maybe cases where better use of valleys might have helped! The extreme climb of Porlock Hill, running westwards from Porlock, reaches 1:4 (25%) gradient. It is said to be the steepest A-road in England and a challenge for cyclists that has defeated the author (for one).[56] An alternative, scenic and hair-pinned route is the toll road than runs from Hawcombe Head in the west, providing outstanding views across Porlock Bay (Chapter 4), built in the 1840s by the Porlock Manor Estate.[57]

Exmoor, due to its terrain, kept packhorses as the main form of transport well into the 18th century, making them likely more commonplace than wheeled vehicles (Figure 5.9). Turnpike

---

[55] Exmoor Fisheries (n.d.) <http://www.exmoorfisheries.co.uk/about-us.html> Accessed April 2020.
[56] Cycling Uphill (2020) <https://cyclinguphill.com/porlock-hill-climb/> Accessed April 2020.
[57] Porlock Manor Estate (2011) <http://porlockmanorestate.org/> Accessed April 2020.

Figure 5.8. Woods in Hawk Coombe above Porlock (the author).

roads were located along the north coast of Somerset as far west as Minehead,[58] while an act of 1765 established a Turnpike Road between Minehead and Bampton, the modern A396. Improvements were made from Minehead onward to Dunster, Crowcombe, Watchet, and Nether Stowey including the modern A39; the core of Exmoor was not involved.[59]

The task of improving the road system of high Exmoor was left to John Knight, who purchased the former Royal Forest in 1818 (Chapter 7). Except for the modern A396 between Dunster and Tiverton, the valleys of Exmoor play only a limited role in defining transport corridors, although they do provide shelter for settlements. Distribution of woodlands is a different matter. The north side of Exmoor, including the National Trust's Holnicote Estate, recorded as a manor in Domesday,[60] and Porlock Manor Estate,[61] are heavily wooded, as is the area immediately inland of Dunster.

[58] Devon turnpike Trusts (2009) <http://www.turnpikes.org.uk/Turnpikes%20in%20Devon.htm?LMCL=i8wzQB> Accessed April 2020.
[59] J. Bentley (1997) Transport, in M. Atkinson (ed.) *Exmoor's Industrial Archaeology*. Exmoor Books, Tiverton, 136-157; Map of Somerset Turnpikes (n.d.) <http://www.turnpikes.org.uk/map%20Somerset%20turnpikes.jpg> Accessed April 2020.
[60] Exmoor National Park (2019) MEM22093 - Holnicote House (Building). <https://www.exmoorher.co.uk/Monument/MEM22093> Accessed April 2020.
[61] Porlock Manor Estate (2011) <http://porlockmanorestate.org/> Accessed April 2020.

Figure 5.9. Packhorse bridge West Luccombe near Porlock, crossing Horner Water (the author).

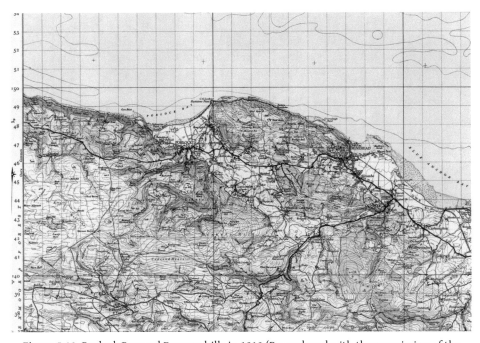

Figure 5.10. Porlock Bay and Exmoor hills in 1919 (Reproduced with the permission of the National Library of Scotland under a Creative Commons License)

The Porlock Manor Estate has been linked to the Blathwayt family since 1686 when William Blathwayt (Secretary of State to King William III) married Mary Wynter. The manor was left to Mary and has since passed down the Wynter-Blathwayt family to the present. The estate is closely linked with the harbour at Porlock Weir (a medieval fish weir, but in effect a harbour), which provided anchorage for coasters carrying timber to South Wales in exchange for coal and limestone for making lime in local kilns.[62]

Scrutiny of Ordnance Survey maps of the area shows almost continuous (deciduous) woodland behind Porlock (Figure 5.10), running up the northwards draining valleys of Hawk Combe and Horner Water. Westward lies more woodland along the coast around Culbone. Around Selworthy there are wooded slopes and to the east much wood and coniferous forestry covers the landscape behind Minehead (the town is located just east of this extract). This port, dating from at least the 14th century, by the 18th century shared much of its trade with South Wales,[63] importing coal, culm for lime kilns (Chapter 4) and livestock. Exports included oak bark for tanning and grain, while smuggling was also important.

By 1830, imports are listed as grain, malt, timber, flour and leather. There was some herring fishing.[64] The next significant harbour to the east (today a marina) is at Watchet, located north-west of the Quantock range. Here the preserved West Somerset Mineral Railway reached Watchet by 1862, initially built to export iron ore from the Brendon Hills to the south. Later it was extended to Dunster and Minehead to the west and south-eastwards to Taunton. In the 19th century, timber grown in the valleys and on the Moors supplied local sawmills, wheelwrights, leather tanners, charcoal makers (for local metal smelting) and a papermill at Watchet, which later used imported pulp from Scandinavia. The railway industry also provided a market for timber products, including sleepers.[65]

By the 20th century, pit props for South Wales from the Dunster Estate were passing through Watchet rather than Minehead. Watchet imported wood pulp and esparto grass for paper–making. Improving road transport and the replacement of coal with oil saw a decline in harbour trade. However, in the mid-1960s a couple of shipping companies revived the harbour by importing timber from Russia and Scandinavia and wine from Portugal and the Mediterranean. Exports included car parts, tractors, and other industrial goods from the Midlands. Eventually Watchet harbour was de-commissioned for commercial traffic in 1999 and became converted to a Marina.[66] Elsewhere on Exmoor, the First Land Utilisation Survey reported that, in the 1930s, while mixed coniferous and deciduous forestry dominated the seaward side of the hills (Figure 5.10), southwards and westwards in the sheltered valleys of the Barle, Exe and Haddon, birch, ash and oak occurred widely with some sycamore and plantations of Douglas fir and a little beech, although this predominated in the hedgerows.[67]

[62] Country Land and Business Association Limited (2016) < https://www.cla.org.uk/your-area/south-west/regional-news/somerset-agm-porlock-manor-estate#> Accessed April 2020.

[63] History of Minehead (n.d.) <https://www.mineheadbay.co.uk/history-of-minehead> accessed May 2021.

[64] J. Bentley (1997) Transport, in M. Atkinson (ed.) *Exmoor's Industrial Archaeology*. Exmoor Books, Tiverton, 136-157.

[65] D. Warren (1997) Miscellaneous Industries, in M. Atkinson (ed.) *Exmoor's Industrial Archaeology*. Exmoor Books, Tiverton, 113-135.

[66] Watchet Museum (n.d.) Maritime History of Watchet <http://watchetmuseum.co.uk/maritime/> Accessed April 2020.

[67] T. Stuart-Menteath (1938) Somerset, in L.D. Stamp (ed.) *The Report of the Land Utilisation Survey of Britain*, part 86. Geographical Publications, London, 62.

The nature and cover of modern woodland, so important to certain valleys in Wessex, will be returned to in chapter 8.

5.7 Case studies: The Vales of White Horse and Pewsey

The Vale of the White Horse is defined on its south side by the chalk escarpment and on the north by the Corallian Ridge. Through it the river Ock drains towards the Thames. Named after the Uffington White Horse, it extends beyond the definition of Wessex used in this book, into Oxfordshire. Prehistoric activity is immediately evidenced by the Ridgeway trackway itself and the Bronze Age horse carved into Whitehorse Hill, which is surmounted by the large Iron Age hillfort of Uffington Castle. There is also a natural mound (called Dragon Hill) that is associated with the legend of St George.[68]

The Vale provides for a natural transport corridor through which both the Wilts and Berks Canal (the eastern section linking Swindon to the Thames at Abingdon) and the Great Western Railway were routed.[69] The Vale is defined on its south side by the chalk escarpment, which provides a strong springline around the junction of the lower (grey) chalk and the Upper Greensand. It otherwise comprises Gault Clay or Kimmeridge Clay, with some Lower Greensand, with the Corallian ridge on the north side. Overlying the (mostly) Gault Clay forming the valley floor is alluvium, often overlying fluvial sands and gravels . The major aquifer is the Chalk, although Abingdon is supplied from the Corallian and there are bores into the Upper Greensand.[70]

The soils are mostly developed on clay, are loamy or clayey in nature and support a rich and varied agriculture. On the Corallian ridge cereals, vegetables, fruit, root crops and other arable may be grown on a range of loamy to sandy soils.[71] A high proportion of fields were enclosed between 1640 and 1750 and the accent was, initially at least, on dairy farming, generally replacing mixed farming in open field communities which produced much wheat.[72] Between 1660 and 1760 there was a dramatic expansion of specialised agriculture, with a range of arable crops including carrots and cabbages, orchards and turnips supplying the London market and adding to a developed mixed agriculture.[73] This basic pattern continues to the present, including orchards on suitable soils.[74]

---

[68] English Heritage (n.d.) Uffington Castle, White Horse and Dragon Hill. <https://www.english-heritage.org.uk/visit/places/uffington-castle-white-horse-and-dragon-hill/> Accessed April 2020.
[69] The Wiltshire, Swindon and Oxfordshire Canal Partnership (n.d.) The Wiltshire and Berkshire Canal <https://www.canalpartnership.org.uk/index.php/13-general/home/1-the-wilts-berks-canal > Accessed 2020.
[70] IGS/Thames Water Authority (1978) Hydrogeological Map of the South West Chilterns and the Berkshire and Marlborough Downs scale 1:100.000. NERC, London; Abingdon Area Archaeological and History Society (2018) Abingdon's Water Supply <https://www.abingdon.gov.uk/abingdons-water-supply> Accessed April 2020.
[71] M. G. Jarvis, R.H. Allen, S.J. Fordham, J. Hazleden, A.J. Moffat and R.G. Sturdy (1984). *Soils and their use in South East England.* Harpenden: Soil Survey of England and Wales. Bulletin 15 with 1: 250,000 scale map and accompanying legend.
[72] J.R. Wordie (1984) The South: Oxfordshire, Buckinghamshire, Berkshire, Wiltshire and Hampshire, in J. Thirsk (ed.) *The Agrarian History of England and Wales vol. V.I. 1640 to 1750: Regional Farming Systems.* Cambridge University Press, Cambridge, 338-339.
[73] J. Thirsk (1997) *Alternative Agriculture: A History* Oxford University Press, Oxford, 62.
[74] Natural England (n.d.) NCA 108: Upper Thames Clay Vale Key Facts & Data <https://www.wildoxfordshire.org.uk/wp-content/uploads/2014/11/108_Upper_Thames_Vale_tcm6-32124.pdf> Accessed April 2020.

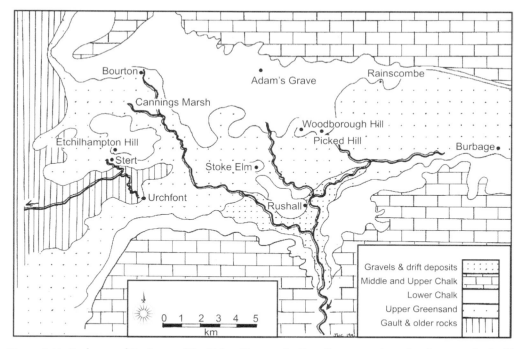

Figure 5.11. Geology and Drainage of the Vale of Pewsey (J. Chandler, 2018, The Vale of Pewsey. (3rd ed))

In many ways, the geology of Pewsey Vale (Figure 5.11) is similar. However, it is drained by two river Avons: one flowing southward, where five tributaries meet to flow through the chalk escarpment, the other from tributaries at Stert and Urchfont westwards towards Bristol. Through downcutting, it likely captured an east-flowing tributary through the Vale. The chalk escarpment on the north side represents the ridge of the Marlborough Downs, that on the south Salisbury Plain (Figure 6.1).

From John Chandler's study of the Vale of Pewsey,[75] the soils which are largely developed on Cretaceous rocks from the Gault to Chalk are potentially fertile, however more problematic soils are developed on the Jurassic Portland stone and Kimmeridge Clay outcrops in the west (Figure 5.11). Historically drainage was a problem, and this is attested by place-names from the Anglo-Saxon period.

The soils are therefore developed on solid geology rocks not unlike those in the Vale of the White Horse and similarly also influenced by drift deposits. These are primarily chalky gravels, brickearth or a clay drift developed from the Gault.[76] However, the nature of these soils is different, comprising coarse, loamy, well-drained calcareous soils, locally waterlogged and suited to both cereals and dairy. Similar soils on the slopes support grassland and woodland with some cereals. Brown earth soils developed directly over chalk, capable of supporting

[75] J. Chandler (2000) *The Vale of Pewsey* (2nd ed). Hobnob Press, Salisbury.
[76] Wiltshire County Council (n.d.) Character Area 9: Vale of Pewsey <http://www.wiltshire.gov.uk/kennet_landscape_character_assessment_part_2_the_character_areas_-_vale_of_pewsey.pdf> Accessed April 2020.

cereals and grass, and suited to stock rearing, including dairying. At the west end of the Vale, developed on clays, are slowly draining seasonally waterlogged soils, fine loamy or clayey in texture, which support winter cereals and short-term grassland, and which are suitable for dairying like those in the Vale of White Horse.[77]

Both field and aerial photography evidence points to there being a strong prehistoric economy, including small rectangular fields defined by low banks, presumably used for arable cropping and dating from the Bronze Age and in the Vale of Pewsey are found impressive burial mounds, field systems and evidence for settlement suggesting the area was densely settled in the late Bronze Age and into the early Iron Age.[78]

Continuing into the Iron Age, a ranch farming system may have operated into the Romano-British period. The evidence comes from the chalklands, for in the bottom of the Vale it has been obliterated through development. These fields are well-preserved north of Bishops Cannings and Allington as well as south of Wilsford. As suggested by the soils and relatively conducive climate, the area likely supported a mixed farming economy in prehistory, for as well as arable fields, animal bones have been found, and certain enclosures look to have been constructed for penning sheep and cattle.[79]

The earliest documentary evidence for settlement comes from Anglo-Saxon charters delineating land holdings by territorial markers. The medieval open fields were largely enclosed by the late 19th century.[80] By William Cobbett's time (1820s), the landscape had developed a familiar tripartite division: downland for rough grazing, arable on the hillsides and meadows providing rich pasture (Chapter 6).[81] More than a century later, two contrasting arable field patterns were recognised, small, isolated areas in the east and large but compact areas in the west. The former occurred on Upper Greensand, the latter on the lower chalk outliers and over river gravels. While the area of arable was noted to have increased by 1940, the Vale was recognised as essentially a pastoral region. Around Market Lavington in the south-west of the Vale, the presence of smallholdings produces market gardening crops.[82]

Today the landscape reflects as much 18th and 19th century enclosure as anything earlier, and is highly agricultural in nature, with woodland and farmland mosaics associated with the Greensand areas. There was early turnpike road improvement in the 18th century, and the Kennett and Avon Canal opened in 1810. In 1862 the Berkshire and Hampshire line from Hungerford was extended via Pewsey and thereby linked Devizes to London. Pewsey's ancient sense of isolation gradually came to an end.[83]

---

[77] D.C. Findlay, G.J.N. Colborne, D.W. Cope, T.R. Harrod, D.V. Hogan, and S.J. Staines (1984). *Soils and their Use in South West England*. Harpenden: Soil Survey of England and Wales, Bulletin 14, 1:250,000 map, legend and memoir.

[78] P.C. Tubb (2011) Late Bronze Age / Early Iron Age transition sites in the Vale of Pewsey : the East Chisenbury midden in its regional context. *Wiltshire Archaeological and Natural History Society Magazine*, 104, 44-61.

[79] J. Chandler (2000) *The Vale of Pewsey* (2nd ed). Hobnob Press, Salisbury, 21-23.

[80] Wiltshire County Council (n.d.) Character Area 9: Vale of Pewsey <http://www.wiltshire.gov.uk/kennet_landscape_character_assessment_part_2_the_character_areas_-_vale_of_pewsey.pdf> Accessed April 2020.

[81] J. Chandler (2000) *The Vale of Pewsey* (2nd ed). Hobnob Press, Salisbury, 19.

[82] A. H. Fry (1940) Wiltshire, in L.D. Stamp (ed.) *The Report of the Land Utilisation Survey of Britain*, part 87. Geographical Publications. London, 190.

[83] J. Chandler (2000) *The Vale of Pewsey* (2nd ed). Hobnob Press, Salisbury, Chapter 6; Pewsey Heritage Centre (2012) Village History <http://www.pewsey-uk.co.uk/pewsey-village-history.html> Accessed May 2021.

5.8 Case studies: Three Clay Vales

The traditional contrast to the chalk country is drawn with the clay vales, where dairying was especially important. The Vale of Marshwood (Figure 6.6) is a small area situated in south-west Dorset. The area lacks nucleated settlements, and comprises isolated farmsteads set among irregular fields, with massive winding hedges. The impact of parliamentary enclosure was relatively light and the earlier field systems have been retained. Place-names commonly end in -hay (for example Bluntshay and Denhay), suggesting personal names from around the time of the Norman Conquest.[84] The Vale contains heavy clay soils derived from the Lower Lias, which includes some calcareous strips, and is defined by encircled, elevated ground caused by the Middle and Upper Lias. The Vale drains south-eastwards by the River Char, which reaches the sea at Charmouth, east of Lyme Regis.

Here too there has been agriculture and settlement since prehistory, with some pre-Neolithic finds, although today many of its small owner-occupied farms are under economic pressure. Ploughing is evident from the medieval period, and arable strips have been identified. Many historic field boundaries have been lost as a result of major field boundary changes since the end of the Second World War, while buried archaeological remains (many unrecorded) are being damaged through continued or increased ploughing, the result of a trend towards drainage and conversion to arable cropping.[85]

Overall, the Vale is considered more fertile than those formed by the Kimmeridge or Oxford Clays, however dairying for milk and cheese making was also important. Hemp was cultivated from the early middle ages into the post-medieval period; it supplied a rope, net, thread and webbing industry at Bridport.[86] Following the arrival of the railways in the second half of the 19th century, farmers ceased to convert their milk to cheese and butter on a large scale, and were able instead to 'export' fresh milk to towns and to London, which had largely lost their dairies due to tighter controls on health, cleanliness and sanitation.[87] Despite many changes, the Vale's pastoral nature remains evident.

The Vales of Blackmore and Upper Stour are next considered. In the NCA classification, they are combined with the Vale of Wardour, which will be considered separately. The Vale of Blackmore is underlain by clays (Oxford and Kimmeridge) as well as Corallian and Upper Greensand. Local gravels may improve soil drainage. The area features in the novels of Thomas Hardy, and there are several stately homes, including Stourhead and the Jacobean Hanford House, now a private school.

In a similar fashion to Marshwood Vale, the Vale of Blackmore (on the border of Somerset and Dorset) is drained by the River Cale, which runs southwards to join the Stour just south of Wincanton. The area lacks nucleated settlements.[88] With the settlement pattern somewhat

---

[84] J. Bettey (1986) *Wessex to AD 1000*. Longman, New York, 37.

[85] NMP Mapping of the Marshwood Vale Dorset AONB (n.d.) Cornwall Archaeological Unit <https://historicengland. org.uk/research/results/reports/6864/NMPMappingoftheMarshwoodValeDorsetAONB> Accessed April 2020.

[86] L. E. Tavener (1940) Dorset in L.D. Stamp (ed.) *The Report of the Land Utilisation Survey of Britain*, part 88. Geographical Publications. London. Geographical Publications. London, 276 and 286; J.H. Bettey (1986) *Wessex to AD 1000*. Longman, New York, 141.

[87] J. H. Bettey (1986) *Wessex to AD 1000*. Longman, New York, 239.

[88] J. H. Bettey (1986) *Wessex to AD 1000*. Longman, New York, 37.

Figure 5.12. Marshwood Vale showing a range of field shapes (photo: James Loveridge).

dispersed, the road pattern reflects the difference between areas of small, irregular fields and narrow lanes that are representative of early woodland clearances.[89]

Prehistoric activity is well attested and there are prominent Iron Age Hillforts on Hambledon Hill and Hod Hill; the Romans conquered set up a fortification on the latter, probably following a conflict with the Durotriges during the conquest of Britannia.[90] There is some systematic, post-medieval enclosure. The Vale is flanked by hills and has a 'rich, tenacious, marshy soil, notable as pastureland, and for the vigorous growth of oaks'.[91] Pasture farming dominates across these landscapes. There was also a royal forest located in the Vale. It appears that arable cultivation south of Sherborne had been abandoned before the Black Death (including that on the Corallian ridge) which may be indicative of a wider pause in population increase even before that calamity.[92] Cattle farming dominated the post-medieval period by far; by the early 19th century, 'in the western part of the county [Dorset] as well as the Vale of Blackmore, the cows were mostly of the Devonshire kind'.[93]

---

[89] Natural England (2014) NCA Profile:133 Blackmore Vale and Vale of Wardour (NE539) <http://publications.naturalengland.org.uk/publication/5858996464386048> Accessed April 2020.
[90] Foundations of Archaeology Project (n.d.) Hod Hill <https://foundationsofarchaeology.wordpress.com/tag/hod-hill-roman-iron-age/> Accessed April 2020.
[91] Vision of Britain (n.d.) Blackmore Vale Dorset <http://www.visionofbritain.org.uk/place/25968> Accessed April 2020.
[92] J.H. Bettey (1986) *Wessex to AD 1000*. Longman, New York, 47 and 85.
[93] J.H. Bettey (1986) *Wessex to AD 1000*. Longman, New York, quotation from William Stevenson 1821 p372

By the inter-war period, the Vale of Blackmore and the Upper Stour valley - subregions 7A (west), 7B (central) and 7C (Vale of the Upper Stour) in Figure 6.6 - are reported as being about two-thirds permanent grass in the west, and of good quality, particularly that developed on the Oxford Clay. Most livestock are dairy cattle with hardly any sheep due to the wet soils. Secondary activities were pig rearing and fruit growing, including cider apples where suitable.[94]

In the central area there are similar heavy clay soils supporting a high area of permanent pasture, but with drier ridges including the Corallian. These permit arable cultivation, particularly straw crops for livestock. In the valley of the Upper Stour, less than 4% arable area is reported, cattle predominated with subsidiary dairying and fruit growing, again virtually no sheep were found.[95]

Essentially, land use patterns described for the 1930s have not altered greatly for:

> The Blackmore Vale, steeped in a long history of pastoral agriculture, is characterised by hedged fields with an abundance of hedgerow trees, many of them veteran. This is productive pastureland that is often waterlogged; it is crossed by streams and several rivers that leave the NCA at all points of the compass.[96]

The NCA refers to mixed farming, and there remain small, scattered broadleaved woodlands. Today the Greensand ridges are important in arable production and urbanisation limited to less than one percent of the land area. Figure 5.13 shows a view from Hambledon Hill hillfort, beyond the settlement a mosaic of small fields, hedges and woods are characteristic of the Vale.

The complex geology of the **Vale of Wardour** reflects and complements at a smaller scale many of the geological exposures to be found on the Jurassic Coast, from Mid-Jurassic clays to the Chalk of the Cretaceous. Stone from Ham Hill, in the Yeovil Scarplands NCA, is often found in high-status buildings.

Some areas, particularly the Greensand terraces and some Upper Greensand dip slopes, were not enclosed until the 18th and 19th centuries. This gave a rectilinear field pattern (Figure 3.6). Around the edge of these areas, large estates were developed in the 16th and 17th centuries, and large landscaped parks were laid out, particularly Wardour, Longleat, Marston Bigot and Stourhead.[97] Such estates often dammed tributaries of the Nadder, which rose in the Chalk and Upper Greensand.[98] Such artificial, and largely ornamental lakes (originally used for fish) may be observed at Compton Chamberlyne, Fovant, Teffont Magna. Swallowcliffe, and between Old Wardour Castle and Donhead St Andrew. One particularly impressive dammed

---

[94]   L.E. Tavener (1940) Dorset in L.D. Stamp (ed.) *The Report of the Land Utilisation Survey of Britain*, part 88. Geographical Publications. London. Geographical Publications. London, 288-289.

[95]   L.E. Tavener (1940) Dorset in L.D. Stamp (ed.) *The Report of the Land Utilisation Survey of Britain*, part 88. Geographical Publications. London. Geographical Publications. London.

[96]   Natural England (2014) NCA Profile:133 Blackmore Vale and Vale of Wardour (NE539) <http://publications. naturalengland.org.uk/publication/5858996464386048> Accessed April 2020.

[97]   Natural England (2014) NCA Profile:133 Blackmore Vale and Vale of Wardour (NE539) <http://publications. naturalengland.org.uk/publication/5858996464386048> Accessed April 2020.

[98]   IGS/WWA (1979). Hydrogeological map of the Chalk and Associated Minor Aquifers of Wessex. scale 1:100,000. NERC, London.

Figure 5.13. View over the Vale of Blackmoor from Hambledon Hill. The settlement beneath the chalk escarpment is Child Okeford (the author).

lake is almost two km in length stretching southwards from Fonthill Bishop. Hydro-electric facilities were later added to this lake.

Fonthill Bishop provides an interesting example of a Wessex manor. The Beckford family who owned it by the 18th century had accumulated extreme wealth derived from their estates in the West Indies. However, the known story starts much earlier.

A field system of some 90ha (230 acres) north of the London to Exeter road on Fonthill Down indicates extensive prehistoric ploughing. In the late 9th century, Athelwulf gifted a 5 hide estate at Fonthill to his wife Athelthryth upon their marriage, which she sold on. The estate passed through several hands, and in 900, Ordlaf granted Fonthill, by then probably about 10 hides, to Denewulf, bishop of Winchester. In 1066 there were 10 hides; in 1086 there was land for 7 ploughs: 5 hides were in the bishop of Winchester's demesne on which there were 2 ploughs and 5 serfs; 8 villeins and 5 bordars shared 3 ploughs. There were 8 acres of meadow, and pasture and woodland were each ½ league long and 3 furlongs broad. The estate was worth £14 having formerly been worth £10. There was also a mill worth 5s. at Fonthill in 1086.[99]

[99] A.P. Baggs, E. Crittall, J. Freeman and J.H. Stevenson (1980) Fonthill Bishop, in D.A. Crowley (ed.) *A History of the County of Wiltshire: Volume 11, Downton Hundred; Elstub and Everleigh Hundred*. Victoria County History, London, 77-82.

From the 13th century, Fonthill Bishop was within the public jurisdiction of the Bishops of Winchester for their hundred or liberty of East Knoyle, and Fonthill was a tithing of that hundred. Sheep and corn husbandry (in the chalkland) was held in common. And in the 13th and 14th centuries, the arable lands belonging to the bishop of Winchester's demesne and those of his tenants were apparently intermingled in the common fields. It remained in Church hands through the Reformation period, and the lordship of the manor was retained by the Ecclesiastical Commissioners although the land was subsequently sold. There was also a customarily held mill in the Middle Ages.

Between the 15th and 17th centuries, cultivation had continued largely in common, regulated through the manor courts and local 'tourns' (bi-annual inspections by the sheriff). The demesne remained the only large farm. In 1539, Fonthill farm was held of the Bishop for £22 12s. a year on leases paid for by substantial fines.[100]

The demesne lands of the manor were leased to farmers from the early 15th century, and passed through several hands. Cultivation continued largely in common with sheep farming apparently profitable. The story of building houses aimed to impress is known to date back to 1533 at Fonthill, when Sir John Mervyn purchased the estate and lived in a house surrounded by a park. In 1539 he was accused of having 'drowned the boundary' between Fonthill Gifford and Fonthill Bishop, likely the result of the original lake being created. The mill was said to be ruined in 1539, and when restored it may have worked until the early 18th century.

The freehold passed with the manor of Fonthill Gifford and was possibly partly merged with land from that manor at this time. Sir James Mervyn was said to have unlawfully fished the Nadder in Fonthill Bishop in 1603-4, while Francis Cottington, in 1722, was denied the right to keep swans. Such are the presumptions of power! The Manor was sub-let for £126 a year in 1724, with some enclosure near the village by 1744.

After 1629, leases passed with the manor of Fonthill Gifford, in the earldom of Castlehaven until the 1630s, in the Cottington family until c.1744, and then in the Beckford family. Most of the parish was in freehold with some leasehold. In 1662 there were named 5 free and 24 customary tenants; no estate grew large until the 18th century. The lords of Fonthill Gifford increased their holdings in Fonthill Bishop by purchases from George Barber and Joseph Bate, probably in the early 17th century.

The Beckford Family, later of Fonthill Bishop, became established in Jamaica in the mid-17th century. Their fortune accumulated based upon a range of activities including wine trading, sugar production, money lending - and the ownership of numerous enslaved Africans. Through their wealth, the Beckfords had also risen politically in the West Indies during the 18th century.

'Alderman' William Beckford (1709-1770), who was twice the Lord Mayor of London as well as MP for that City, purchased the Fonthill estate in 1745 and started a re-building phase that, despite a setback caused by a fire in 1755, enabled him to display his considerable wealth. His son, William Beckford (1760-1844) was a writer, collector of books, art and other objects as

---

[100] In medieval England the term 'fine' referred to a sum of money paid voluntarily to some other individual in return for the grant of some right, benefit, or property.

Figure 5.14. Symbol of power (i): The Fonthill Estate Archway (a) with detail of face from top of the arch (b) (the author).

well as being reclusive. As a boy of ten he inherited one of the greatest individual fortunes in England and as Lord Byron titled him, 'England's wealthiest son'.

From around 1750 the Beckfords steadily bought up the remaining freeholds and copyholds. In 1800 William Beckford's leasehold, freehold, and copyhold estate included nearly the whole parish and, in 1822, a lease was sold with Fonthill Gifford by William Beckford (d. 1844). The transfer was a complicated affair, however by 1823 John Farquhar possessed the estate. Farquhar sold some of the land and assigned the lands south of the road (with Fonthill Abbey) to his nephew George Mortimer.

The Beckfords' activities had made enclosure more practicable. By an agreement, possibly dating to *c.*1760, the arable lands around the village were enclosed. Common husbandry ceased after William Beckford bought a farm in 1796, and after all the tenantry lands were merged in that farm and a small exchange was made with the rector. By 1837, when there were 871 acres of arable in the parish, some 70 acres of downland pasture had been ploughed and there was an appreciable amount of woodland.[101]

The classic model for the growing wealth of particular families in England during the seventeenth, eighteenth and nineteenth centuries involved the accumulation of land through expropriation of peasants by enclosure, resulting in a class of landless labourers who might also enter labour markets as industrial workers. There were other means – shortcuts to becoming wealthy landowners – notably via banking, trading and industry.

Elsewhere, mercantile dynasties dominated cities such as Amsterdam, Ghent or Venice. It seems a trait of aspiring English aristocrats of the seventeenth and eighteenth centuries to require a country seat, not content with a high position as a merchant or financier living in a wealthy area of a city. Examples in England of wealth gained in the financial sector enabling the purchase of large tracts of land include the brilliant economist and stockbroker David Ricardo (1772-1823), who purchased the Gatcombe estate in Gloucestershire.[102] The banking Hoare family purchased Stourton manor in 1717, and created the Palladian house of 'Stourhead',[103] while the Huguenot ancestors of the Lords of Radnor became first successful merchants, then bankers and acquired Longford Castle.[104]

While one cannot claim that the operation of the domestic economy in general terms worked for the common good, slavery was and remains an abomination. It was the cornerstone of the notorious triangular trade involving Africa (providing labour from enslaved Africans), the West Indies (providing such commodities as cotton, tobacco and sugar) and Britain (providing manufactures to Africa). Wessex was central to this exploitation which fuelled the growing economic might of the port of Bristol and the development of Bath as a playground for the decadent wealthy.

---

[101] A.P Baggs, E. Crittall, J. Freeman and J.H. Stevenson (1980) Fonthill Bishop, in D.A. Crowley (ed.) *A History of the County of Wiltshire: Volume 11, Downton Hundred; Elstub and Everleigh Hundred.* Victoria County History, London, 77-82.
[102] J. Hamlin (n.d.) David Ricardo, 1772-1823. <https://www.d.umn.edu/cla/faculty/jhamlin/4111/Ricardo/David%20 Ricardo.htm> Accessed April 2020.
[103] Anglotopia (2020) Stourhead <https://www.anglotopia.net/british-history/great-british-houses-stourhead-everything-need-know-great-house-wiltshire/> Accessed April 2020.
[104] The Daily Telegraph Obituaries (2008) The Earl of Radnor <https://www.telegraph.co.uk/news/obituaries/2560301/The-Earl-of-Radnor.html> Accessed April 2020.

Figure 5.15. Symbols of power (ii): The Fonthill Lake and gauge-board (a) and dam including hydroelectric facility dating from 2011 (b) (the author).

The life of the Bristol merchant Edward Colston (1636-1721) is especially illustrative. On the one hand, through being a member of the Royal African Company he was responsible for the enslavement and deaths of many thousands of west Africans transported to the West Indies, on the other he has been celebrated for his philanthropy which benefitted his home City.[105] Still being remembered with gratitude in the nineteenth century, Colston's 1895 bronze statue was pulled from its plinth during a demonstration in June 2020 and thrown into the harbour from where the slave ships departed.[106] It is worth noting that the folk song 'Captain *Coulson*' while very maritime in character, has no connection with the merchant Colston.

What is generally ignored is how slavery in the colonies financed lifestyles and investment in England. British mercantile ventures did produce families that benefitted specifically from slavery in cities such as Glasgow and Liverpool. Inevitably the relationship with the developing financial sector in such cities as Bristol and London was strong.[107]

The eccentric and sexually ambiguous William Beckford proceeded to build the ill-fated Fonthill Abbey to house his fantastic collection, commissioning the architect James Wyatt in 1796. It is evident that Beckford himself was well-aware of controversy around the institution of slavery, but probably defended it (and his own interests) on the basis that Africans were probably better of under the control of the white man![108] Construction of the Abbey seems to have marked the beginning of the end of Beckford's wealth; he sold the estate to John Farquhar, a gunpowder contractor from Bengal, India. The theme of wealth accumulation through the colonies was to continue!

Within two years of this sale, a large part of Fonthill Abbey collapsed and Farquhar tried to sell all his land but died intestate in 1826.[109] Perhaps this was a lucky escape for Beckford, although other factors in the decline in the family's fortunes arose from falling income from the sugar plantations due to poor management and the abolition of the transatlantic slave trade in 1807. However, the tale does tell us of wealth earned through slavery becoming invested in Wessex.[110]

The Beckford Tower in Bath is another story, but it reflects William's lingering desire to become immortalised in stone. He evidently tired of being a country gentleman and moved to Bath, following the sale of the Fonthill Estate.[111]

---

[105] M. Dresser (2007) *Slavery Obscured: The Social History of the Slave Trade in an English Provincial Port*. Redcliffe Press, Bristol, Introduction.

[106] BBC (2020) 'Edward Colston: Bristol slave trader statue 'was an affront''. <https://www.bbc.co.uk/news/uk-england-bristol-52962356> Accessed June 2020.

[107] C. Hall, N. Draper, K. McClelland, K. Donnington and R. Lang (2014) *Legacies of British Slave Ownership: Colonial Slavery and the Formation of Victorian Britain*. Cambridge University Press, Cambridge, 86; K. Morgan (1999) *Edward Colston and Bristol*. The Bristol Branch of the Historical Association, Local History Pamphlets, Bristol. <http://www.bris.ac.uk/Depts/History/bristolrecordsociety/publications/bha096.pdf> Accessed April 2020.

[108] L. Klein (2018) William Thomas Beckford Between dalliance and duty in C. Dakers (ed.) *Fonthill Rediscovered*. UCL Press, London, Chapter 15.

[109] The Fonthill Estate (n.d.) History <https://www.fonthill.co.uk/history> Accessed April 2020.

[110] A. Frost (2014) Big Spenders: The Beckfords and Slavery. BBC Wiltshire. http://www.bbc.co.uk/wiltshire/content/articles/2007/03/06/abolition_fonthill_abbey_feature.shtml

[111] D. Ross (n.d) Beckford's Tower. Britain Express <https://www.britainexpress.com/attractions.htm?attraction=2443> Accessed April 2020.

Aside from the abbey venture, the Alderman and his son William Jr. had made other marks. Figure 5.14 shows the (Grade 1 listed) Fonthill Archway (with flanking walls), dating from the 1756, which operated as a gatehouse to the Estate.[112] Closer inspection of faces carved high up in the arch on both sides is suggestive of African features, cruelly recalling the original source of the family's wealth. Figure 5.15 shows the lake, an integral part of the landscaped park, and dam. In the 1820s this enabled a textile mill to be powered (built by John Farquhar), which was not economically successful and was removed for aesthetic purposes. Later in the 19th century, waterwheels powered pumps to lift water to the lakes and were later superseded by electric pumps. Today there is a hydro-electric facility.

Semley Manor is in deep 'cheese country', although known historically for butter production from clay soils supporting good pasture. There are both Kimmeridge and Gault clays in the area, with Upper Greensand hills in the south and drift deposits in the valleys. It was probably part of the estate given by King Edwy to Wilton Abbey in 955, and it belonged to that abbey until the Dissolution when it passed to the Crown. Later, the manor was held by the Arundells of Wardour, who remained Roman Catholic; there were several legal challenges and sequestrations of the title in the succeeding centuries.

The proportion of arable remained low. In the mid-14th century some, and by the late 16th century most, arable lay in closes. Between 1599 and 1769, c. 500 acres of pasture in the parish were enclosed. Of the lowland pasture for cattle, nearly all that west of the Warminster to Shaftesbury road and nearly all that in the east had been enclosed by 1769. The area remained deeply agricultural. In 1831, of 145 families living in the parish, 127 were employed in agriculture and 14 in trades, crafts, or manufacturing. Small dairy farms remained characteristic of the parish in the late 19th and early 20th century, although the trend was towards larger holdings. In 1839, as indeed in 1985, there remained 300 acres of common pasture for cattle, which was allocated in strips. Until 1922 use of the common was regulated by Semley Manor court. Thereafter a common master was appointed annually.[113] The railway arrived in 1859.

Semley, located on the London and South West railway line linking Sherborne and Salisbury, had a wholesale milk depot. As in the Vale of Marshwood, the arrival of the railway meant that most farmers stopped converting their milk into butter or cheese; it was easier and more profitable to supply milk to towns including London.[114] The numbers of depots in the south Wiltshire and Dorset grew and in 1928 at a factory in Semley milk was pasteurised and stored and cheese was produced.[115] A fair proportion of Woodland remains, much of it located on the hilly ground. Figure 5.17 shows Oysters Coppice, managed by the Wiltshire Wildlife Trust.

In the 1930s, the area over the outcropping Kimmeridge Clay in the west of the Vale of Wardour was described as 'not suitable to arable land'. Portland beds in the central Vale did support arable agriculture, especially around Tisbury, while in the east of the Vale there was

---

[112] Historic England (1987) Nos. 65 and 66 (The archway with flanking walls). <https://historicengland.org.uk/listing/the-list/list-entry/1318805> Accessed April 2020.
[113] J. Freeman and J.H. Stevenson (1987) Semley, in D.A. Crowley (ed.) *A History of the County of Wiltshire: Volume 13, South-West Wiltshire: Chalke and Dunworth Hundreds*, 66-79.
[114] J.H. Bettey (1986) *Wessex to AD 1000*. Longman, New York, 239.
[115] J. Freeman and J.H. Stevenson (1987) Semley, in D.A. Crowley (ed.) *A History of the County of Wiltshire: Volume 13, South-West Wiltshire: Chalke and Dunworth Hundreds*, 66-79.

Figure 5.16. Common Pasture strip form wide verges to road, Semley Common (the author).

Figure 5.17. Oysters Coppice, Semley (the author)

less arable over the Purbeck beds; the narrow belt of Gault Clay carries no arable land.[116] There was arable on the southern margin of the Vale on the Upper Greensand and, in general, this land use situation pertains today. Bordered on the south by the Chalk escarpment, parklands including Wardour Castle, Fonthill Abbey and Dinton add character to the modern Vale of Wardour. A trend is apparent of converting pasture to arable and field boundary removal on the productive soils of the Greensand terrace around the Vale of Wardour, towards Warminster and on limestone ridges.[117]

5.9 Conclusions: Valley landscapes

The *structure* of the landscapes is defined by the erosion of successions of hard and soft strata, producing complicated systems of valleys separated by plateaus and interfluve ridges. Frequently the valley bottoms are not only occupied by flat floodplains of modern alluvium, but also display (particularly in the chalklands) river terraces, generally wind-blown loess ('brickearth') found over Pleistocene gravelly material (Chapter 2). These have had a profound impact on the economic development of such valleys, providing materials for brick making (Chapter 6) and forming fertile soil for the development of agriculture and horticulture. The most notable exception is Exmoor, where the valleys may or may not contain any significant alluvial material (where above the point of alluviation there is a simple V-shape valley profile).

The *functions* of the vales of Wessex are therefore central to our understanding of the economic roles of the region. Not only do they harbour a range of land uses in the wider vales, but they also permit settlement and assist communication. In most areas where there is little opportunity for development of significant routeways, the preference is often to travel across upland areas between practical crossing points in the river valleys. This is observable not only in the chalklands, but also across Exmoor and in the Mendips. Vales provide most areas of cattle husbandry and economically viable woodland and forestry in Wessex. Their *value* therefore, in an economic sense, is complimentary to many upland areas, while they also have provided opportunities for elites to develop prestige landscapes around country houses.

The *scale* of vales runs from the tiny gullies of Exmoor to some considerable wide valleys, such as that of the Hampshire Avon, associated with the chalklands and with the Hampshire basin as well as within the Jurassic strata of Dorset, Gloucestershire and elsewhere. While agricultural *change* has been relatively conservative in recent decades, valleys experienced enclosure, plantation and more subtle changes, from cheese production in certain areas (especially Wiltshire and Dorset) to dairy or beef production from cattle prior to the First World War (see also Chapter 6). Where the valleys provided useful transport corridors, notably for turnpikes, canals and railways, this infrastructure facilitated many changes.

---

[116] A. H. Fry (1940) Wiltshire, in L.D. Stamp (ed.) *The Report of the Land Utilisation Survey of Britain*, part 87. Geographical Publications. London, 190.
[117] Natural England (2014) NCA Profile:133 Blackmore Vale and Vale of Wardour (NE539) <http://publications.naturalengland.org.uk/publication/5858996464386048> Accessed April 2020.

Chapter 6

More than just calcium carbonate and grass?

'Rock of Ages, cleft for me,
Let me hide myself in Thee;
Let the water and the blood,
From Thy riven side which flowed,
Be of sin the double cure,
Cleanse me from its guilt and power.'

Words by the Revd Augustus Toplady (1740 –1778) probably composed
whilst he was sheltering in Burrington Coombe in the Mendip Hills.

6.1 Limestone country

It may seem unbelievable that the activities of microscopic marine organisms have such a profound influence. Limestone thus produced typically displays an undulating land surface made from porous and fissured rocks which store, transmit and yield water sufficiently to enable river flow and wetland integrity at a regional scale. They also influence soils in terms of both chemical (i.e. relatively high pH) and physical conditions that produce a solid, generally shallow and gradually dissolvable bedrock. Eventually this led to religious inspiration for an 18th century clergyman.

Revd Toplady was apparently sheltering in the hard, Carboniferous Limestone that also produces familiar landscapes from the Peak District, Yorkshire Dales, to the Gower Peninsula in South Wales and is home to another clergyman – Father Ted, and his colleagues on the fictitious 'Craggy Island' (aka the Burren, Co. Clare in Ireland). Following deposition of $CaCO_3$, sometimes liquids rich in magnesium carbonate ($MgCO_3$) may permeate limestone, forming the mineral dolomite ($CaMg(CO_3)_2$). To qualify as 'limestone', deposits must have over 50% $CaCO_3$, and true limestones contain over 90% calcite (a stable form of calcium carbonate). Chalk, on the other hand, is widespread in southern and eastern England and is merely a soft and pure form of limestone that is predominantly calcium carbonate ($CaCO_3$). Limestone, like most sedimentary rocks, contains impurities such as sand grains, silt, and clay. Much of the Carboniferous Limestone and parts of the Chalk of England contain less than one per cent impurities.[1]

In our region, limestones including chalk result from carbonate deposition between 359my (the oldest date for the Shirehampton Formation of the Carboniferous Limestone) and 65.5 my (the youngest date for the Portsdown Chalk Formation of the Chalk). In-between in age, the Jurassic period (201.3 to 145 my) displays limestone bands within the Lias, the Great and Inferior Oolite, and the Corallian, Portland and Purbeck limestones.[2] Most are of economic significance, particularly for building (see Chapter 2). So how do limestones form?

---

[1]   British Geological Survey (2019) The composition of limestone <https://www.bgs.ac.uk/discoveringGeology/geologyOfBritain/limestoneLandscapes/whatIsLimestone/composition.html> Accessed April 2020.
[2]   Mostly from information provided in publications of the British Geological Survey.

Limestones most commonly form in clear, warm, shallow marine waters. Evaporation played a role in sediment formation from the Purbeck of the upper Jurassic and lower Cretaceous. Visible shells or visible fragments of marine creatures, including echinoderms (such as sea urchins and sea lilies), bivalves, brachiopods, molluscs' fecal debris and corals, would never account for the original volume of calcium carbonate that constitutes limestone formations. The bulk of the material is formed by microscopic marine algae. Limestone can also form from deposition by calcium carbonate saturated waters (such as speleothems in caves) or as evaporite deposits from hot environments.

Where limestones do not display fossil remains that are obvious to the naked eye, the bulk composition is from coccoliths (calcareous skeletons of microscopic, single-celled, photosynthetic algae called coccolithophores) and foraminifera (single-celled organisms termed protists, with a hard shell).[3] The Chalk is largely formed from coccoliths.

While fossils in building stone are often regarded as a flaw by masons, to palaeontologists they are of key interest. Limestone is renowned for containing embedded fossils, and many of these had calcium carbonate shells. Because of this, the Portland Stone used in buildings throughout the country, including Sir Christopher Wren's St Paul's Cathedral, was chosen from beds containing relatively few fossils.

The term 'oolite' is puzzling to the non-geologist. The Jurassic contains the Inferior and Great Oolite Groups that form the Cotswold Hills. The term derives from Greek for 'egg stone', because many limestones in the Jurassic have the appearance of minute spheres (typically 0.5-1.0mm) resulting from the gentle rolling around of small sand grains in the calcium carbonate mud, caused by gentle currents.[4] Where the diameter of the spheres comprising the rock exceeds 2.0mm, then the term 'pisolite' is used. In the Inferior Oolite of the Cheltenham area, such limestones form the 'pea grit'.

Limestone country commonly exhibits 'karst topography', which results from the dissolution of the limestone by rainwater (which is slightly acid), giving rise to features including caves, gorges, underground rivers, sink holes and limestone pavements. These features are well known from Carboniferous Limestone areas (the Burren and at a small scale, the Mendips included), and in Wessex they are found in the Mendips and Avon Gorge. Carboniferous Limestone also contains mineral veins, a source for ores of lead and zinc.

Chalk exhibits karstic features, but due to its softer nature, these are not always so apparent as in harder limestones. Chalk and Limestone areas provide water that may be harnessed for waterpower, a significant factor in their historic importance, alongside habitat provision and agriculture. Soils developed over the chalk are variable, but are typically of 'rendzina' type, that is thin, with organic rich or brown, sometimes grey 'A-horizons' over fragmented white chalk that can support both grassland and productive arable, although they are prone to serious soil erosion.[5]

---

[3] <https://www.bgs.ac.uk/discoveringGeology/geologyOfBritain/limestoneLandscapes/whatIsLimestone/howFormed.html> Accessed April 2020.
[4] I. Geddes (2003) *Hidden Depths*. Ex-Libris Press, Bradford-on-Avon, 17.
[5] D.C. Findlay, G.J.N Colborne, D.W. Cope, T.R. Harrod, D.V. Hogan, and S.J. Staines (1984) *Soils and their Use in South West England*. Harpenden: Soil Survey of England and Wales, Bulletin no. 14, 1:250,000 map, legend and memoir, 75-81, 213-217.

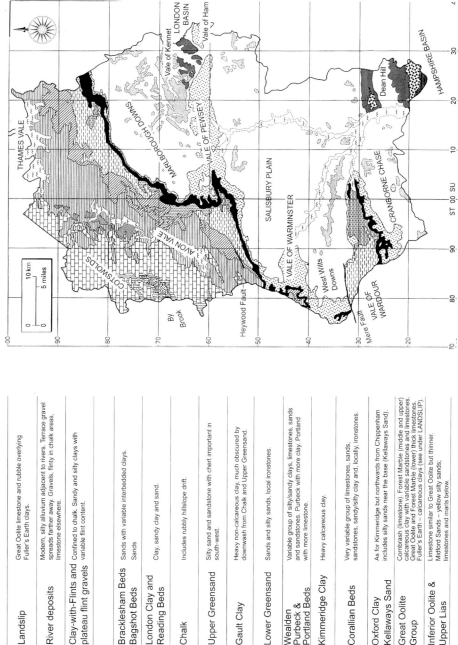

Figure 6.1. Geology of Wilshire (I. Geddes 2003. *Hidden Depths* Ex-Libris Press Bradford-on-Avon 17)

Calcium rich limestone and chalk soils support species rich habitats, including calcareous grasslands. The soil pH is above 6.5, generally as humic gleys and brown rendzinas. Conversion to intensive grass production and arable is highly destructive to the habitat, and lowland calcareous grassland constitutes only around 2.6% of UK semi-natural grasslands. Grasslands are the product of grazing and generally date from prehistory, representing a plagioclimax or semi-natural seral stage of vegetation community development (Chapter 1). Because the grasslands have generally been changed by agriculture and consequential soil erosion, we talk of them being an outcome of secondary succession. This is in contrast with primary succession on unaltered geological substrates that would occur on (for example) sand dunes of salt marshes.[6]

6.2 Landscapes: Chalk and cheese

Figure 6.1 is a geological map of Wiltshire that illustrates the origins of 'chalk and cheese' clearly (Chapter 5). Firstly, the dominance of the Chalk in the County is clear from earliest times and including the development of the Stonehenge landscape.[7] This is unlike (for example) Somerset, in Wiltshire Jurassic limestones are restricted to the north-west of the county (Cotswold Hills) and the Chalk of the Marlborough Downs, Salisbury Plain and Cranbourne Chase. Between the two outcrops is extensive clay country ('Avon Vale') and in the south-west this reaches eastwards into the Chalk outcrop. This is the result of gentle anticlinal folding, bringing strata older than the Chalk to the surface:

> It is not widely known that the saying 'as different as chalk from cheese' originated in Wiltshire. It refers to the division of the county into two distinct but unequal parts. Approximately two-thirds, in the south and east are chalk country, characterised by rolling downs. Just a few decades ago nearly all this land was unploughed. Now most is cultivated – the downs became a granary for wheat and barley. The lowlands of the north-west are, by contrast, sheltered country, meadow and dairy land – a land of milk and cheese.[8]

There is a land-use distinction between different kinds of grassland. The National Character Assessment Profiles (Chapter 1 and Figure 1.2) provide for a combination of landscape, biodiversity, geodiversity, history, and cultural and economic activity, with boundaries following 'natural' lines rather than administrative boundaries. The main examples are summarised below and are precis of the original documents online.[9]

6.2.1 Chalk: NCA Profile 116. Berkshire and Marlborough Downs

Vast arable fields stretch across the sparsely settled, rolling Chalk hills, and included is the Vale of Pewsey (Figure 6.1 and mostly areas 5a-c on Figure 6.2). Extensive views from the escarpment are punctuated by historic landmarks including chalk-cut horse figures, beech

6   D. Blakesley and P. Buckley (2016) *Grassland Restoration and Management*. Pelagic Publishing, Exeter, Chapter 1.
7   A. Lawson (2007) *Chalkland: An Archaeology of Stonehenge and Its Region*. Hobnob Press, Salisbury.
8   T.M. Lewis (2012) Wiltshire Countywide & General Information. <https://www.wiltshire-opc.org.uk/genealogy/index.php/research/county-wide-information> Accessed April 2019.
9   Natural England (2014). National Character Area profiles: data for local decision making. < https://www.gov.uk/government/publications/national-character-area-profiles-data-for-local-decision-making/national-character-area-profiles >Accessed April 2020.

Figure 6.2. The Land Use regions of Wiltshire
(A. H. Fry, 1940. *The Land of Britain: Wiltshire*, public domain).
Key to small areas not marked on map:
2a Arable Islands of the upper Thames Valley
4 Bromham Market and Gardening Region
5b Southern Arable Foreland
5d (i) Savernake Grass-Forest Region
5d (ii)West Savernake Clay-with-Flints Grin-Clover Region
5e Kennet valley
5f Bedwyn Grass-Forest Region
6b Upper Greensand Market Gardening Region

clumps and ancient monuments. Historic routeways, including the Ridgeway provide public access. Writers and artists have been inspired by this landscape, which has also attracted historians and antiquarians, such as John Aubrey, who described Avebury stone circle. Most of the area is included in the North Wessex Downs Area of Outstanding Natural Beauty. Farmland habitat supports the brown hare, the harvest mouse, rare arable plants and farmland birds, including the stone curlew. Along the escarpment and steep slopes, limited tracts of hanging woodland and species-rich chalk grassland can be found. The historic hunting forest of Savernake presents the largest concentration of woodland, with much of it ancient.

6.2.2 Chalk: NCA Profile 130. Hampshire Downs

Most of the area is an elevated, open, rolling landscape dominated by large arable fields with low hedgerows on thin chalk soils, scattered woodland blocks (mostly on Clay-with-Flint caps) and shelterbelts. To the east, hedgerows are often overgrown, and there are larger blocks of woodland. That a fifth of the area is within the North Wessex Downs Area of Outstanding Natural Beauty and six per cent in the South Downs National Park is due to the scenic quality of the landscape. Flower- and invertebrate-rich remnants of calcareous grassland remain mostly along the northern scarp and on isolated commons throughout.

6.2.3 Chalk: NCA Profile 132. Salisbury Plain and West Wiltshire Downs

Salisbury Plain is a sparsely settled, predominantly agricultural area with a strong sense of remoteness and openness, and is an important military training area (Figure 6.2). The dominant element in the landscape is the gently rolling chalk downland. The Plain is designated as both a Special Protection Area and Special Area of Conservation (SAC) under European legislation, notably for the populations of stone curlew and hen harrier, and for the chalk grassland habitat, comprising one of the largest remaining areas of calcareous grassland in north-western Europe. Much of the natural environment is also protected through designation as a Site of Special Scientific Interest for its populations of rare bumblebee species, and many rare birds, plants and invertebrates. Around one third of this NCA is also designated as Area of Outstanding Natural Beauty.

6.2.4 Chalk: NCA Profile 134. Dorset Downs and Cranborne Chase

The Dorset Downs and Cranborne Chase NCA lies across the counties of Dorset, Wiltshire and Hampshire, from east of Bridport to the outskirts of Salisbury. Dorchester and Blandford Forum are the largest settlements in an otherwise sparsely settled rural area important in food production. This agricultural area is characterised by large, open fields of pasture and arable, punctuated by blocks of woodland, all draped over the undulating chalk topography, with dense assemblages of prehistoric sites revealing some 8,000 years of human activity. The area also provides drinking water and maintaining river flow from the chalk aquifer is a further essential service. The area's outstanding landscape is recognised in the designation of two Areas of Outstanding Natural Beauty (AONB): the Dorset AONB and the Cranborne Chase and West Wiltshire Downs AONB.

6.2.5 NCA Profile 133. The Blackmore Vale and Vale of Wardour

The Blackmore Vale and Vale of Wardour NCA comprises a large expanse of lowland clay vale and Upper Greensand terraces and hills, together with an area extending northwards from Penselwood around the edge of the Salisbury Plain and West Wiltshire Downs NCA. The Blackmore Vale is steeped in a long history of pastoral agriculture and characterised by hedged fields with an abundance of hedgerow trees, many of them veteran. This is productive pastureland and it is often waterlogged (Figure 6.2). The elevated, drier and fertile Upper Greensand terraces and hills are characterised by arable agriculture and display some impressive stately homes with their associated gardens, parks, plantations and woodlands. Rural settlement is mixed, with small villages, hamlets and isolated farmsteads set in landscapes of medieval and later enclosed fields. The road pattern reflects the difference between areas of small, irregular fields and narrow lanes that are representative of early (probably prehistoric) clearances, and later, systematic, post-medieval enclosure.

6.2.6 NCA Profile 118. Bristol, Avon Valleys and Ridges

The Bristol, Avon Valleys and Ridges NCA encompasses the City of Bristol with its historic port, and the surrounding area including the Chew and Yeo valleys, Keynsham, Clevedon, Portishead and parts of the Cotswolds and Mendip Hills AONB. The area is characterised by alternating ridges and broad valleys, with some steep, wooded slopes and open rolling farmland. It is flanked by the Somerset Levels and Moors and the Mendip Hills to the south, the Cotswold Hills to the east and the Severn and Avon Vales to the west, which largely separates it from the Severn Estuary except for a small stretch of coastline between Clevedon and Portishead. It has a complex geology including the dramatic Avon Gorge, and there are many designated exposures and rich [Carboniferous] fossil beds. Species-rich grasslands and ancient woodlands are a feature of the area, with ancient woodland and limestone grassland areas. Although the urban area (over 21%) is significant, much of the surrounding rural landscape is farmed. The area is rich in history, from evidence of Neolithic activity through the Roman port at Sea Mills, to the more recent industrial history of mines and mills and the wealth of the port at Bristol. The Chew Valley Lake supplies water to Bristol and is designated a Special Protection Area (SPA) for its internationally important numbers of shoveler ducks and nationally important numbers of gadwall, tufted duck and teal.

6.2.7 NCA Profile 140. Yeovil Scarplands

The Yeovil Scarplands run in an arc from the Mendip Hills around the southern edge of the mid-Somerset Hills and the Somerset Levels and Moors to the fringes of the Blackdowns. This remote, rural landscape comprises a series of broad ridges and steep scarps separating sheltered clay vales. Less than five per cent of the area is urban, largely the industrial town of Yeovil in the south of the area. The area has a long history of settlement from the Neolithic through Roman villas to remnant medieval open fields and there are many Listed Buildings. The area is known for its collection of fine manor houses and associated parklands. The geology includes a variety of limestones and sandstones giving distinctive local settlement character. Foremost among these is the Ham Hill stone. Approximately 85% of the NCA is farmed and, away from the towns, some places are intensely rural.

There is a tradition of growing soft fruit and vegetables as well as remnant orchards. The south-west of the area features arable systems with fragmented grassland assemblages and ancient woodland. The rest of the NCA is mainly pastoral in nature, though in some of the clay vales between the scarps mixed farming brings a variation of character.

6.2.8 Limestone: NCA Profile 107. The Cotswold Hills

The Cotswolds comprise predominantly oolitic Jurassic Limestone belt that stretches from the Dorset coast to Lincolnshire. The dominant pattern of the Cotswold landscape is of a steep scarp crowned by a high, open 'wold'.[10] These form the beginning of a long and rolling dip-slope, cut by a series of wooded valleys. The scarp provides a backdrop to the towns and expansive views across the Severn and Avon Vales to the west. Smaller towns and villages nestle at the scarp foot springline, in the valley bottoms and on the gentler valley sides with scattered hamlets and isolated farmsteads on the higher ground.

Limestone imparts strong character to the buildings and walls. The distinctive character is reflected in its partial designation as the Cotswolds AONB. Nationally important beech woods feature in the landscape and are a notable feature on the scarp edge and in some incised valleys. Mixed oak woodlands are concentrated on the upper slopes of valleys and on the flat high wold tops. Woodlands can contain a wide and notable range of calcicole [calcium loving] shrubs and ground flora. Parkland and estates are characteristic, farming is mixed, with much of the high wold dominated by arable on thin, brashy (crumbly) soils that are prone to erosion. Pasture is predominant in the valleys, and on steeper slopes and more clayey soils. Meadows and treelined watercourses are found along the valley bottoms.

6.2.9 Limestone: NCA Profile 141. The Mendip Hills

The Mendip Hills rise abruptly from the flat landscape of the Somerset Levels and Moors to the south. This Carboniferous Limestone ridge, with its more weather-resistant sandstone summits, illustrates the classic features of a karst landscape, the result of the response of the soluble limestone to water and weathering, creating surface features, complex underground cave and river systems, gorges, dry valleys, surface depressions, swallets (disappearing streams), sink holes and fast-flowing springs. Such natural features have interacted with human influences to result in complex ritual, industrial and agricultural landscapes extending from the prehistoric period to modern times (Figure 6.3). This is a rural area, and around 53% of the area lies within the Mendip Hills AONB. There are four Special Area of Conservation designations and two National Nature Reserves. The concentration of 29 geological and mixed-interest Sites of Special Scientific Interest demonstrates the geological importance of this relatively small NCA. The Mendip aquifer supplies water to Cheddar, Blagdon and Chew Valley lakes, which provide much of the water supply for the Bristol area. The aquifer also possibly supplies the hot springs in Bath[11], following deep percolation in the crust. The NCA provides many recreational opportunities and is particularly of interest to cavers and potholers. Cheddar Gorge and Wookey Hole are tourist attractions.

---

[10]   The term 'wold' originally implying a high area with wood covering, later a piece of high uncultivated land in a similar sense to 'moor'.

[11]   R.W. Gallois (2006) The Geology of the hot Springs at Bath, Somerset. *Geoscience in south-west England* 11, 168-173.

6.2.10 NCA Profile 142. The Somerset Levels and Moors

The Somerset Levels and Moors NCA is a flat landscape extending across parts of the north and centre of the historical county of Somerset, reaching from Clevedon near Bristol in the north to Glastonbury in the east and Ilchester and Langport in the south. The Somerset Levels and Moors NCA is dissected by the Mendip Hills NCA and the Mid Somerset Hills NCA, notably the limestone ridge of the Polden Hills. The western boundary is formed by Bridgwater Bay and the Bristol Channel beyond. The landscape blends almost seamlessly into the Vale of Taunton in the south-west and into the Yeovil scarplands to the south. This is a landscape of rivers and wetlands, artificially drained, irrigated and modified to allow productive farming.

The coastal (marine alluvial) Levels were once mostly salt marsh and the meandering 'rhynes' (artificial drainage ditches) and irregular field patterns follow the former courses of creeks and rivers. They contrast with the open, often treeless, landscape of the inland (peat) moors and their chequerboard-like pattern of rectilinear fields, ditches, rhynes, drains and engineered rivers, and roads. Today, the Levels and Moors have many similarities, but their histories are quite distinct. The Levels landscape was probably established by the time of the Norman Conquest, while the Moors remained an open waste until enclosure and drainage between 1750 and 1850. Much of the NCA lies below the level of high spring tides in the Bristol Channel. The biodiversity of the area is of national and international importance, reflected in the designation of 13% of the NCA as Sites of Special Scientific Interest.

Figure 6.3. Enclosure walls of Carboniferous Limestone on the Mendip hills (the author).

6.3 Sheep may safely graze

Prior to 2006, the Lord Chancellor was both Speaker for the House of Lords and head of the judiciary. He sat on the 'Woolsack' which signifies the central importance of the wool trade to the economy of England in the middle ages. This cushion originated as a bale of wool, commanded thus to be used by Edward III in the 14th century. Today the Woolsack remains the seat of the Lord Speaker of the House of Lords.

The Woolsack is a large, wool-stuffed cushion or seat, covered with red cloth. Whilst it has neither a back nor arms, in the centre of this cushion is a back-rest. The Lords' Mace is placed on the rear part of the Woolsack. In 1938 the Woolsack was found to be stuffed with horsehair. To remedy this blunder it was subsequently re-stuffed with wool from all over the British Commonwealth.[12]

While this may be about as far as British constitutional reform is likely to go in one step, the historic resonance is clear. Although the English economy moved towards cloth manufacture and 'finished goods' in the later Middle Ages and Tudor period (increasing prosperity as industry moved up the value chain), the historic resonance of wool as an economic basis of the realm is clear.

Textile manufacture in Wessex, in a similar manner to East Anglia, remained important into the modern era (see Chapter 2), yet in a similar fashion, it failed to industrialise on the scale found elsewhere. The Chalk country and the limestone areas were largely the source of wool and this determined the direction of the rural landscape on Downs, Cotswold Hills, Mendip Hills and Purbeck alike. Sheep are grazed in most upland areas of Wessex.

Domestication of animals inevitably dates from prehistory; in prehistoric Britain domestic livestock included sheep, goats, cattle, and pigs. During the Neolithic sheep husbandry formed part of a developing agriculture. The relict sheep varieties 'mouflon' (*Ovis orientalis orientalis*) and, as prehistoric farming developed, Soay sheep (*Ovis aries*) are varieties derived from wild sheep (*Ovis orientalis*). Soay sheep have been demonstrated to eat leaf fodder in winter in preference to grass, unlike modern breeds, suggesting a direct link to woodland.[13] By the Iron Age, sheep were a significant part of the pastoral economy of Wessex, providing wool, meat and ewes milk, with the archaeological record pointing to the importance of woollen textiles.[14]

Grazing flocks and herds would have provided incentives to enclose areas of land from the Neolithic onward, particularly to protect fields growing grain.[15] By the 11th century AD, there is clear evidence of the economic importance of sheep in Somerset and Dorset on demesne land, providing among other products, dung for corn-land. There were large flocks, mainly on royal manors on the Chalk of Dorset. The lower, lusher grassland (for example on the Vale of the White Horse) supported dairy farming (including cheese making). Goats and horses

---

[12] Hansard (1965) House of Lords 3rd August 1965. *Hansard* Vol. 269, 115-7. <https://api.parliament.uk/historic-hansard/lords/1965/aug/03/the-woolsack > Accessed May 2021.
[13] P.J. Reynolds (1987) *Ancient Farming*. Shire Books, Aylesbury, 24, 27, 45.
[14] B. Cunliffe (1993) *Wessex to AD 1000*. Longman, London, 186-187.
[15] B. Cunliffe (2013) *Britain Begins*. Oxford University Press, Oxford, Chapter 5.

were widely distributed as was boar.[16] Between 1300 and 1500, sheep flocks greatly increased, notably those owned by monasteries and the Bishops of Winchester and Salisbury; manorial demesne flocks could exceed 1,000 sheep. Where grassland has persisted, it plays an essential part in preserving the archaeological record, for tree roots can disturb the ground up to several metres in area, while ploughing mixes and damages the topsoil. Subsoiling furthermore takes the damage deeper still.

The 14th century was catastrophic. Rising population in Britain ceased around 1300,[17] and the period 1315 to 1322 experienced unusually bad weather, specifically heavy rain and floods which caused poor harvests, animal diseases – especially among sheep – and famine.[18] The Black Death (plague) first hit England in 1348, possibly brought by a ship from Bordeaux that landed at Melcombe Regis (part of modern Weymouth in Dorset),[19] or alternatively it may have entered via the Port of Bristol. However, by 1375 there had been a total of four visitations of the pestilence.[20] The Black Death is variously estimated to have killed between 30% and 40% of the population of England.[21] The impact on the European economy and society was profound.

The feudal system was already facing change by the 14th century with a cash-based economy replacing feudal labour and military obligations. The disruption caused by de-population allowed many peasants to escape feudal bondage, working for other manors or escaping to the relative freedom of the towns, which had themselves suffered serious de-population. The response was a legal attempt to re-enforce feudal labour relations. Labourers demanded higher wages which impacted on the profits of the landed classes, who appealed to the government. One response, the *Statute of Labourers*, was issued by Edward III in 1351 and directed against the rise in prices and wages. It failed.[22]

England was involved in the Hundred Years' War. By the 1370s, this had left the treasury empty; the barons were tired of paying for the war and there was inflation. The government response was to raise a Poll (head) Tax. In 1377, John of Gaunt imposed this to cover the cost of the war. The tax was to be paid by peasants and landowners alike. The immediate impact was successful so that it was repeated twice more. The first tax was 4d from every adult (individuals of (14 years of more); by 1381 it reached 12d per adult. Spreading the burden of taxation was popular with the barons, but predictably not with the peasants who could ill afford it, and many attempted to hide from tax collectors.[23]

The response to this taxation burden on the peasantry and others was the Peasants Revolt of 1381. There were outbreaks around England, but mostly in East Anglia, London, Essex, and Kent. During this event, the Archbishop of Canterbury, Simon Sudbury, and Sir Robert Hales

[16] J.H. Bettey (1986) *Wessex to AD 1000*. Longman London 21-22.
[17] E. Mason (2000) Portrait of Britain: AD 1200. *History Today* 50(5).
[18] J.H. Bettey (1986) *Wessex to AD 1000*. Longman, London, Chapter 3.
[19] O.J. Benedictow (2005) The Black Death: The Greatest Catastrophe Ever. *History Today* 55 (3).
[20] O.J. Benedictow (2005) The Black Death: The Greatest Catastrophe Ever. *History Today* 55 (3).
[21] V. Masson (n.d). The Black Death. Historic UK <https://www.historic-uk.com/HistoryUK/HistoryofEngland/The-Black-Death/> Accessed, April 2020.
[22] S. Kreis (2006) The Statute of Labourers (1351). *The History Guide* <http://www.historyguide.org/ancient/statute.html> Accessed April 2019.
[23] The Peasants' Revolt 1381 (n.d.) <https://www.marxists.org/history/england/peasants-revolt/story.htm> Accessed April 2020.

(both held responsible for the Poll Tax) were killed, and the King agreed to meet the rebel leader Wat Tyler. The rising was hardly an immediate success, the leaders Wat Tyler and the radical priest John Ball (among others) were killed; Tyler by the Lord Mayor of London on the spot, while Ball was later hanged. Promises made by the young King Richard II in Smithfield regarding abolition of feudal bondage and wages were apparently not kept.

However, the barons were perturbed by the revolt and in the longer term several things were achieved. Specifically, Parliament gave up trying to control wages, the Poll Tax was at an end, and with feudal services increasingly difficult to enforce, more peasants became free men. In short it represented the final breakdown of the feudal system. A drift to the towns continued, limiting the rural labour supply. The labour shortage and overall reduction in the population meant an end to assarting economically marginal land, while some arable was deliberately grassed over. In response, a free yeomanry emerged. Actually, there had already been pressure on land with elements of food insecurity, particularly for landless men dependent on wage-labour, during the years prior to the pandemic.[24]

David Hall can write:

> Medieval population was at its maximum in the early decades of the fourteenth century. Thereafter, for well over a century, it declined because of continual havoc caused by disease and economic recession. By the fifteenth century, there was no longer sufficient population to maintain a labour-intensive agricultural system. Lords attracted villeins from nearby manors by offering them money for work rendered, so freeing them from feudal obligations of work service for the lord, and permanent residence in their home manor.[25]

Problems of disease and poor harvests saw the end of a medieval boom period. In Wessex (as elsewhere in Britain) the changes impacted on population; settlement; agriculture, with a reduced arable land area; towns; trade; and the prosperity of the monasteries, who as major landowners were hit by the drop in population and its effect on the manorial economy. However, demand for English cloth in the 15th century benefitted the ports of Southampton and Bristol, while textile producing towns managed to buck the trend of economic recession.[26] Often sheep-runs replaced populated countryside, greatly changing rural dynamics. The labour shortage meant that many strips and strip lynchets within the open field landscapes were left as permanent grass, called leys, generally in areas of poor soils and difficult topography. Figure 6.4 shows modern-day grassed-over strip lynchets. By way of illustration, the area under arable on the Bishop of Winchester's demesne land of his manor of Downton in south Wiltshire in the 15th century was less than a quarter of that in the early 13th century; marginal land, waterlogged clay lands and chalk downland reverted to grass on an impressive scale.[27]

With a switch to sheep husbandry across England, the ancient and lonely occupation of shepherd on the chalklands became important once more. Shepherds managed flocks and

---

[24] J.H. Bettey (1986) *Wessex to AD 1000*. Longman, London, Chapter 3.
[25] D. Hall (1982) *Medieval Fields*. Shire Books, Aylesbury, 37.
[26] J.H. Bettey (1986) *Wessex to AD 1000*. Longman, London, Chapter 3.
[27] J.H. Bettey (1986) *Wessex to AD 1000*. Longman, London, Chapter 3.

Figure 6.4. Grassed strip lynchets on Purbeck, near Worth Matravers, Dorset (the author).

managed grazing; flocks in Wessex were common on the downlands and would graze on arable land. A high degree of co-operation occurred in the valleys, exemplified by the daily spring movement of sheep flocks between arable and water meadows (Chapter 4).

Shepherds were far fewer in number than arable-based peasants who could prove difficult to manage (from the lord of the manor's point of view) and lower numbers of farm holdings were more easily regulated in manorial courts[28]. The expansion of grazing land in chalk and limestone areas had benefits for the landlord which might compensate for the fading feudal bonds of the villeins and their fellow travellers, whose greater personal mobility would furthermore give stimulus to urbanisation across England.

Into the 17th century, the chalkland arable remained unenclosed, emphasizing the importance of sheep flocks for fertilizing the fields. Indeed, the rapid decline of sheep husbandry only really occurred in response to the agricultural depression from 1879 and continued into the mid-20th century. When the decline in folding sheep on arable and water meadow operation continued apace.[29]

As commonly occurs during prolonged agricultural recession, arable acreage declined. In Wiltshire this was from over 400,000 (*c*.160,000ha) acres to around 150,000 acres (*c*.60,000

---

[28] T. Williamson and L. Bellemy (1987) *Property and Landscape: A Social History of Land Ownership and the English Countryside*. George Philip, London, Chapter 4.
[29] T. Williamson and L. Bellemy (1987) *Property and Landscape: A Social History of Land Ownership and the English Countryside*. George Philip, London, 108-109, 121-122, 259.

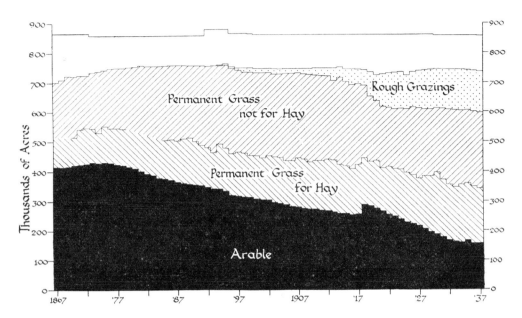

Figure 6.5. Changes in Land Use in Wiltshire 1867-1937. The blank portion in the upper part of the graph represents land unused for crops and grass (A. H. Fry, 1940. *The Land of Britain: Wiltshire*, public domain).

ha), shown in Figure 6.5. While the farmed area remained approximately steady, the area under hay increased, and permanent grass that is not cut increased until the First World War when there was a spike in arable area. The balance was met by a substantial increase in rough (unimproved) grazing from the 1890s, although questions were raised concerning the classification of downland pasture in the agricultural census. For the same period in Dorset, arable fell from slightly above 200,000 acres (*c*.81,000ha) acres to around 100,000 acres (*c*.40,500 ha) with a comparable and compensatory increase in pasture.[30]

The compensatory change in the pastoral economy more than doubled dairy cattle in the county.[31] Nationally before the First World War, there was a switch in all but the most inaccessible regions of England away from cheese production, caused by a combination of imported cheeses and better transport.[32] The railways enabled rapid transport of fresh milk to towns and cities, and Wessex was no exception. A comparable change was recorded for the same period in Dorset, with massive declines (typically more than halving) in the area planted with wheat, barley, oats, turnips and swedes, potatoes, beans and vetches. Only areas under cabbage or mangolds remained relatively stable. While there was once more compensation in the area under pasture, sheep again declined from around 500,000 to nearer 200,000; total

[30] L.E. Tavener (1940) Dorset in L.D. Stamp (ed.) *The Report of the Land Utilisation Survey of Britain*, part 88. Geographical Publications. London. Geographical Publications. London, 253.
[31] A.H. Fry (1940) Wiltshire, in L.D. Stamp (ed.) *The Report of the Land Utilisation Survey of Britain*, part 87. Geographical Publications. London,177, 186, 192-194.
[32] D. Taylor (1987) Growth and Structural Change in the English Dairy Industry, c. 1860-1930. *Agricultural History Review* 35 (1), 47-65.

Figure 6.6. Land Use Regions of Dorset. (L.E. Tavener, 1940. *The Land of Britain: Dorset*, public domain).
Key to small areas (letters) not marked on map:

3A Small hills, pipe clay pits and mines, boggy valleys and stunted plantations. Soils largely
  developed on Bagshot sands

3B Better drained soils than around Poole Harbour (3A), pinewoods and heath with poor pasture
  and clay workings

3C Apart from the water meadows in and cultivation in the Piddle valley, heathlands developed on
  plateau gravels. Military training areas.

3D Flat heathlands and commons north of Poole Harbour

5A Complex ridged area on clays with pasture and some arable

5B Oolitic ridge, many thin calcareous soils, complex topography

5C Light and fertile soils on Middle and Upper Lias

5D Vale of Marshwood, heavy clays soils developed over the Lower Lias

5E 'Devon Edge', flat-topped greensand outliers, some arable.

7A Western part of the Vale of Blackmore, claylands including Oxford Clay, important for dairying.

7B Central portion of the Vale of Blackmore, broad low, wet claylands with drier Corallian ridge

7C Vale of Shaftesbury dominated by pasture, many cattle

cattle including dairy increased, while pigs and horses remained relatively stable. The blame can be squarely pinned upon a combination of bad harvests with no compensatory rise in grain prices due to competition from the United States where wheat production trebled between 1860 and 1880. Farmers were also hit by increased importation of foreign meat.[33] For Somerset, the arable area also fell 1866-1936, rough grazing more than doubled 1894-1936 and permanent grass increased, with most of this cut for hay.[34] The land use regions of Dorset in the 1930s are shown in Figure 6.6, demonstrating the importance of the central chalk uplands region but with a varied agriculture surrounding the Dorset Downs,

6.4 Grass and grassland on chalk and limestone

Chalk grassland stretches across southern England and the outcrop of the Chalk extends northwards into East Anglia and the Yorkshire Coast. Typically, in one square metre there will be up to 45 species of flowering (and often aromatic) plants. Figure 6.7 shows the distribution of calcareous grassland developed on chalk and limestone.[35] Lowland chalk grassland provides one of Western Europe's most diverse plant communities. Similar habitats are provided on other limestones, particularly those from the Jurassic and Carboniferous periods, and the habitat is typically on dry valley slopes. The definition applies when the altitude is less than 250m, although both the Cotswold and Mendip hills locally exceed this elevation.

The Chalk Country and associated river valleys areas have been important since wildwood clearance and the evolution of agriculture. The downs have been grazed since the first farmers from Europe 6,000 years ago. Neolithic farmers cleared the natural forest using flint-bladed tools, providing timber and agricultural land. A combination of sheep grazing over several hundred years, soil erosion from arable plots and a soil deficient in most plant nutrients allowed the development of the short, springy grassland.

Ploughing of Chalk Grassland become significant in the 18th century. Daniel Defoe, writing in the 1720s, was impressed by the farming on downland, the increase in size of the mobile sheep flocks and how farmers had been able to extend their arable land over parts of the downland of Wessex. Sheep dung was especially efficacious on newly ploughed land with thin soils and that was too far from the farmsteads in the valley to cart manure.[36] There was significant intensification from 1940 and more than 80% of chalk grassland has been lost since the Second World War. This is because land use moved away from extensive animal grazing. Intensive farming with the use of herbicides and fertilisers changes the nature of the soil such that the traditional chalk grassland species cannot grow.[37]

Broadly speaking, grassland is a widespread habitat across the UK. In terms of area, however most has been 'improved' by cultivation, fertilisers, herbicides and reseeding, so that only 2%

[33] L.E. Tavener (1940) Dorset in L.D. Stamp (ed.) *The Report of the Land Utilisation Survey of Britain*, part 88. Geographical Publications. London. Geographical Publications. London, 253-255.

[34] T. Stuart-Menteath (1938) Somerset in L.D. Stamp (ed.) *The Report of the Land Utilisation Survey of Britain*, part 86. Geographical Publications. London. Geographical Publications. London, 138.

[35] Joint Nature Conservation Committee (2006) Common Standards Monitoring for Designated Sites 2006: Report <http://jncc.defra.gov.uk/page-3562 > Accessed April 2019.

[36] J.H. Bettey (2007). The Floated Water Meadows of Wessex: A Triumph of English Agriculture, in H. Cook and T. Williamson (eds.) *Water Meadows* Windgather Press, Macclesfield, Chapter 2.

[37] National Trust (n.d.). What's special about chalk grassland? <https://www.nationaltrust.org.uk/features/whats-special-about-chalk-grassland > Accessed April 2020.

Figure 6.7. Lowland calcareous grassland (DEFRA).

of grasslands now have a high diversity of species. The Joint Nature Conservation Committee (JNCC) has estimated that 97% of lowland meadow was lost in England and Wales between the 1930s and 1980s with most remaining meadows displaying poor wildlife value.[38] The change is compounded by the fact that native plants within arable farmland struggle to survive when herbicide and nitrogen fertiliser are applied. There has also been reduction in hedgerows and non-cropped areas impacting birds, small mammals, butterflies, and other insects.

These grasslands are found on lime-rich (pH >7), nutrient- poor and dry soils derived from, or overlaying, chalk or limestone bedrocks, especially low in available phosphorous. The large areas of chalk and limestone grassland have been lost and the disappearance of these areas has led to major declines in wildlife, including wildflowers, bumblebees, butterflies, reptiles, amphibians and farmland birds.[39] Lowland, semi-natural grassland has been the focus of conservation efforts in recent decades as it is of great importance for biodiversity. Much that remains is now designated as protected areas, including 70% of lowland calcareous grassland.

[38] D.B. Hayhow *et al.* (2016). State of Nature 2016. The State of Nature partnership. <https://www.bnhc.org.uk/wp-content/uploads/2016/01/state-of-nature-report-2016-uk-2.pdf> Accessed April 2020.
[39] Farm Wildlife (2019) < https://farmwildlife.info/how-to-do-it/existing-wildlife-habitats/chalk-and-limestone-grassland/> Accessed April 2020.

More work needs to be done from a conservation point of view, although there are success stories. For one, the marsh fritillary (*Euphydryas aurinia*) butterfly in Dorset is recovering due to the promotion of appropriate land management since 1980 on Wessex downland sites, achieved largely through payments to farmers under agri-environment agreements in the South Wessex Downs Environmentally Sensitive Area and its successor Higher Level Stewardship.[40]

Current estimates suggest that up to 41,000ha of lowland chalk grassland remain in the UK, mostly in the south of England, where common flowers like small scabious (*Scabiosa columbaria*) can be found alongside nationally rare plants such as late spider orchid (*Ophrys fuciflora*) and pasqueflower (*Pulsatilla vulgaris*). Before 1940 lowland chalk grassland was widespread, but it declined, and by 1984, 80% of sheep-grazed lowland chalk and limestone grassland had disappeared. The main reason for this decline is the reduction in grazing and traditional management due to changes in land use to intensive agriculture and afforestation.[41] Believe it or not, on account of its diversity of plant and insect life, chalk grassland is sometimes called Europe's tropical rainforest.[42]

An example of the land-use on chalkland in the 1930s is illustrated in Figure 6.8.[43] This is dominated by the area acquired by the military from 1897 to 1920, and today constitutes the largest area of chalk grassland in North-West Europe, including 40% of the remaining area in the UK. 20,000 ha are designated as SSSI, Special Conservation Areas and Special Protection Areas for birds and there are some 2,300 ancient monuments.[44]

Arable areas, blank on Figure 6.8, are restricted to the valley margins and river terraces. This is a legacy of times of limited mechanisation, when sheep could be led off the water meadows daily in winter and spring to fertilise the nearby arable. The distinction is made between hill pasture (rough grazing), permanent pasture and 'improved pasture'. Productivity was improved through fertilisation and, like the permanent pasture, it would have been mown. Improved pasture is close to valleys, as is the arable. While animal agriculture is practiced on Salisbury Plain, other areas had already been ploughed. Today the Salisbury Plain area is an island in a county of otherwise intensified agriculture.

Restoration of chalk and limestone grassland is an important aspect of semi-natural habitat restoration. It requires care, monitoring and restoration to a stable grazed community and may take several years. The process will have occured following previous recessions in agriculture as in the 14th and 19th centuries, and in the present-day there are seed-mixes available to speed the process. For example, the North Wessex Downs AONB has a restoration strategy as

[40] F. Burns *et. al.* (2013). State of Nature report. The State of Nature partnership <https://ww2.rspb.org.uk/Images/stateofnature_tcm9-345839.pdf> Accessed. April 2020.
[41] The South Downs National Park (n.d.) Chalk Grassland. <https://learning.southdowns.gov.uk/wp-content/uploads/sites/2/2015/11/EB-slides-on-chalk-grassland.pdf> Acc. April 2020.
[42] National Trust (n.d.) What's Special about Chalk Grassland? <https://www.nationaltrust.org.uk/features/whats-special-about-chalk-grassland> Accessed April 2020.
[43] A.H. Fry (1940) Wiltshire, in L.D. Stamp (ed.) *The Report of the Land Utilisation Survey of Britain*, part 87. Geographical Publications. London, 181.
[44] The University of Bristol (n.d.). Salisbury Plain Army Training Estate, Wiltshire <http://www.bristol.ac.uk/history/militarylandscapes/sites/britain/salisbury/> Accessed April 2020.

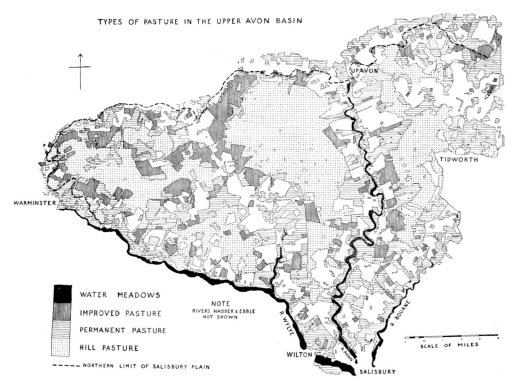

Figure 6.8. The Grasslands of Salisbury Plain in the 1930s (A. H. Fry, 1940. *The Land of Britain: Wiltshire*, p.181, public domain).

part of its Biodiversity Action Plan (BAP).[45] Seed mixes for chalk and limestone areas form part of a strategy for restoration including recolonisation from core areas of surviving habitat.

Chalk grasslands have been managed by sheep grazing. Sheep are the ideal animals to manage chalk grasslands since they graze the sward between the flower stalks, leaving the nectar sources for the insects and the seeds for future years, although burning is also a valuable method of destroying the dense mattress of dead grass leaves so often found at the base of un-grazed chalk grassland, because grass litter is broken down only slowly and tends to accumulate and smother many low-growing species. This has been used in the Cotswold Hills to control the coarse grasses.[46]

The daily transfer of sheep from water meadows to arable land at night largely ceased by the 19th century (see Chapter 4). Its origins are to be found in earlier similar practices of folding sheep on arable to fertilise soils and the use of 'brookland' to provide the early grass

[45] Plantlife (n.d.) Calcareous grasslands. <http://www.magnificentmeadows.org.uk/assets/pdfs/Calcareous_Grasslands.pdf> Accessed April 2020; Biological Records Centre (2005) Chalk Grassland Strategy Report <http://www.northwessexdowns.org.uk/uploads/File_Management/Publications/Biodiversity/NWDAONBChalkGrasslandStrategy.pdf> Accessed April 2020.
[46] A. Fowler and V. Brown (n.d.) Site Management and climate. Natural England Research Report 76, Chapter 9. <https://publications.naturalengland.org.uk/publication/59049 > Accessed April 2019.

for animals when South Downland pastures were not yielding sufficient grass for the sheep.[47] Sheep were transferred ('folded') at night from chalk grassland downland onto arable land where their dung and urine acted as a fertiliser. This helped the transfer of nutrients from the pasture, while a decline in this practice could have led to an increase in nutrients. The effect may have been strong in areas where sheep lie up at night, such as against walls, fences and scrub. Other animals including horses may also form 'latrines' in similar areas and these will become dominated by fast growing, nutrient demanding species.[48] The impact is to reduce sward biodiversity.

Sheep graze from roughly dawn to dusk. They eat on and off during daylight, stopping frequently to chew the cud. Pasture is preferred (grass, forbs, and legumes) with night-time grazing only when there is a problem in securing food. Feed must have sufficient nutrients to meet the sheep's needs during shorter days when there are fewer hours of daylight, so supplementation with grain is commonplace, and winter feed historically has been supplemented by certain root crops, particularly turnips. Because sheep are gregarious, they eat more (and better) when they are in a flock.[49]

Grazing and animal movement works to maintain the habitat through the physical removal of potential woody species and through constant nutrient depletion where there has been no agricultural application of fertilisers:

> In the daytime, shepherd boys would take the sheep onto the hills to graze. At night, they would be brought down to some of the uphill fields where sheep folds were pitched during the day by the Head Shepherd and his men. A fold was simply an area of fodder fenced off by hurdles. It was often arranged so that the outside row of hurdles in the morning fold became the back row of the new fold in the evening. Swedes and turnips were grown in some of the uphill fields to feed the sheep in the mornings and, of course, the sheep helped to manure the soil.[50]

Under-grazing presents problems because it causes loss of grassland habitats, obscuring landscape features and ultimately rendering the land unsuitable for grazing (Figure 6.9). Today chalk scrub occupies much of the area that was once open 'sheep-walk' grassland. This is the result of a decline in mixed farming, reduced sheep grazing on (for example) the South Downs and fluctuations in rabbit populations.

Scrub has always been a component of the downland ecosystem, providing shaded and protected habitat for insects, birds, and reptiles, depending on stage of scrub development. Grazing (ideally by cattle and ponies) can be used following cutting of scrub to browse and prevent regrowth, with a balance sought with scrub grazing before dense thickets develop. Deciduous chalk scrub was kept at bay in part by rabbits, but due to the onset of myxomatosis, their numbers decreased, reducing their impact.

---

[47] P. Brandon (1988) *The South Downs*, Phillimore, Bognor Regis, Chapter 5.
[48] A. Crofts and R.G. Jefferson (eds) (2007). *Lowland Grassland Management Handbook* Natural England, Peterborough, Chapter 5.
[49] P. Simmons and C. Ekarius (2009) *Storey's Guide to Raising Sheep* (4th ed.) Storey Publishing, North Adams MA, 179, 187.
[50] Buriton Heritage (2020) <http://buriton.org.uk/history/the-days-of-sheep-rearing-on-the-downs/> Accessed April 2020.

Figure 6.9. Scrub meets archaeology: Iron Age Woolbury Ring, Stockbridge, Hampshire (the author).

Juniper scrub occurs only in the West Sussex and the Hampshire part of the South Downs at a very few sites and appears to be mostly senescent or diseased, as there is very little regeneration.[51] On Cranborne Chase, historically grazed by deer, the removal of trees, of scrub and ploughing was forbidden until 1829, thereafter land clearance and arable agriculture followed. However, while scrub invasion is identified as a problem, calcareous scrub is recognised as a transitional semi-natural habitat, and it may be retained where there is a potential for serious surface runoff.[52] It is probably a truism that while uncontrolled scrub invasion is problematic, especially for species diversity, when properly managed (for example by rotational cutting) it has value in maintaining a landscape mosaic of semi-natural habitats.[53]

On the chalk ridge that crosses the 'Isle' of Purbeck, from Swanage in the east to West Lulworth in the west, maintaining open grassland for biodiversity is a matter of balance. Grazing, on this occasion by traditional breeds (including the North Devon and Soay sheep), can be carefully

[51] Natural England (2013) South Downs Natural Area Profile. DOI: 10.1.1.559.4533.
[52] Natural England (2013) National Character Area Profile 134 Dorset Downs and Cranbourne Chase <https://publications.naturalengland.org.uk/publication/5846213517639680?category=587130> Accessed April 2020.
[53] S.J. Gough and R.J. Fuller (1998) <https://www.bto.org/sites/default/files/shared_documents/publications/research-reports/1998/rr194.pdf > Accessed April 2020.

Figure 6.10. Scrub invasion following excavation 35 years previously of a late Neolithic henge monument (c.2,800BC) at Wyke Down, Cranborne Chase, Dorset (the author).

controlled to prevent invasion, in this case by gorse and coarse grasses. The simple objective is therefore to restore chalk grassland, and this restoration is proving successful.[54]

Chalk and limestone grassland is biodiverse not only because of its management by grazing, but also because it is a low nutrient system. The soils supporting it are thin (generally described as 'rendzinas') and display a thin, humose (organic-matter-rich) calcareous topsoil (no more that 20cm deep) of the Icknield series that is common on the scarp slopes.[55]

Figure 6.9 shows scrub encroachment on an archaeological site at Down Farm, Cranborne Chase, near Sixpenny Handley, in Dorset. Such sites provide opportunities to study the establishment of vegetation following archaeological investigations. If scrub invasion on spoil gives an idea of the rate of scrub invasion, where the site has been dug to chalk bedrock there is a parallel opportunity under poor, calcareous conditions to observe natural recolonisation of chalk grassland.[56]

---

[54] Dorset Biodiversity Partnership March (2010) Dorset Biodiversity Strategy mid-term review Summary <https://dorsetlnp.org.uk/wp-content/uploads/2019/01/Dorset-Biodiversity-Strategy-Review-2010.pdf> Accessed April 2020.
[55] Cranfield University (2019). *The Soils Guide.* <www.landis.org.uk> Accessed April 2020.
[56] Martin Green (2019) pers. comm.

Human interference of one kind or another, including grazing, means such grasslands support a plagioclimax community, that is a semi-natural habitat caused by grazing. The diversity of plant species implies consistent nutrient depletion, otherwise fast-growing species of grass or other plants take over at the expense of herbaceous species. While calcium carbonate levels are extremely high, generally nutrient levels are severely lacking in these skeletal soils that are undergoing recolonisation over time.[57]

Many ground beetles and different types of bee can be found amongst rarer insects such as the phantom hoverfly (*Doros profuges*) and the even more marvellously named wart-biter bush cricket (*Decticus verrucivorus*). Rare butterflies such as the Silver spotted skipper (*Hesperia comma*) and Adonis blue (*Polyommatus bellargus*), and rare moths such as the Gothic moth (*Naenia typica*) and Four-spotted moth (*Tyta luctuosa*) are also to be found on chalk grassland. Threatened birds such as the stone curlew (*Burhinus oedicnemus*) and skylark (*Alauda arvensis*) rely on chalk grassland, where reptiles such as adders and slow worms can bask. While grazing maintains the plagioclimax, under-grazing leads to the development of scrub cover, and over-grazing can prevent plants from flowering and seeding. Other threats come from building development and recreational use.[58]

6.5 Drovers

While the occupations of a shepherd, dairy or cattle farmer, or arable cultivator have existed from prehistory to the present, and there still are individuals described as 'drowners' and marshmen (who collect reed for thatching), drovers are no more. The arrival of the railway may have stimulated some rural occupations, but the drovers once vital to the Wessex economy for the movement of animals over long distances, lost out. The Drovers' House in Stockbridge, Hampshire has a memorial; picked out in flint are the Welsh words:

> Gwair tymherus porta flasus cwrw da a gwalcysurus
> Fine hay, sweet pasture, good ale and a comfortable bed.[59]

The placename 'Stockbridge' could be misleading. It was a river crossing on the main route from Gloucester and South Wales to Portsmouth, where the drover's found accommodation, good pasture and a place to cross the river Test. But, in fact, the '*stocc*' part likely comes from the Old English for a tree trunk and hence probably relates to construction, rather than referring to animals, despite the importance of the town to droving![60]

This implies two basic points. First that there was a network of drovers' roads throughout the region, and second that there was a mobile workforce that travelled with their animals. Comparatively, little is known of their way of life, something all too common to nomadic and semi-nomadic people.

---

[57] RSPB (n.d.) The Wiltshire Chalk Grassland Project <http://ww2.rspb.org.uk/Images/The%20Wiltshire%20Chalk%20Grassland%20Project_tcm9-132715.pdf> Accessed April 2020.
[58] National Trust (n.d.) What's Special about Chalk Grassland? <https://www.nationaltrust.org.uk/features/whats-special-about-chalk-grassland> Accessed April 2020.
[59] J.H. Bettey (1986) *Wessex to AD 1000*. Longman, London, 150.
[60] Old English translator (n.d.) < https://www.oldenglishtranslator.co.uk/> Accessed April 2019.

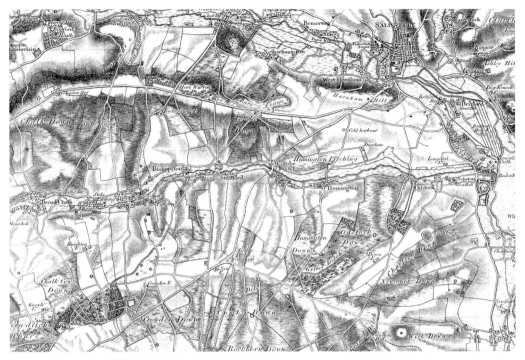

Figure 6.11. The Ebble ('Chalke') valley in Wiltshire between Broad Chalk (west) and Nunton (east) Ordnance Survey First Series 1811-1817. Width of extract c. 15km. (© Cassini Publishing Ltd)

What, on the other hand, was its economic importance? Annual cattle movements in the 1660s into the port of Minehead from Ireland were measured in thousands, sheep in tens of thousands.[61] In England today, former drovers' roads may be metaled, or more usually constitute byways or bridleways.

Figure 6.11 shows an example southwest of Salisbury (see Chapter 5). Drover roads include the Shaston or Old Shaftsbury Drove that runs on the ridge from Shaftsbury, via Compton Hut in the north-west of the figure, eastwards via the racecourse to Harnham hill and hence Salisbury. From the southwest corner, Ox Drove runs past Knoyle Farm eastwards to Combe Common. It forks allowing northward travel to Coombe Bisset (centre of extract) as the Old Blandford road, or east-southeast towards the lower Avon valley in the southeast.

Croucheston Drove runs north-south linking the part of modern Bishopstone of that name (adjacent to Flamestone) through a dry valley with the Ox Drove close to the wooded area. Other examples include 'Drove lane' that links Combe Bisset northwards, then northwestwards, to the Shafts Drove opposite the Race Course, and Pennings Drove that joins Combe Bisset running south to the crossroads just west of Humington Down, then southwest past Pennings Farm (marked but not named in the figure) to join the continuation of the Ox Drove near to the crossing of the prehistoric 'Grims Ditch'.

---

[61] J.H. Bettey (1983) Livestock Trade in the West Country during the Seventeenth Century. *Proceedings of the Somerset Archaeological and Natural History Society* 127, 123.

Mapping regional scale drovers' roads shows a network that links such routes with Wales, the Midlands, Wessex and the Thames valley.[62] Many became metaled, but where the original roads have survived (for example as bridleways) they are generally wider than other tracks that merely accommodated packhorses, allowing the passage of large numbers of livestock. The typical width of a drove road is 12-15m, sometimes appreciably more.

The occupation certainly left its mark on the Wessex landscape where the roads may be understood as linking areas of pasture. While the daily movement of animals between pasture of meadow and arable is well-attested for sheep-corn systems (for example between downland and arable, or water meadows and arable), drovers operated at a regional, or even national scale.

Identification is easy where a tell-tale road name survives, for example Old Shaston and Ox Drove, while others may take a little investigative work. Where Welsh drovers set up temporary communities on common land a short distance from an established village, nearby tracks may be called Welsh or Welshman's Road while Cow, Bullock or Ox Lane, is likely to be the drover's route itself.[63]

The origins of many drovers' roads are likely ancient, as may be inferred from the known antiquity of animal movement in Britain[64] and the presence of ancient trackways, especially across the chalklands[65]. Although one study on Pewsey Vale found no evidence of regional-scale transhumance at the Bronze Age-Iron Age transition[66], it is nevertheless probable that a transhumant economy was developing at the time.[67] Droving was important in the middle ages, and transhumance in southwest England, including western Wessex, was probably practised in the early medieval period.[68] However, growth in demand for meat was notable in the early 17th century, as part of a general rise in living standards.

Generally, 'enclosure roads' were created when arable or common land was enclosed and were wide enough to assist the passage of many animals. Cattle and sheep could be imported from Wales and Ireland via Somerset ports, notably (but not exclusively) Minehead. There was also a lucrative onward trade for fattened livestock into the Thames valley and London markets. Fat cattle from Somerset were driven to markets or fairs, sheep bred and fattened on Somerset pastures were purchased by arable farmers on the chalklands of Wiltshire and Devon for folding on corn or fallow land and there was significant trade in fattened cattle from Dorset to London.[69]

---

[62] K. Harris (2019) Drovers' Roads <http://www.swanbournehistory.co.uk/drover-roads> Accessed April 2020.

[63] British Landscape Club (2013) Drove Roads <http://www.britishlandscape.org/reading-the-landscape/files/20043 6b595d3c6c0dd53e93fd057f88f-7.htm> Accessed April 2020.

[64] B. Cunliffe (2013) *Britain Begins*. Oxford University Press, Oxford, 146.

[65] Dorset Council (n.d.) Wessex Ridgeway <https://www.dorsetcouncil.gov.uk/sport-leisure/walking/walking-in-west-dorset/wessex-ridgeway.aspx > Accessed April 2020.

[66] P.C. Tubb (2009). The Bronze Age-Iron Age transition in the Vale of Pewsey, Wiltshire unpub. PhD thesis University of Bristol, 287.

[67] A. Tullett (2010) Information Highways – Wessex Linear Ditches and the Transmission of Community, in M. Sterry, A. Tullett and N. Ray (eds) *In Search of the Iron Age: Proceedings of the Iron Age Research Student Seminar 2008, University of Leicester*. Leicester Archaeology Monographs 18. University of Leicester Press, Leicester, 111-126.

[68] S.T. Turner (2004) Christianity and the Landscape of Early Medieval South-West Britain. Unpublished PhD thesis, University of York. <http://etheses.whiterose.ac.uk/9848/> 86, 287 Accessed April 2020.

[69] J. H. Bettey (1983) Livestock Trade in the West Country during the Seventeenth Century. *Proceedings of the Somerset Archaeological and Natural History Society* 127, 123-128.

Figure 6.12. a. Formerly turnpiked Ox drove south of Grovely Woods, Wilton.
b. mid-18th century limestone milestone. (the author).

Trading at Taunton fair during the 1620s and 1630s included livestock sold to Somerset farmers by Welsh drovers; at Whitedown Fair, which was held on a hilltop between Crewkerne and Chard in Somerset, there are records of sales of cattle from Ireland. The fair at Magdalen Hill outside Winchester was noted for horses, and in the 1620s more than half the horses sold there were colts from Somerset, most of them having been bred on the marshlands; others came from the pasture lands of north Wiltshire. Animal products were also involved, for fairs at Devizes and Marlborough were famous for cheese and tripe was observed to be consumed at a fair in 1634 at Woodbury Hill, Dorset.[70] Prior to the railway, the Napoleonic war period saw the price of oxen rising, giving a grand finale to this form of transhumance that was prehistoric in origins.[71]

Drovers' roads would be natural candidates for improvement, being largely across upland areas and integrated into the working landscape. One chalkland example is Ox Drive which runs south of Groverly woods between Chickslade and Wilton; it was turnpiked relatively early. The indistinct carved inscription on a limestone pillar milestone located immediately south of Grovely Woods near Wilton (Figure 6.12) reads: 'VI / Miles From / SARUM / 1750'.[72] The turnpike road is an unmetaled restricted byway. There are other milestones along the track marked by the Ordnance Survey.

The drove is likely of great antiquity. A Roman road connected the Mendips (important for the lead mining) to Old Sarum, which ran through Grovely Woods. The Ox Drove crosses this today and in historic times was used for driving cattle to Wilton and Salisbury markets.[73]

6.6 Mendip: agriculture, mining, hunting and manufacture

Introduced in this Chapter as NCA 141, the Mendip Hills provide another insightful regional case study. The hills are located southwest of Bristol and Bath and constitute a plateau of Carboniferous Limestone with a core of Devonian Old Red Sandstone. Figure 6.13 shows, diagrammatically, the roads, caves and archaeological sites on Mendip, a small area measuring approximately 200km^2 and only 30 km in length, reaching up to 325mAOD. The western, highest area is designated an AONB.

While the soils include thin rendzinas over limestone, deep, fine silty-loess derived fertile soils are also present which are generally too wet (and hence fragile) for arable agriculture. These form-acidic soils over Old Red Sandstone. The natural vegetation includes limestone grassland, notably within Cheddar Gorge, and limestone woodland supporting small-leaved lime (*Tilia cordata*), pendunculate oak (*Quercus robur*) and wych elm (*Ulmus glabra*). On Old Red Sandstone, acid heathland has become established, and there is 'limestone heathland' where

---

[70] J.H. Bettey (2004) West-Country markets and fairs in the seventeeth century: some documentary evidence *The Local Historian* 34(4), 227-234. Bettey (1986) *Wessex to AD 1000*. Longman, London, 148.

[71] Local Drove Roads (n.d.) Drovers' House Stockbridge <http://www.localdroveroads.co.uk/drovers-house-stockbridge/> Accessed April 2020.

[72] Historic Milestone (2003) Milestone. <https://historicengland.org.uk/images-books/photos/item/IOE01/11138/10> Accessed April 2020.

[73] E. Crittall (1965) Dinton, in E. Crittall (ed.) *A History of the County of Wiltshire: Volume 8, Warminster, Westbury and Whorwellsdown Hundreds*. Victoria County History, London, 25-34.

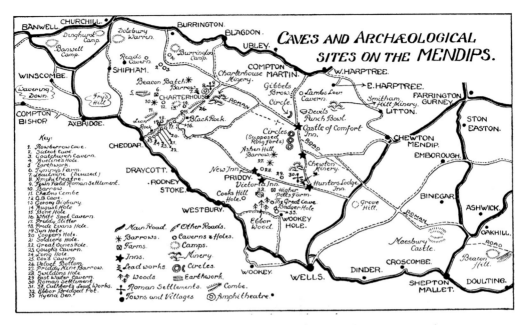

Figure 6.13. The caves and archaeological sites on Mendip (A.W. Coysh, E.J. Mason and V. Waite, 1977, The Mendips. Robert Hayle, London, p 21).

soils have become acidified by leaching. There is also lead, zinc and cadmium contamination, associated with former mineries, which affects semi-natural habitat composition.[74]

Figure 6.13 gives an immediate indication of the economic importance of the Mendips. Its cartoon-like quality belies an historic and landscape quality that is disproportionate to its size in a purely geographical sense, yet it manages to capture much of the local complexity. The approximately Westnorthwest-Eastsoutheast trend of the landscape reflects the overall shape of the plateau. This in turn is the outcome of the intersection of two periclinal folds (elongated dome-like structures) in the Carboniferous Limestone and other strata which together with the syncline (downfold) beneath and a series of major faults (most prominent on the south side) creates an impressive massif.[75]

The north side (above 'Winscombe' on the map) presents a strong east-west ridge of Carboniferous Limestone with Old Red Sandstone on the south side, and a terrestrial tropical weathering deposit traditionally known as 'Dolomitic Conglomerate' below the limestone ridge (both north and south) that was deposited in 'wadi-like' valleys cut at right angles to structural trends of the periclinal folds.[76] This landscape was once buried beneath Jurassic

---

[74] R.D. Russell (1995) Agriculture on Mendip. *Proceedings of the Bristol Naturalists Society* 55, 79-86; K.C. Allen (1995) Botanical walks and wanderings on the Mendip Hills. *Proceedings of the Bristol Naturalists Society* 55, 3-18.

[75] M.J. Simms (1995) The Geological history of the Mendip Mills and their margins. *Proceedings of the Bristol Naturalists Society* 55, 95-112.

[76] E. Jamieson (2015) *The Historic Landscape of the Mendip Hills*. Historic England, Swindon, 5.

Figure 6.14. Aveline's Hole (ST 476105 8670) in the Mendip Hills,
probably Britain's oldest cemetery, formed in steeply dipping
Carboniferous Limestone (the author)

and Cretaceous sediments, typically limestones; the recession of these beds' scarp fronts eastwards re-exposed the Triassic and Palaeozoic land-surface.[77]

The hillfort of Dolebury Warren occupies the heights of this ridge on the north side, while the famous Burrington Combe (of 'Rock of Ages Cleft for Me' fame) is part of a 'karst' landscape cut in Carboniferous Limestone. 'Karst' refers to a suite of landforms created by the action of water on limestone. Where there is no gorge, the routes up to Mendip from the north side are especially steep. The Revd. Toplady, while sheltering in Burrington Combe and contemplating whatever sins he may have committed, must have realised something about the interaction of water and limestone. Indeed, a century and a half later, the Bristol Waterworks Company created the nearby Blagdon Lake (Formerly the Yeo Reservoir) and, fifty years after that, the Chew Valley Lake to store water to supply Bristol.[78]

[77] R. Bradshaw (1987) The Geology and Evolution of the Avon Gorge. *Proceedings of the Bristol Naturalists Society* 47, 45-64.
[78] A.W. Coysh, E.J. Mason and V. Waite (1977) *The Mendips*. Robert Hale, London, Chapter 16.

Turning to prehistory, the Mendips are famous for cave archaeology including hunter-gather remains at Ebbor George, Cheddar George and the immediate Wookey Hole area.[79] At Aveline's Hole, a cave in Burrington Combe (Figure 6.14), yet another clergyman, the geologist Revd. William Buckland investigated an accidental find in 1797. It is known that Pleistocene period Palaeolithic people (including Neanderthals) hunted over Mendip, for example from the 'hyena den' at Wookey Hole, south side of Figure 6.13. However, the archaeology of Aveline represents a significant early Mesolithic find. This period started with climate warming from about 8,000 BC. The sea level rose forming the Bristol channel and submerging the Somerset Levels and Moors. Regionally, tundra vegetation was replaced with temperate species. On the Mendips these consisted of open birch-pine woodland with grassland becoming deciduous woodland including alder, hazel, and oak by the later Mesolithic. There were likely Mesolithic huts established near Priddy.[80]

Hunting bands utilising the Yeo Valley and the Failand Ridge (near to the modern mouth of the Bristol Avon to the north) may have made a transitory site at Aveline's Hole.[81] Pre-agricultural communities would have had access to rich hunting environments: upland trees sheltering fauna for hunting in the valleys and gorges and lowland marshlands. The poor recording and curating of the early finds from Aveline's hole made interpretation difficult, but it may represent an early funerary site including perhaps 50 individuals, many disarticulated and buried with fossil ammonites, red ochre and perforated animal teeth, possibly surviving from body adornment. Radiocarbon dates originally suggested that the burials date between *c.* 8400 and *c.* 8200 BC, although subsequent analysis suggests that some human remains from Aveline's Hole are in fact Early Neolithic.[82] DNA analysis of the crania in question suggests a considerable influx of migrants associated with Neolithic farmers.[83] These hunter gathers must have had some religious beliefs associated with death.[84]

Few scientific revelations spark the imagination as did those around Cheddar Man, dated to around 10,000 years ago (*c.* 8000 BC), the time of the Late Upper Palaeolithic to Early Mesolithic. On the south side of Mendip (location 23 on Figure 6.13) is Gough's Cave in the side of the lower part of Cheddar Gorge, the vicinity being presented as a tourist destination. However, the skeleton of an unfortunate young man was found in 1903 with the skull showing injury. This time (fortunately) the original bones were removed to the British Museum for safe keeping.[85] His head injury has led to speculation around how he might have died, but that was nothing compared with his genes.

The analysis of Mitochondrial DNA ('mtDNA') shows that he appears to have identifiable living relatives. mtDNA is passed from mother to daughter unchanged but cannot be passed beyond one generation of male-line of descent. Media interest in genetics and archaeology

[79] E. Jamieson (2015) *The Historic Landscape of the Mendip Hills*. Historic England, Swindon, 14.
[80] E. Jamieson (2015) *The Historic Landscape of the Mendip Hills*. Historic England, Swindon, 32-33.
[81] B. Cunliffe (1993) *Wessex to AD 1000*. Longman, London, 19.
[82] R.J. Schulting, T. Booth, S. Brace, Y. Diekmann, M. Thomas, I. Barnes, C. Meiklejohn, J. Babb, C. Budd, S. Charlton, H. van der Plicht, G. Mullan and L. Wilson (2019) Aveline's Hole: An unexpected twist in the tale. *University of Bristol Spelaeological Society Proceedings* 28(1), 9-63.
[83] R.J. Schulting *et al.* (2019) Aveline's Hole: An unexpected twist in the tale. *University of Bristol Spelaeological Society Proceedings* 28(1), 9-63.
[84] B. Cunliffe (2013) *Britain Begins*. Oxford University Press, Oxford, 124.
[85] K. Lotzof (2018) Cheddar Man: Mesolithic Britain's blue-eyed boy. <https://www.nhm.ac.uk/discover/cheddar-man-mesolithic-britain-blue-eyed-boy.html?gclid=CjwKCAjw-vjqBRA6EiwAe8TCk2bEjW282sEQ> Accessed May 2021.

was raised when Cheddar history teacher Adrian Targett was found to possess these genetic similarities, tests being prompted by knowledge of the antiquity of Mr Targett's matrilineal line in the vicinity.[86] His ancestor would have been a hunter gatherer living in a relatively rich environment in the vicinity of Mendip,[87] making something like 500 generations ago. Media interest during 1997 caused much amusement, including one student commenting that hitherto his teacher had had no nickname(!) and his late wife Katie stated that she now had an explanation as to why he liked his steak rare.[88]

Such a finding does not sit readily with recent research suggesting a large-scale changes to the population of northwest Europe. There were extensive migrants in Britain associated with the arrival of agriculture around 4000BC.[89] There is also suggested an extensive replacement associated with the Beaker Culture period 2500 to 1800 BC.[90]

Investigation of Cheddar Man's genome suggests that he had dark skin with blue eyes.[91] Light skins, typical of modern northern Europeans, may only have arisen after the time of Cheddar man. Blue or green eyes may not have been unusual in early times, but brown eyes may have come to Britain with migrants bringing agriculture. Cheddar Man was likely lactose intolerant, a property that changed over time in European adults following the adoption of agriculture.[92] Whatever the case, the Cheddar study tells us a proportion of the population remained in place from before Britain became an island. Human evolutionary change may have continued alongside technological advancement since the end of the Pleistocene.[93]

The spread of sites shown in Figure 6.13 shows much evidence of economic development beyond that of pre-agricultural hunter-gatherers. Agriculture arrived in the area after 4,200 BC,[94] but in line with the cliché 'dead men tell no tales', funerary evidence outstrips evidence for settlement and agriculture. In the Neolithic, the activity on Mendip may have been limited when compared with the adjacent lowlands, however as newer evidence comes to light a picture of a diverse Neolithic and Bronze economy may yet emerge.[95]

---

[86] L. Barham, P. Priestly and A. Targett (1999) *In Search of Cheddar Man.* Tempus Books, Stroud, Chapter 6.

[87] E. Blinkhorn and N. Milner (2013) Mesolithic Research and Conservation Framework 2013 <https://archaeologydataservice.ac.uk/researchframeworks/mesolithic/wiki/Meso_Res_Cons_Framework> Accessed June 2020.

[88] C. Arthur (1997) The family link that reaches back 300 generations to a Cheddar cave. *Independent* 8 March. <https://www.independent.co.uk/news/the-family-link-that-reaches-back-300-generations-to-a-cheddar-cave-1271542.html> Accessed April 2020.

[89] J. McNish (2018) The Beaker People: A New Population for Ancient Britain. <https://www.nhm.ac.uk/discover/news/2018/february/the-beaker-people-a-new-population-for-ancient-britain.html> Accessed Aug 2021.

[90] I. Olalde, S. Brace, M. Allentoft *et al.* (2018) The Beaker phenomenon and the genomic transformation of northwest Europe. *Nature* 555, 190–196. https://doi.org/10.1038/nature25738

[91] S. Gibbens (2018) Britain's Dark-Skinned, Blue-Eyed Ancestor Explained. <https://www.nationalgeographic.com/news/2018/02/ancient-face-cheddar-man-reconstructed-dna-spd/> Accessed June 2020.

[92] C. Gamba *et. al.*, (2014) Genome flux and stasis in a five millennium transect of European prehistory. *Nature Communications* 5, 5257. DOI: 10.1038/ncomms6257; K. Lotzof (2018) Cheddar Man: Mesolithic Britain's blue-eyed boy. <https://www.nhm.ac.uk/discover/cheddar-man-mesolithic-britain-blue-eyed-boy.html> Accessed May 2021.

[93] Cheddar Man: Britain's Black Heritage <https://www.youtube.com/watch?v=lWDWVDu01P0> Accessed June 2020.

[94] B. Cunliffe (2013) *Britain Begins.* Oxford University Press, Oxford, 142.

[95] P. Ellis (1992) *Mendip Hills An Archaeological Survey of the Area of Outstanding Natural Beauty.* Somerset Historic Environment Publications, Taunton.

Early Neolithic activity is demonstrated by long barrows on the western side of Mendip; eight likely locations have been proposed including at Barrow Batch and near to Priddy.[96] The prehistoric economy of Mendip indicates a classic shift from hunter-gathering to a settled society which, during the Neolithic, constructed monuments, presumed to be funerary in purpose (long barrows are contemporary in Wessex with 'causewayed enclosures') with social or religiously purposed henges dating typically a millennium later. The day-to-day lives of the early farmers have been revealed through the usual pottery fragments and stone axes as well as human skeletal remains from Chelm's Combe, immediately north of Cheddar. These were taken from a rock shelter that has since been quarried away.[97]

Near to Nempnett Thrubwell (north of the mapped area of Figure 6.13) there once stood the chambered long barrow of Fairy Toot, where the merriment of fairy folk was said to be heard.[98] Four late Neolithic 'circles' (Priddy Circles) in the middle of the map date from (2800-2000 BC) and are enclosures located towards the centre of the Mendip Plateau.[99] These are considered to be henge monumentsbut have characteristics which distinguish them from other henges presumed made for ritual or ceremonial purpose. The Priddy Circles were constructed as roughly circular or oval shaped enclosures comprising a flat area 180m-200m in diameter, enclosed by a bank and, unusually, an external ditch. There is a possible ovoid barrow in the north east quadrant of one of the circles.[100] Additionally, four bowl barrows are to be found within the northernmost of the Priddy Circles, and another just to its west.[101]

Evidence of Beaker Culture is found at Gorsey Bigbury (a henge monument near to Charterhouse) where there may have been livestock enclosures associated with initial Beaker period burials, however the nature of the supposed settlement remains unclear.

A Bronze Age cremation cemetery at Burrington Combe should indicate an adjacent settlement and may also indicate nearby settlements on the hills themselves.[102] The Beacon Batch Barrows (Figure 6.13 near to Charterhouse) are 'ring ditch' round barrows of the Bronze Age and represent a significant funerary landscape, deliberately located at high points on the Mendips.[103] Round barrows and a circular earthwork at Beacon Hill Wood (in the southeast corner of the map) have been found to contain cremated human remains.[104] Off the map to the north-east, is the late Neolithic/ early Bronze Age stone circle and cove of Stanton Drew.[105]

---

[96] J. Lewis (2008) The long arows and long mounds of West Mendip. *Proceedings of the Bristol Spelaeological Society* 24(3), 187-206.

[97] A.W. Coysh, E.J. Mason and V. Waite (1977) *The Mendips*. Robert Hale, London, 26-27.

[98] A.W. Coysh, E.J. Mason and V. Waite (1977) *The Mendips*. Robert Hale, London.

[99] E. Jamieson (2015) *The Historic Landscape of the Mendip Hills*. Historic England, Swindon, 47.

[100] Historic England (1997) Three of the Priddy Circles and one barrow, 400m west of Castle Farm <https://historicengland.org.uk/listing/the-list/list-entry/1015498> Accessed April 2020

[101] Historic England (1997) Priddy Circle and barrow cemetery 400m north of Castle of Comfort Inn <https://historicengland.org.uk/listing/the-list/list-entry/1015501> Accessed April 2020

[102] P. Ellis (1992) *Mendip Hills An Archaeological Survey of the Area of Outstanding Natural Beauty*. Somerset Historic Environment Publications, Taunton, 17-18.

[103] Discovering Black Down (n.d.) Beacon Batch Barrow Cemetery <http://www.discoveringblackdown.org.uk/beacon-batch-barrow-cemetery> Accessed April 2020.

[104] E. Jamieson (2015) *The Historic Landscape of the Mendip Hills*. Historic England, Swindon, 67-8.

[105] English Heritage (n.d.) Stanton Drew Circles and Cove <https://www.english-heritage.org.uk/visit/places/stanton-drew-circles-and-cove/> Accessed April 2020.

Evidence of agriculture from the Bronze Age (in the form of field systems) is limited, but a possible example of a trackway associated with two Bronze Age enclosures has been reported from just southeast of Cheddar.[106]

Long barrows and Iron Age hillforts alike punctuate the skyline and may be associated with making public statements. The former may be about early farmers establishing the territory of a clan, while the developed economy of the Iron Age was tribal in nature, and we may imagine chiefs displaying their prominence, although in practice these defensible spaces, where settled within, may represent proto-urbanisation where goods, processing and other services were exchanged.

By the late Bronze Age, large areas of woodland were being cleared to make way for blocks of planned fields, it is imagined by a community composed of family units living in single or grouped roundhouses enclosed by banks and ditches. The importance of land division is evident from linear ditch systems created at this time. Metal objects (weapons, tools and ornaments) have been found in hoards, buried possibly for religious purposes, and indicate that prestige goods were available for exchange between social networks.[107]

Exchange goods are often limited in type (e.g. amber, copper, glass beads and marine shells) but the society in question ascribes high status or value to them. Such prestige items underlined the political importance of local elites and the systems of exchange that they directed and controlled.[108] It becomes easy to see how social hierarchies emerged (who controlled the agricultural landscape) as well as particular forms of production, including mining and manufacture of high-end goods.

New field systems on the Mendips dating from the Bronze Age and Iron Age have been identified from the aerial photography; these are in general on more productively marginal land. During the first millennium BC, iron working became commonplace in Britain.

Iron age activity was evidently significant, and Figure 6.13 shows 'camps' or 'hillforts' at Banwell Camp, Dolebury Warren, Burrington Camp (a hillside enclosure) and Maesbury Castle.

Evidence of Iron Age settlement is strong for both upland and lowland areas in the vicinity of Mendip. By the middle Iron Age, there was probable population growth, maybe suggesting new technologies were combatting adverse climatic impacts on the agricultural system.[109]

Based on work by Barry Cunliffe, a model for the Early Iron Age has been suggested where enclosures represent the centres controlling upland pastoral land use on behalf of a community based in the lowland. In the Middle Iron Age, there was a marked reduction in the number of enclosures to a few heavily defended sites, each with their own territory. In the Early Iron Age some form of collection and redistribution of livestock pastured on the hills may have occurred, later to be replaced by the collection and processing of food within

---

[106] P. Ellis (1992) *Mendip Hills An Archaeological Survey of the Area of Outstanding Natural Beauty*. Somerset Historic Environment Publications, Taunton, 19.

[107] E. Jamieson (2015) *The Historic Landscape of the Mendip Hills*. Historic England, Swindon, 73.

[108] Archaeology Wordsmith (n.d.) <https://archaeologywordsmith.com/search.php?q=prestige%20goods> Accessed April 2020.

[109] E. Jamieson (2015) *The Historic Landscape of the Mendip Hills*. Historic England, Swindon, 79-80.

the major hillforts. The idea comes from investigations at Danebury Hillfort in Hampshire but might be applicable to Mendip. The collapse of this system may have occurred in the first century BC.[110]

The extant developed defences of Dolebury Warren are late Iron Age (4th to 3rd century BC), although there may have been an earlier monument on the site. At Pitchers Enclosure (near to West Harptree) there are hut circles and other earthwork enclosures which may have been occupied. At Chew Park (North of West Harptree) there were open farmsteads dating from the middle of the 1st millennium BC as well as a Roman villa site now submerged by the lake. Nearby at Worberry Gate, occupation of an Iron Age rural site continued into the Roman period. Mendip lead was exploited from the Bronze Age into the modern period, but continuity of sites is unclear in the archaeological record.[111]

The Romano-British period (AD 43 to c AD 410) saw Roman rule imposed on the *Dobunni*. The period saw both intensification of agriculture and rapid development of industry, especially lead and possibly silver extraction from the Mendips. The significance of mining may be signified by the presence of an early Roman fort at Charterhouse-on-Mendip, a nearby amphitheatre and the discovery of a stamped lead ingot from AD 49 at Wookey Hole.[112]

The Mendip plateau had long been accessible, and this must have incentivised mining. In Figure 6.13 there is a Roman road running Westnorthwest-Eastsoutheast across the plateau, which was linked with another spur running northeast by Compton Martin towards the Chew Valley Lake, beneath which lies the remains of a Roman villa. It may be that British lead, as already mined in Derbyshire and on Mendip was one spur to the Roman invasion of Britain, successfully completed during the reign of Emperor Claudius. The Romans mining industry was sustained in Britain from about AD 49 for about 350 years. Mendip alone gives plenty of evidence that what was to become the Province of *Britannia* had a developed economy, ripe for conquest, administration and taxation. Lead mining was particularly centred around Charterhouse-on Mendip.[113]

Post-Roman mining continued to extract galena ore, lead sulphide (PbS). Since then, other centres developed at East Harptree, Chewton and nearby St Cuthberts. Lead mining continued in Mendip until its closure during the 19th century. Today, due to re-working of the spoil, reclamation for agriculture and re-colonisation by vegetation of the hummocky or 'gruffy ground' left by miners, there is little remaining impact of this once important industry.[114]

Iron mining is not so readily associated with Mendip, but the mineral haematite (Fe_2O_3) is found throughout the hills, other ores being pyrite (FeS_2), goethite (FeO.OH), and a mixture of hydrated iron oxides known as 'limonite'. Here haematite and goethite commonly occur

[110] P. Ellis (1992) *Mendip Hills An Archaeological Survey of the Area of Outstanding Natural Beauty*. Somerset Historic Environment Publications, Taunton, 21-22.
[111] P. Ellis (1992) *Mendip Hills An Archaeological Survey of the Area of Outstanding Natural Beauty*. Somerset Historic Environment Publications, Taunton, 32.
[112] E. Jamieson (2015) *The Historic Landscape of the Mendip Hills*. Historic England, Swindon, Chapter 5.
[113] Historic England (2015) Charterhouse on Mendip Roman lead mining settlement <https://www.pastscape.org.uk/hob.aspx?hob_id=1519564> Accessed April 2020.
[114] Northern Mine Research Society (n.d.) Mendip Mines <https://www.nmrs.org.uk/mines-map/metal/mendip-mines/#targetText=Mendip%20Mines,of%20zinc%20ores%2C%20especially%20calamine> Accessed April 2020.

together as either massive, granular, often siliceous masses or amorphous earthy varieties known as ochre. Much as iron attracted interest in the Brendon hills (see Chapter 7), so Mendip ironmasters such as the Fussell family of Mells were productive in iron-making from the mid-18th century until the collapse of English agriculture from 1880; agricultural machinery that was their bread and butter.[115]

The early medieval period (c. AD 400–1086) on Mendip commenced with economic decline following the end of Roman rule, with probable settlement abandonment, although Cheddar was one area of continuity. West Somerset was likely part of the West Saxon kingdom by AD 680. Place name evidence suggests Anglo-Saxon settlement started along springlines at the foot of the plateau, with woodland prominent along its slopes. In terms of field patterns, there was probably initially continuity from the Romano-British period, although by the 10th century there open field systems predominated around the plateau. Significantly, there was an important royal estate of the Kings of Wessex at Cheddar by the 9th century, its location likely influenced by upland hunting.[116]

A date for afforestation of the Royal Forest of Mendip is not forthcoming but the removal of the legal status (disafforestation) *effectively* occurred early on in 1338. Anglo-Saxon kings evidently hunted on their demesnes, and there were officers appointed for the royal sport as shown by the *Life of St Dunstan* (see Chapter 3). The land over Mendip was not the best for agriculture in the area, providing one factor behind its designation as of a royal hunting forest.[117] Cheddar was a royal manor in 1086. After the fourteenth century, Mendip reverted to common moorland with private woodland, but in 1633 Charles I formally abolished the Royal Forest in a way that benefitted the Crown and landowners, but not commoners. There was much encroachment on the forest and wood pasture was converted to grassland; common land on Mendip was enclosed by Act of Parliament in 1830.[118]

Another significant Mendip industry was paper-making. Historically this was at Stoke Bottom (Figure 6.14) just east of Ashwick, and today the industry is still associated with Wookey Hole (from where the river Axe emerges) and with St Cuthberts Mill (Wells) in the south. There were mills north of Mendip at North Wick near Chew Magna, at Compton Martin and at Cheddar. Papermaking requires 'quick streams and clear water'; a clean water supply being particularly important, even more so if high-end (white) papermaking is undertaken, of the kind used for bank notes, books, and letters.

The domestic paper industry would replace the importation of paper for printing before 1495, but activity around Mendip tended to be later. Here, papermaking was largely an 18th or early 19th century operation, although the Wookey Hole mill dates from before 1610; after 1860 many small mills went out of business. Associated with French Huguenot refugees, paper making became liable to an excise duty imposed from 1712 to 1860, suggesting the attention of government to a profitable enterprise. The demise of small, high quality manufacturers

---

[115] R. Atthill (1964) *Old Mendip.* David and Charles, London, Chapter 6; BGS (2017) Minerals and mines <https://www.bgs.ac.uk/mendips/minerals/Mins_Mines_6.htm> Accessed April 2020.

[116] E. Jamieson (2015) *The Historic Landscape of the Mendip Hills.* Historic England, Swindon, Chapter 5.

[117] H. Cook (2018) *New Forest: The Forging of a Landscape.* Windgather Press, Oxford, Chapter 7.

[118] M. Richardson (2003) An Archaeological Assessment of Cheddar. English Heritage. <https://www.somersetheritage.org.uk/downloads/eus/Somerset_EUS_Cheddar.pdf> Accessed April 2020; O. Rackham (1988) Woods, Hedges and Forests, in M. Aston (ed.) *The Medieval Landscape of Somerset.* Somerset County Council, Taunton, Chapter 1.

was due to industrialisation, which enabled the handling of larger volumes of rags and wood as inputs. Steam technology, using coal, replaced local waterpower, and the industry would become located elsewhere.[119]

In the 19th century, with lead being produced on the Mendip plateau, and groundwater emerging from caves within this largely karstic topography, the provision of clean water was compromised. A landmark test case (*Hodgkinson v Ennor*, 1863) saw judgement in favour of the plaintiff, a paper manufacturer (Hodgkinson), upheld against pollution of water issuing from Wookey Hole arising from the St Cuthbert's leadworks at Priddy. Groundwater contamination here was considered to affect *underground river* supplies. It would be some time before there was a sufficient legal framework to protect groundwater that flowed through permeable strata as distinct from in an 'underground river' identifiable in karst landscapes.[120]

Agriculture on Mendip has long been dominated by grass production. Although the geology is complicated, the soils (mostly well-drained brown earths) are generally acidic, despite the prevalence of limestone, due to the Old Red Sandstone, Triassic formations and presence of aeolian (wind-blown) silty drift across the plateau, producing deep and fertile soils of the Nordrach soil series.[121]

In Tudor times, ploughland was sown for sheep and enclosure began. Dairy came to replace sheep over time, and today the region produces milk for yoghurt, cheese, and whey for pigs. There has been little re-introduction of arable agriculture due to the climate. There has been some destruction of natural grassland and removal of some boundaries, while the legacy of mining has caused localised zinc toxicity on grass and other crops.[122]

Systematic enclosure of Mendip overall occurred between 1770 and 1870, accounting for about 17.7% of the total land enclosed in Somerset at this time.

The pre-enclosure landscape would have been desolate, with little deep soil and protruding rocks, suited to grazing sheep and young cattle. Areas of Nordrach soil association were prime candidates for enclosure and improvement. These comprise well-drained fine silty over clayey soils, stone-less or with chert stones, and are often deep, derived from wind-blown loess over carboniferous limestone.[123] The surface in parts was still covered with heath, fern, and furze, but there were few remaining trees. Alongside land improvement there were improvements to the road system and enclosure using limestone walls with a few white thorn hedges, creating new farmsteads.[124]

In modern biological conservation terms, the Mendips display semi-natural dry grassland and scrubland facies on calcareous substrates (*Festuco-Brometalia*), including important orchid sites. Found on the coastal and inland outcrops of Carboniferous Limestone, they are of

---

[119] R. Atthill (1964) *Old Mendip*. David and Charles, London, Chapter 5.

[120] H.F. Cook (1995) Groundwater development in England. *Environment and History* 5, 75-96.

[121] E. Jamieson (2015) *The Historic Landscape of the Mendip Hills*. Historic England, Swindon, 8-9.

[122] R.D. Russell (1995) Agriculture on Mendip. *Proceedings of the Bristol Naturalists Society* 55, 79-86.

[123] Landis (2020) 0581a NORDRACH. <https://www.landis.org.uk/services/soilsguide/mapunit.cfm?mu=58101> Accessed May 2020.

[124] M. Williams (1971) The Enclosure and Reclamation of the Mendip Hills. *Agricultural History Review* 19(1), 65-81; Jamieson E. (2015) *The Historic Landscape of the Mendip Hills*. Historic England, Swindon, 9.

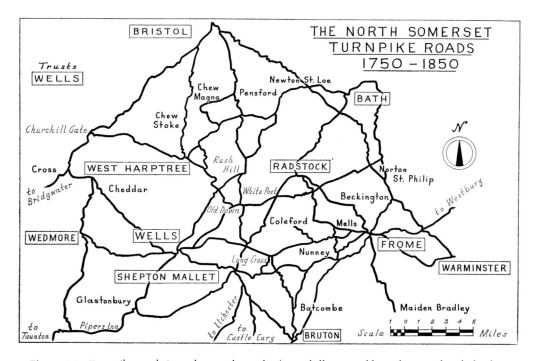

Figure 6.15. Turnpike roads in and around Mendip (R. Atthill, 1964, *Old Mendip*. David and Charles, London, p.107, public domain).

European significance, designated under the EU's European Habitats Directive (92/43/EEC), known as the Directive on the Conservation of Natural Habitats and of Wild Fauna and Flora.[125]

The coastal headland and inland hills support the largest area of *Festuca ovina–Carlina vulgaris* grassland in England, including two sub-types (*Carex humilis* and *Trinia glauca* sub-communities) known from no other site in the UK. Areas of short-turf *Festuca ovina–Avenula pratensis* grassland are also found inland. The JNCC's National Vegetation Classification community CG1 (*Festuca ovina-Carlina vulgaris* grassland) is one of the calcicolous (lime-loving) grassland communities in the (British) National Vegetation Classification system. It is a short-sward lowland grassland community associated with heavy grazing and is regarded as the south-west coastal counterpart of "typical" chalk grassland (CG2).[126]

Overall, there are exceptional areas that support a number of rare and scarce vascular plants typical of the oceanic southern temperate and even 'Mediterranean' elements of the British flora. These include white rock-rose (*Helianthemum apenninum*), Somerset hair-grass (*Koeleria vallesiana*) and honewort (*Trinia glauca*). Transitions to limestone heath (European dry heaths) are situated on flatter terrain.[127] Today, on Black Down (2km southwest of Burrington, Figure

[125] JNCC (n.d.) Semi-natural dry grasslands and scrubland facies <https://sac.jncc.gov.uk/habitat/H6210/> Accessed April 2020.
[126] JNCC (n.d.) Mendip Limestone Grasslands <https://sac.jncc.gov.uk/site/UK0030203> Accessed April 2020.
[127] JNCC (n.d.) Mendip Limestone Grasslands <https://sac.jncc.gov.uk/site/UK0030203> Accessed April 2020.

6.13) poorly drained soils at the highest part of Mendip (maximum 325m on the adjacent Beacon Hill) are developed on Old Red Sndstone and support heather and heathland grasses.[128]

Figure 6.15 shows the final major stage in the improvement of road transport in the Mendip area. Although the first Turnpike Trust (Bath) dated 1707-8, the maturity of the new road network was not achieved until around 1800.[129] It is worth noting that few turnpikes actually cross the Mendip plateau, and none at the western end. Most notable is that from Wells running northeastwards through the combe below Prior's hill to the summit and thence to Chewton Mendip (Figure 6.13) and on to Bath. A second from the Cathedral City of Wells also runs northeastwards towards Old Down (Figure 6.15). Both these turnpikes cross the Roman Road that traverses Mendip running Westnorthwest-Eastsoutheast, which itself was not turnpiked. The reminder of the network crossing Mendip lies east of here and links Shepton Mallet, Mells and Radstock, a significant coal mining centre. The mid-18th C, development of the road system would have gone together with the development of the coalfield.

6.7 Conclusions: Landscapes and grasslands

The *structure* of the original limestone and chalk grassland landscapes is, by definition, that of a grass sward of a height that is dependent on grazing regime, season, and climate. The most biodiverse of these valued habitats display many plant species per square metre, the result of low soil nutrient conditions and persistent grazing. The *functions* of such landscapes are not just grazing for livestock, but also provision of a diverse habitat for plants, animals and invertebrates. There are also considerable mineral resources.

The *value* of chalk and limestone landscapes lies not only in the economic sphere, with production of milk, meat, grain etc., but in the ecological and societal goods that they provide, in the provision of open landscapes and in the protection of the archaeological resource. Other economic value resides in mining and manufactures as well as historically as recreation for elites.

The *scale* of chalk, limestone and other grassland areas depends on changes of land-use. Humans have been active since prehistory as shown by archaeological and skeletal finds as well as monuments. Ecological islands of diverse grasslands are found where grassland around a meadow or pasture has been destroyed; regional scale survivals are most desirable but chalkland and limestone soil ploughing has fragmented many formerly extensive habitats. *Change*, it follows, is due to obvious destruction (by ploughing, development, or forestry plantation) and more subtly through fertilisation, sward improvement, and under-grazing enabling scrub invasion. Industries have come and gone, including metalliferous extraction and smelting; in this we can see the interaction of economic sectors. The destruction of chalk and limestone grasslands together with their enormous biodiversity is of great concern. These grasslands, furthermore, are often of European significance. The task of restoration is greatly assisted by the availability of specific seed mixes, making it possible to start the slow process of recreating chalk and limestone grassland.

---

[128] E. Jamieson (2015) *The Historic Landscape of the Mendip Hills.* Historic England, Swindon, 10.
[129] R. Atthill (1964) *Old Mendip.* David and Charles, London, Chapter 8.

Chapter 7

Heathland and upland moorland

7.1 Heathland, moorland and Clym Yeobright

No contemporary ecologist or landscape historian would pretend that the heaths and moors of Wessex are anything other than the outcome of human intervention, making them semi-natural habitats *par excellence*. They present a plagioclimax plant community; other examples are chalk grassland (see Chapter 6) and coppice woodland (see Chapter 8). As is commonplace in landscape studies, there are devils hiding in the detail. The actual distinction between heathland and upland moorland is not so clear. For one thing, lowland heaths (commonplace in Dorset) differ from upland moorlands (such as those of Exmoor), which are found characteristically in steeper and often wetter terrains and at altitudes of more than 250m OD, providing a topographic distinction.[1] In general terms, mean annual rainfall tends to be higher over moors than for lowland heath, a function of both elevation and their frequent location in western Britain.

Unhelpfully, the term 'moor' may also be used for areas at elevations close to sea level in both Dorset and Somerset, adding to the impression there is no *clear* differentiation to be found in terms of ecologies of areas above or below 250m. A more detailed consideration of moorland will be given below when Exmoor is discussed. The main areas described in this Chapter are the Dorset Heaths (NCA 135), heaths within the New Forest (NCA 131) and upland Exmoor (NCA 145).[2] Figure 4.5 shows the context of Exmoor within Somerset.

In general, heathland is characterised by the presence of *Ericaceous* plants such as heathers, acid loving grasses and where hydrology permits, their more aquatic associates. Dominated by low, woody shrubs, larger trees are rare or absent. If not managed by vegetation removal, heathland is prone to loss of habitat from gorse, birch, or conifer invasion.[3]

Both heathland and moorland form highly evocative and characteristic landscapes in Wessex. Heath is developed in lowland areas, on dry sandy soils and on wet peaty soils, often developed on acidic, nutrient poor sand and gravels. Characterised by the presence of a range of dwarf-shrubs including species of heather (*Calluna vulgaris* and *Erica* spp.) and gorse (*Ulex* spp.), bilberry or blueberry (*Vaccinium myrtillus*, *Genista* spp.) and including common broom (*Cytisus scopartius*). Lowland heathland is a priority for nature conservation because it has declined greatly since 1800, so that in England it is estimated that only one sixth of the heathland present in 1800 remains.

[1] C.H. Gimingham (1992) *The Lowland Heathland Management Book*. English Nature Science book no.8, English Nature, Peterborough, 15.
[2] Natural England (2014) National Character Area Profiles: data for local decision making <https://www.gov.uk/government/publications/national-character-area-profiles-data-for-local-decision-making/national-character-area-profiles> Accessed May 2020
[3] C.H. Gimingham (1992) *The Lowland Heathland Management Book*. English Nature Science book no.8, English Nature, Peterborough.

Heathland is particularly important for reptiles such as the rare sand lizard (*Lacerta agilis*) and smooth snake (*Coronella austriaca*). Rare bird species are also found, for example it is the primary habitat for the nightjar (*Caprimulgus europaeus*) and Dartford warbler (*Sylvia undata*). Many scarce and threatened invertebrates and plants are also found on lowland heathland. The UK supports around 20% of the lowland heathland in Europe.[4]

Across Europe, lowland heathland is valued for its biodiversity, yet it may have declined in area by up to 80% due to land use change (generally for agriculture and forestry) as well as inappropriate management.[5] The term is broad, for it refers to a mosaic of wet, damp and dry habitats, characterised by attractively flowering dwarf shrubs, which came into existence during or after the Neolithic period.[6] Earlier origins attributed to small clearances by Mesolithic people have also been suggested on account of buried podzol soil profiles beneath a Mesolithic dwelling at Oakhanger in Hampshire[7]. Reliable data about habitat extent and fragmentation is available going back to the 1930s, and modern approaches to conservation and restoration are being activated at the landscape scale rather than merely protecting individual isolated areas.[8] Management strongly implies a return to traditional land management practices.

Figure 7.1 shows heathland between Abbots Well and Fritham in the New Forest. Heathers dominate the foreground, there is also grazed acid grassland, bracken and gorse. Wood pasture, seen in the middle distance, has been a likely component of heathland since at least Anglo-Saxon times. Over-grazing of what was typically a land cover of common land, caused loss of tree cover.[9]

New Forest heathland areas are extensive with dry heath, wet heath and dry grassland generally found to be in a 'favourable condition'.[10] There remains a danger of scrub invasion where grazing is insufficient both within the forest and on the adjacent commons, or where there is damage to heathland ecosystems from over-grazing. Heathland requires appropriate management.

Historically, programmes of 'rolling inclosure' associated with the creation of hardwood plantation, which date from around 1700, were legally challenged in the 19th century on account of their impact on the open forest areas and hence conflict with the interests of commoners. The New Forest actually presents a mosaic of heathland, woodland, commercial forestry, lawns and wetlands, of which the heaths are an integral part of a complex landscape.[11]

---

[4]   The Joint Nature Conservation Committee (2015) UK Lowland Heathland Habitat Types & Characteristics 7<http://jncc.defra.gov.uk/page-5939> Accessed May 2020.
[5]   R.F. Pywell, W.R. Meek, N.R. Webb, P.D. Putwain and J.M. Bullock (2011) Long-term heathland restoration on former grassland: The results of a 17-year experiment. *Biological Conservation* 144, 1602–1609.
[6]   English Nature (2002) Lowland heathland: a cultural and endangered landscape. English Nature, Peterborough < https://publications.naturalengland.org.uk/publication/81012> Accessed May 2020.
[7]   O. Rackham (1986) *The History of the Countryside*. Weidenfeld and Nicholson, London, Chapter 13; G.W. Dimbleby (1962) *The Development of British Heathlands and Their Soils*, Oxford Forestry Memoir no.23, Oxford University Press, Oxford.
[8]   A. Diaz, S.A. Keith, J.M. Bullock, D.A.P. Hooftman and A.C. Newton (2013) Conservation implications of long-term changes detected in a lowland heath plant metacommunity. *Biological Conservation* 167 325-333. https://doi.org/10.1016/j.biocon.2013.08.018
[9]   O. Rackham (1986) *The History of the Countryside*. Weidenfeld and Nicholson, London, Chapter 13.
[10]   E. Cantarello, R. Green and D. Westerhoff (2010) The condition of New Forest habitats: an overview in A.C. Newton (ed.) *Biodiversity in the New Forest*. Pisces Publications, Newbury, 124-131.
[11]   H. Cook (2018) *New Forest: The Forging of a Landscape*. Windgather Press, Oxford, Chapter 7.

Figure 7.1. Heathland in the New Forest, Hampshire. Note the heather, grassy ride and gravel path with Wood pasture (the author).

Otherwise in Wessex, the major areas of lowland heath are in Dorset, for example on the Isle of Purbeck. Aspects of their causation by historic management, fragmentation and habitat loss, restoration and present-day management challenges will be discussed. We may consider how they are culturally as well as ecologically important in the landscape through a lengthy quotation from the famed (and gloomy) author, Thomas Hardy. Here, as if offered on a plate, is an introduction to lowland heathland in Wessex:

> A Saturday afternoon in November was approaching the time of twilight, and the vast tract of unenclosed wild known as Egdon Heath embrowned itself moment by moment. Overhead the hollow stretch of whitish cloud shutting out the sky was as a tent which had the whole heath for its floor.....

> ....his [Clym Yeobright's] toys had been the flint knives and arrow-heads which he found there, wondering why stones should "grow" to such odd shapes; his flowers, the purple bells and yellow furze: his animal kingdom, the snakes and croppers; his society, its human haunters.......To many persons this Egdon was a place which had slipped out of its century generations ago, to intrude as an uncouth object into this. It was an obsolete thing, and few cared to study it.

How could this be otherwise in the days of square fields, plashed hedges, and meadows watered on a plan so rectangular that on a fine day they looked like silver gridirons? The farmer, in his ride, who could smile at artificial grasses, look with solicitude at the coming corn, and sigh with sadness at the fly-eaten turnips, bestowed upon the distant upland of heath nothing better than a frown. But as for Yeobright, when he looked from the heights on his way he could not help indulging in a barbarous satisfaction at observing that, in some of the attempts at reclamation from the waste, tillage, after holding on for a year or two, had receded again in despair, the ferns and furze-tufts stubbornly reasserting themselves.[12]

The contrast between the ordered and intensively farmed landscape and heathland, where improvement had proven futile, is clear. Hardy's picture mirrors prehistoric, typically Bronze Age, efforts to turn heath into productive farmland.[13] The word 'heath' has hardly changed in meaning since the Old-English *hæð* was used. Place-names containing this element and its derivatives (especially 'hoth', for example Hothfield common in Kent) can be readily recognised on maps. The term 'heathen' apparently means heath dweller, or dweller on uncultivated land. The term was never meant to be complimentary, referring to outsiders and those beyond the social, even religious norms of the person providing the label. This observation the stresses social as well as economic marginalisation of a heath's occupants or users, as well as providing landscape description.

Significant in the Hardy extract is the recognition of human activity associated with heaths. Here as elsewhere furze (gorse) is cut. Its uses include fuel, broom manufacture and even winter fodder for horses and cattle. In heaths in general there are sand and gravel extraction sites, an economic activity complementary to the poor economic returns from agricultural activity, which reflects the poverty of the soil in the first place.

By talking of flint implements Hardy refers to ancient human occupation on the heath as well as less than successful efforts of restoration. In existence since prehistory, it may be there was an incentive to maintain them during the middle ages and, since the twelfth century, landowners with a grant of free warren could maintain warrens on common land irrespective of commoners' grazing rights.[14] Actually, it is likely that the Romans originally introduced rabbits to Britain.[15] Any notion that heaths are entirely 'natural' is fallacious and Hardy clearly appreciates something of vegetation succession.

Egdon Heath does not exist as a real place, but it stands for the many heathland areas in Wessex, and yet it is worth noting that the (real) Iron Age hillfort of Eggardon Hill is on chalk, and like the fictitious heath, it commands views of the surrounding countryside. Interestingly, Hardy echoes agricultural efforts to reclaim the heathlands from 'the waste'. The reference to turnips and enclosure allude to the period of agricultural improvement and there is clearly a contrast drawn with corn fields, water meadows and improved grassland.

---

[12] T. Hardy (1878) *The Return of the Native*. Smith, Elder & Co, London.
[13] H. Cook (2018) *New Forest: The Forging of a Landscape*. Windgather Press, Oxford, Chapter 3.
[14] O. Rackham (1986) *The History of the Countryside*. Weidenfeld and Nicholson, London, 292.
[15] Current Archaeology (2019) Roman Rabbit discovered at Fishbourne. https://archaeology.co.uk/articles/news/roman-rabbit-discovered-at-fishbourne.htm> Accessed Aug 2021.

The educated Clym Yeobright clearly takes satisfaction in the failure of earlier improvers to incorporate heaths into an intensively farmed landscape. In penning this passage, Hardy therefore had an inkling of previous failed efforts to restore poor soils that formed heaths in his day, and in reality, reclamation for agriculture can be dated from prehistory to the 12th century. For agriculture, a former heath requires investment, first to turn it to farmland, then to meet the recurrent costs of applying fertilisers and irrigation as poor sandy soils have poor retention of water and nutrients. Ecosystem loss should be considered against a potential agriculture that requires high investment.

7.2 Heathland ecosystems and their management

So, what makes a 'heath ecosystem'? First, it is semi-natural, a plagioclimax ecosystem (see Chapter 1), a response to activities including arable agriculture (in the past this may have been only temporary due to the poor soils being rapidly depleted of the few nutrients they may hold), burning of vegetation (intentional or otherwise), cutting vegetation for animal or human use, digging peat for domestic or industrial purposes and grazing at an intensity to keep the vegetation from regenerating to scrub. While over-grazing can produce acid grassland given time,[16] in some instances the sheep grazed on heaths were folded over adjacent arable fields where their dung and urine fertilised the arable soil. This daily pattern of transhumance is comparable with that of the 'sheep-corn system' of water meadows (Chapter 4).[17] Whatever else, traditional management of heathland tends to increase the predominance of species of heather.[18]

Figure 7.2 demonstrates the life cycle of heather. In the **Pioneer phase**, the heather develops from seed, forming small pyramid shaped plants. The height is usually less than 10-15 cm. Short swards caused by mowing, burning or grazing can be included as 'pseudo-pioneer' within the cycle. In the **Building phase**, the heather forms a closed canopy up to 40 cm in height. This is followed by the **Mature phase** when the heather plants become woody, with thick stems, and display fewer green shoots. The heather canopy now begins to open, allowing other plant species such as mosses to increase in cover. Taller vegetation (60-100 cm) provides shelter and cover for animal and bird species, but when too high there can be decline in habitat quality.

Finally, in the **Degenerate phase** the central branches of heather plants tend to die off, creating gaps in the centre of the bush in which heather seedlings may sometimes establish. If unmanaged the heath vegetation may be invaded by bracken, then scrub, and finally mature woodland, but this can be prevented earlier in the cycle including burning the sward. *Calluna*, for example, is a relatively fire-resistant shrub and burning the above-ground sward during the building phase will not only release nutrients to the air, thereby helping to maintain a low nutrient ecosystem[19], but will permit reversion to the pioneering phase thereby maintaining the *Calluna*'s dominance.[20]

[16] D. Blakesley and P. Buckley (2016) *Grassland Restoration and Management*. Pelagic Publishing Exeter, Chapter 1.
[17] C.H. Gimingham (1992) *The Lowland Heathland Management Book*. English Nature Science book no.8, English Nature, Peterborough, Chapter 8.
[18] C.H. Gimingham (1992) *The Lowland Heathland Management Book*. English Nature Science book no.8, English Nature, Peterborough, Chapter 8.
[19] C.C. Evans and S.E. Allen (1971). Nutrient losses in smoke produced during heather burning. *Oikos* 22, 149-154.
[20] C.H. Gimingham (1972) *Ecology of Heathlands*. Chapman and Hall, London, 199, 203.

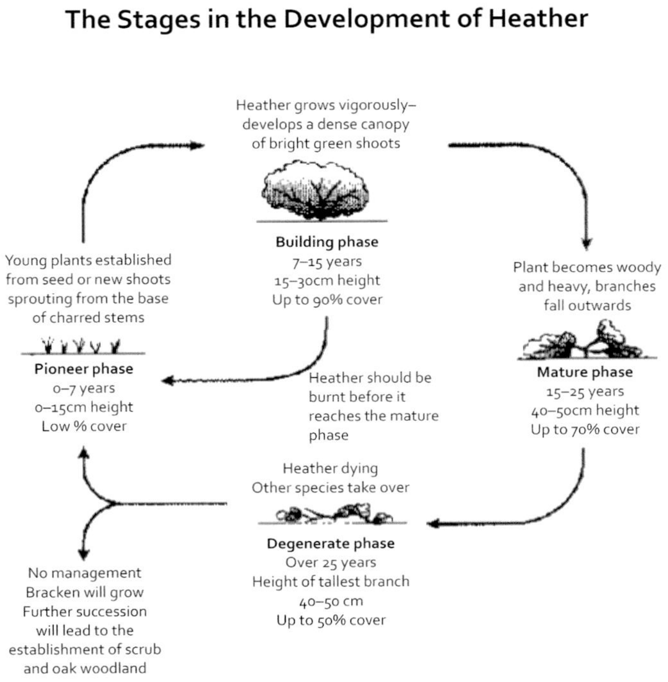

The Stages in the Development of Heather

Heather grows vigorously– develops a dense canopy of bright green shoots

Building phase
7–15 years
15–30cm height
Up to 90% cover

Young plants established from seed or new shoots sprouting from the base of charred stems

Plant becomes woody and heavy, branches fall outwards

Pioneer phase
0–7 years
0–15cm height
Low % cover

Heather should be burnt before it reaches the mature phase

Mature phase
15–25 years
40–50cm height
Up to 70% cover

Heather dying
Other species take over

No management
Bracken will grow
Further succession will lead to the establishment of scrub and oak woodland

Degenerate phase
Over 25 years
Height of tallest branch
40–50 cm
Up to 50% cover

Figure 7.2. Characteristic life cycle of heather (from 'Slideshare' based on C.H. Gimingham (1972) *Ecology of Heathlands* Springer and JNCC (2004) Common Standards Monitoring Guidance for Lowland Heathland).

Furze (gorse) and ling heather (*Calluna vulgaris*) were widely used for fuel, as a low-grade thatch. Bracken was also used for fuel and for litter for livestock, for thatch and as a source of potash in glassmaking once burned. Alongside cutting and grazing, burning may be accidental, or it may be deliberately used in order increase stocking density in 'swaling'. This is a term for the controlled burning of moorland and is used on Exmoor. Swaling is a traditional form of management to encourage regeneration of vegetation. Sections of moor are carefully burned between 1st October and 15th April on a rotational cycle.[21] However, burning is regarded as destructive for certain species of heathland plants and animals.[22]

The Joint Nature Conservation Committee (JNCC) classifies the heathlands of Dorset and Hampshire as south-western oceanic dry heath, recognising the increasingly mild and oceanic climate towards the south-west of England.

Here *Ulex minor-Agrostis curtisii* heath occurs, although heather frequently dominates this vegetation, particularly where there has been no recent burning or heavy grazing. Dwarf gorse can occur variably in abundance, while both bell heather (*E. cinerea*) becomes commonplace

---

[21] Exmoor National Park (2020) Swaling <https://www.exmoor-nationalpark.gov.uk/living-and-working/info-for-farmers-and-land-managers/swaling> Accessed May 2020.
[22] O. Rackham (1986) *The History of the Countryside*. Weidenfeld and Nicholson, London, 295-6.

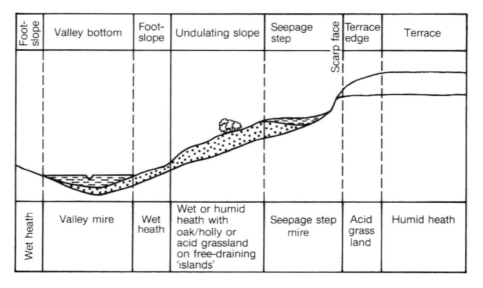

Figure 7.3. Gradation of vegetation types from terrace to valley bottom in the New Forest C.R. Tubbs, 2001, *The New Forest: History, Ecology and Conservation*. New Forest Ninth Century Trust, p.246).

following burning, and cross-leaved heath (*E. tetralix*) can also be prominent on waterlogged gleyed soils[23]. Bristle bent and purple moor-grass are characteristic grasses, especially after burning. Eastern Dorset forms the boundary where dwarf gorse (*Ulex minor*) tends to replace the western gorse (*U. gallii*).[24]

The stereotype of heath becoming established on dry sandy and gravelly soils that are permeable yet yield little in terms of fertility is only part of the picture. Actually, lowland heaths are commonly associated with wet heath areas, ponds that may be ephemeral or mires that are more-or-less permanent. By way of illustration, the classic book on the New Forest and its management by Colin Tubbs provides an illustration of a gradation of wetland types from valley mire to terrace. It is immediately clear from Figure 7.3 that the interplay of topography, hydrology and resulting soils supports a range of habitats and hence vegetation types.

Figure 7.4 shows a simplified heathland food web. The range of habitats available provides for a range of rare species including the Dartford warbler (*Sylvia undata*), smooth snake (*Coronella austriaca*) and more. The importance of heather at the base of the food web is self-evident.

Where soils in heathland areas are sandy and permeable, dry heath may result. However the inclusion of hard pan in subsoils or clay horizons in the bedrock or as iron pans deposited in the subsoil will typically impede drainage, giving areas of 'wet heath' (see Figure 7.3). Soils

---

[23] 'Gleyed' soil refers to where groundwater affects the soil profile for long enough, the soil colour tends to be greenish-grey, bluish-grey or just plain grey due to the reduction of iron compounds in the soil that would otherwise be reddish, yellow or orange.
[24] JNCC (2015) UK Lowland Heathland Habitat Types & Characteristics <http://jncc.defra.gov.uk/page-5939> Accessed May 2020, note 22.

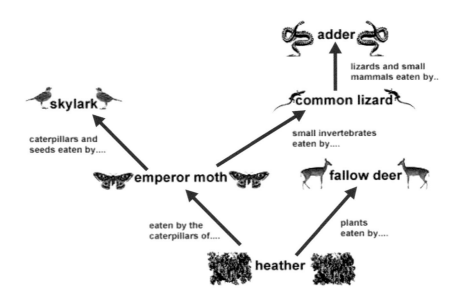

Figure 7.4. A simplified heathland food web.

may be only slowly permeable, or seasonally or permanently waterlogged. They are, however, overwhelmingly acid in nature,[25] specifically:

> 'The New Forest is a major area of semi-natural soils affected to the greater extent only by prehistoric and historic changes in vegetation and cultivation practices, primarily woodland to heathland. As a result, the Forest displays continuous sequences of soil with all their interrelated facets, such as soil water regimes and interactions with plant communities, intact. This is the one place in the southern UK where the interaction of soils and their geological substrates can be seen in relation to vegetation, and to valley mires in particular'.[26]

Like the vegetation community, the soils may also be regarded as 'semi-natural'. Podzols usually occur on high gravel terraces with good drainage and display bleached flint pebbles at the surface and, in many cases, iron pan (with manganese) around a metre below the surface. Podzols are identified by having a bleached ash-like horizon close to the surface that is depleted in nutrients, organic matter, iron and aluminium. Surface-water gley soils, on the other hand, are the most extensive soil type within the New Forest. They are often blue-grey and pyritic at depth where un-weathered, forming on the Barton Clay and elsewhere where

---

[25] R. Allen (2005) Soil and water in the New Forest and the valley Mires <http://www.soilandwater.co.uk/index.php?id=33&click=0> Accessed May 2020.
[26] R. Allen (2005) Soil and water in the New Forest and the valley Mires <http://www.soilandwater.co.uk/index.php?id=33&click=0> Accessed May 2020.

there is clay in the underlying geology. Groundwater gley soils are common on river alluvium of flood plains and will display grey or mottled subsoils due to groundwater fluctuation.[27]

On the plateau, gravels and river gravel soils are mapped as the Bolderwood association (including podzols). These give rise to heathland, gravel extraction and deciduous woodland. These slowly permeable soils display slight seasonal waterlogging. They are very acid coarse loamy over clayey with a bleached subsurface horizon with local humose or peaty surface horizons. Holidays Hill association soils occur where acidic and podzolic soils dominate on Tertiary or Cretaceous sand, loam or clay. These are again slowly permeable soils that display slight seasonal waterlogging and also acid sandy over clayey soils, also with local humose or peaty surface horizons that give rise to wet lowland heath habitats with some coniferous plantation and limited cultivation.

Drift deposits over Mesozoic or Tertiary clays and loam give rise to Wickham 3 and Wickham 4 association soils, slowly permeable and seasonally waterlogged coarse and fine loamy horizons over clayey soils giving way to a range of agricultural, woodland uses and also wet lowland heath uses on Wickham 3.[28]

Upland moorland in our region is found in Exmoor Forest, to be described below, and elsewhere in the northern and western regions of Britain. Here the topography varies greatly in height but is commonly between 250 and 500 MOD and experiences considerably higher rainfall than Dorset, Somerset or Hampshire (Introduction).

7.3 Origins and benefits of lowland heaths

Lowland heaths are important and valued because they typically support a range of birds, reptiles and invertebrates that are not found on moorlands.[29] In Hampshire, the clearance, colonisation and cultivation of the New Forest area was later than for the surrounding downland. Although heaths may date from the Neolithic, those of the New Forest and Poole Basin in Dorset likely began to be formed during the Bronze Age through land being cleared for agriculture and settlement.[30] Pollen diagrams for England generally show a succession from woodland to more open habitats including *Ericaceous* plants.[31] This has been postulated because brown forest soils can be identified beneath woodland today.

Podzols have developed beneath acid grassland resulting from forest clearance[32] and overall it is likely that podzolisation in the New Forest (as elsewhere) occurred from the Bronze Age.[33] Population pressures would have forced reclamation of poorer soils, most likely after 1500

---

[27] L.M. West (2018) Geology of the New Forest National Park <http://www.southampton.ac.uk/~imw/New-Forest-Geology-Guide.htm> Accessed May 2020.

[28] M.G. Jarvis, R.H. Allen, S.J. Fordham, J Hazleden, A.J. Moffat and, R.G. Sturdey (1984) *Soils and their use in South East England.* Harpenden: Soil Survey of England and Wales. Bulletin no. 15.

[29] Natural England and RSPB (2014) Climate Change Adaptation Manual <https://publications.naturalengland.org.uk/publication/5629923804839936> Accessed May 2020, Chapter 17.

[30] N.R. Webb and L. E. Haskins (1980) An ecological survey of heathlands in the Poole Basin, Dorset, England. *Biological Conservation* 17(4) 281-296 DOI: 10.1016/0006-3207(80)90028-2

[31] C.H. Gimingham (1972) *Ecology of Heathlands.* Chapman and Hall, London, 21.

[32] C.R. Tubbs (2001) *New Forest, the history, ecology and conservation.* New Forest Ninth Centenary Trust, Lyndhurst, 168-169.

[33] H. Cook (2018) *New Forest: The Forging of a Landscape.* Windgather Press, Oxford, 51.

BC.[34] The style of clearance may have been similar to landnam (Danish for land-take) at first, whereby land is converted for agriculture on a temporary basis, then allowed to scrub over such that in time the fertility may be recovered.[35]

Direct market economic benefits are limited and have been so for a long time due to poor soils, although the value of moorlands for shooting provides an exception. Heathlands are valued for the ecosystem services they provide:[36]

- Grazing (difficult to assess because of its use as a conservation tool)
- Timber, heather and other vegetation cutting
- Venison, wool and other animal products
- Freshwater provision
- Waste detoxification
- Climate regulation
- Renewable energy
- Recreation, tourism and aesthetic considerations

Recreation is best assessed through willingness to pay; health and well-being through medical bills avoided. Then there are aesthetic issues that may be ranked according to preference and educational uses. However, habitat and soil protection are not quantifiable. In the past, from a purely economic point of view, heathlands and moorlands were a case of 'making the best of a bad job', and generally used as rough grazing, while agricultural improvers would enclose and improve areas where soil quality, topography and climate - with particular attention given to shelter and elevation - were deemed to make it a worthwhile investment. We can imagine a kind of economic selection based on land capability and productivity, leaving these areas we have come to value so much being historically exploited to within their carrying capacity. Where this was not the case, scrub invasion would follow, and this may develop towards mature secondary woodland.[37]

7.4 The decline and fragmentation of heathlands

It is undeniable that a growing population had (and has) to be fed. Even before the drive for agricultural productivity started in 1940, the loss of semi-natural habitats to agricultural endeavours had been phenomenal. Lowland acidic heath has declined by 78% in area since A.D. 1830, and by 40% from the 1950 level.[38] Typical replacement land cover includes scrub, acid grassland or, with suitable inputs, agriculturally productive grass and arable.

---

[34] H. Cook (2018) *New Forest: The Forging of a Landscape*. Windgather Press, Oxford, Chapter 3; C.R. Tubbs (2001) *New Forest, the history, ecology and conservation*. New Forest Ninth Centenary Trust, Lyndhurst, Chapter 4.

[35] C.H. Gimingham (1972) *Ecology of Heathlands*. Chapman and Hall, London, 24.

[36] A. Newton and J. Cordingley (n.d) Evaluation of the scientific rationale for a landscape-scale approach to heathland conservation and management. Centre for Conservation Ecology and Environmental Science, School of Applied Sciences, Bournemouth University.and Dorset wildlife Trust <https://www.dorsetwildlifetrust.org.uk/hres/ Scientific-rationale.pdf > Accessed September 2018.

[37] J.C. Underhill-Day (2005) *A Literature Review of Urban Effects on Lowland Heaths and Their Wildlife*, English Nature Research Report, 623. English Nature, Peterborough.

[38] D.A. Ratcliffe (1984) Post-mediaeval and recent changes in British vegetation: The culmination of human influence. *New Phytologist* 98, 73-100. <https://doi.org/10.1111/j.1469-8137.1984.tb06099.x> Accessed May 2020.

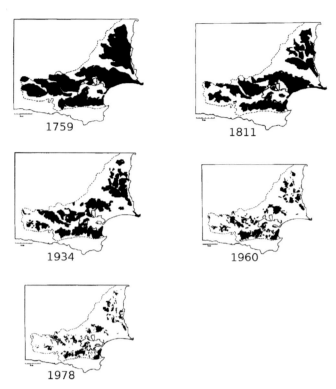

Figure 7.5. The decline of heathland in Dorset (Webb, N.R. and
Haskins, L.E., 1980, An ecological survey of heathlands in the Poole
Basin, Dorset, England in 1978. *Biological Conservation*, 17).

Today, around 20% of the international area of lowland heath is to be found in England, totalling around 58,000ha.[39] This habitat is distributed across western and north-western Europe with appreciable areas in south-western England, including Dorset.[40] Activities generally causing loss and fragmentation of habitat are attributed to the cutting of vegetation (for fuel and fodder), grazing, scrub invasion, burning, forestry, agriculture and urban development, all potentially affecting gene flow, migration and dispersal of heathland species.[41]

Changes would have taken place in heathland areas of the Poole Basin since their origin in the late Bronze Age. Figure 7.5 shows that from the mid-18th century when there were about 40,000ha of heathland, the area declined to about 6100ha by the 1970s. A 2005 publication

[39] JNCC (2015) UK Lowland Heathland Habitat Types & Characteristics <http://jncc.defra.gov.uk/page-5939> Accessed May 2020, 5941.
[40] A. Newton and J. Cordingly (n.d.) (n.d) Evaluation of the scientific rationale for a landscape-scale approach to heathland conservation and management. Centre for Conservation Ecology and Environmental Science, School of Applied Sciences, Bournemouth University.and Dorset wildlife Trust <https://www.dorsetwildlifetrust.org.uk/hres/Scientific-rationale.pdf > Accessed September 2018.
[41] J.C. Underhill-Day (2005) *A Literature Review of Urban Effects on Lowland Heaths and Their Wildlife*, English Nature Research Report, 623. English Nature, Peterborough.

shows basically the same pattern of decline continuing, although the areas were subdivided by habitat into grassland, dry heath, humid/wet heath, mire, scrub and wetland.[42]

Current management approaches aim to avoid succession to woodland from the typical dwarf shrub community. For example, it is likely that grazing alone is insufficient to prevent an increase in grassland from ericoid shrub cover; other management options include burning, cutting and selective herbicide application.[43] Yet a paucity of information, small studies and the general heterogeneity of heathland sites can lead to unclear, if not downright contradictory evidence available to those who manage heathlands. Site specific experiments and monitoring remain the best way forward. And the climate is changing.

Lowland heaths in the south are expected to be of medium climate sensitivity.[44] All areas of the UK are projected to get warmer, more so in summer than in winter, with changes in summer mean temperatures greatest in southern England. Overall, annual rainfall may not change very much, but average winter rainfall will increase and average summer rainfall decrease. Warmer temperatures in summer will increase evapotranspiration, enhancing summer drought conditions. There is also likely to be an increase in the proportion of rain falling in heavy storm events. It is probable that continuing increase in global greenhouse gas emissions will drive further change.

Possible changes to lowland heaths include:

- Dwarf shrub may become less dominant as other more competitive plants become established
- Increased nutrient cycling and insect herbivory could cause grasses to become dominant over dwarf shrubs leading to a shift from dry heath to acid grassland
- Increased length of growing season and activity period of key species may result in a reduced window of opportunity to conduct winter management, such as controlled burning and cutting
- Hotter summers, causing increased evapotranspiration leading to drying sites and a change in the balance of species, particularly on wet heathlands
- Increased risk of wildfire including damage to peat soils
- An increase in unmanaged access could lead to more erosion on access routes and irreversible damage to vegetation
- Increased disturbance of ground nesting birds
- Climate change may impact on the amount of carbon stored or emitted
- Rare species (e.g. Dartford Warbler) may benefit from a warmer climate
- Bracken could have a competitive advantage over slower growing heather
- Drought may alter community composition and loss of wet heath threatening wet heathland species such as *E. tetralix*
- Wetter winters may increase surface runoff and nitrogen deposition

---

[42] J.C. Underhill-Day (2005) *A Literature Review of Urban Effects on Lowland Heaths and Their Wildlife*, English Nature Research Report, 623. English Nature, Peterborough.
[43] A.C. Newton, G.B. Stewart, G. .Myers, A. Diaz, S. Lake and J.M. Bullock (2009) Impacts of grazing on lowland heathland in north-west Europe. *Biological Conservation* 142 (5), 935-947.
[44] Natural England and RSPB(2014) Climate Change Adaptation Manual <https://publications.naturalengland.org.uk/publication/5629923804839936> Accessed May 2020.

- The addition of Nitrogen may increase the sensitivity of heather to drought, frost, and heather beetle outbreaks
- Key animal species currently at the northern end of their range (e.g. the smooth snake and sand lizard) may benefit as the climate becomes milder

7.5 Hartland Moor SSSI, Dorset

Hartland Moor[45] is adjacent to, and drains into, Poole Harbour (south side), near to Wareham in Dorset. It is all below 35m OD and includes prehistoric remains suggesting long-term human activity. Poor, infertile soils are formed from the Bagshot sands and clays.

The 'Moor' (it is better described as a heath) comprises a spectacular mosaic of open lowland heathland and bogs, which supports insects, reptiles and birds that are rarely found elsewhere alongside many specialised plants and animals. The area comprises both dry heath and valley mire and is managed by Natural England and the National Trust. Its significance is demonstrated by its various designations as an Area of Outstanding Natural Beauty, National Nature Reserve and Site of Special Scientific Interest. It is managed jointly with the adjacent Stoborough Heath National Nature Reserve.

Typical plants on the SSSI comprise ling, cross-leaved heath (*Erica tetralix*), bell heather (*E. cinera*), bog asphodel (*Narthecium ossifragum),* white beak sedge (*Rhynchospora alba*) and western gorse (*U. gallii*). Rare plants include Dorset heath (*E. ciliaris*), marsh gentian (*Gentiana pneumonanthe*) and bog orchid (*Hammarbya paludosa*). Rare heathland insects include heath grasshoppers (*Chorthippus vagans*) and large marsh grasshoppers (*Stethophyma grossum*) with bird species Dartford warbler (*Sylvia undata*), hobby (*Falco subbute*), meadow pipit (*Anthus pratensis*), stonechat (*Saxicola rubicola*), nightjar (*Caprimulgus europaeus*) and hen harrier (*Circus cyaneus*).

The valley mire system (which is roughly Y-shaped) runs west to east, forking to the west where vegetational differences are apparent between the two arms. The northern arm is acidic with a community dominated by the bog mosses *Sphagnum* spp., including an abundance of the rare *S. pulchrum*, and by pools containing the scarce bog sedge (*Carex limosa*) and the rare bog orchid (*Hammarbya paludosa*).

The southern arm, by contrast, is relatively base rich with vegetation dominated by tussocky black bog-rush (*Schoenus nigricans*). At the confluence of these streams are intermediate base-status conditions where broad-leaved cotton-grass (*Eriophorum latifolium*) may be found. Eastwards the bog becomes more acidic with a rich flora that includes the rare brown beak-sedge (*Rhynchospora fusca*) and marsh gentian (*Gentiana pneumonanthe*). To the south are areas of reed (*Phragmites australis*) and there is a carr community with common sallow (*Salix cinerea*) and bog myrtle (*Myrica gale*). All six species of British reptile are found, including sand lizard and smooth snake.[46]

---

[45]   Natural England (2008). Dorset's National Nature Reserves <https://www.gov.uk/government/publications/dorsets-national-nature-reserves/dorsets-national-nature-reserves#hartland-moor> Accessed May 2020.
[46]   Hartland Moor SSSI designation (n.d.) <https://designatedsites.naturalengland.org.uk/PDFsForWeb/Citation/1002993.pdf> Accessed May 2020.

Where productive agricultural grassland replaced heathland, several means of restoration have been tried.[47] In 1989 a large-scale experiment was begun on Hartland Moor to reverse the impacts of land reclamation for agriculture as grazing land. Treatments imposed were: (i) natural regeneration; (ii) herbicide application to facilitate regeneration; (iii) cultivation and application of seed-rich heathland vegetation; (iv) soil removal and incorporation of heathland topsoil; and (v) heathland translocation.[48]

Nitrogen compounds introduced to agricultural grassland are leached out of reclaimed heathland over a few years although they may be replenished by atmospheric input.[49] However, residual phosphate in soils may remain available to plants for decades and is not easily removed. One approach that was tried is to render the Phosphorous inert within the soil using treatments such as sulphurous soil amendments and burning.[50]

After 17 years, the pH of the unamended agricultural soils remained significantly higher than the amended areas. Natural re-vegetation was slow due to a limited supply of seed (the result was an acid grassland community), and overall the different treatments resulted in variable vegetation communities. The most effective practical and long-term sustainable method of heathland restoration was the application of seed-bearing vegetation as a part of routine management, although the methods employed require further development.

7.6 The development of Exmoor Forest for agriculture

Ecologically speaking, lowland heath grades into upland moorland without enclosure. In practice this distinction may be spatially defined at the upward limit of agricultural enclosure and improvement, beyond which moorland remains. On Exmoor, enormous efforts in pursuit of economic development were made but were of limited success due to the nature of the land. In contrast to Thomas Hardy's gloomy heath, the poet Robert Southey (1774-1843) could write of the evocative landscape of Exmoor:

'The inland walks are striking: the hills dark, and dells woody and watery, winding up them in ways of sequestered coolness'[51]

Exmoor does present a range of attractive landscapes and (provided it is not raining or misty) appears as open, wild and airy. Just like lowland heaths, humans have been at work shaping the landscape of Exmoor for millennia.

---

[47] R.F. Pywell, W.R. Meek, N.R. Webb, P.D. Putwain ,and J.M. Bullock (2011) Long-term heathland restoration on former grassland: The results of a 17-year experiment. *Biological Conservation* 144 1602-1609.
[48] UK Biodiversity Action Plan Priority Habitat Descriptions Lowland Dry Acid Grassland (2008) <http://jncc.defra. gov.uk/pdf/UKBAP_BAPHabitats-26-LowlandDryAcidGrass.pdf > Accessed May 2020.
[49] M. Herrmann, J. Pust and R. Pott (2005) Leaching of nitrate and ammonium in heathland and forest ecosystems in Northwest Germany under the influence of enhanced nitrogen deposition. *Plant and Soil* 273: 129. https://doi.org/10.1007/s11104-004-7246-x
[50] M. Tibbett and A. Diaz (2005). Are Sulfurous Soil Amendments (S0, Fe(II)SO4, Fe(III)SO4) an Effective Tool in the Restoration of Heathland and Acidic Grassland after Four Decades of Rock Phosphate Fertilisation? *Restoration Ecology* 13 (1), 83-9. https://doi.org/10.1111/j.1526-100X.2005.00010.x
[51] Exmoor National Park (2020) Robert Southey. <http://www.exmoor-nationalpark.gov.uk/Whats-Special/culture/ literary-links/robert-southey> Accessed May 2020.

Figure 7.6. Exmoor landscape near to Porlock, Somerset (the author).

Figure 7.6 shows such a landscape. While there is a proliferation of *Juncus* near to the stream set in a deep gulley, the hills on either side are moorland in rough grazing. The sale of the royal hunting forest after 1818[52] led to Exmoor becoming what amounts to a heavily invested experiment in land improvement.

Clear evidence for human presence dates from the Mesolithic; Neolithic activity may only inferred, because contexts are often lacking and dating poor or absent. Bronze Age farmers left cairns, barrows and cist burials and limited pottery finds; there is evidence of settlement from the Iron Age.[53] Later medieval field systems may have their origins in the Bronze age.[54]

As with the New Forest and the Dorset heaths, human activity cleared the forests leading to the development of characteristic vegetation. Upland heather moorland is of high conservation value across Britain and can be an important source of winter fodder for hill sheep as well as serving to maintain grouse populations. However, over the last 40 years, the

---

[52] Exmoor National Park (2020) The Historic Monument Record for Exmoor National Park <https://www.exmoorher. co.uk/hbsmr-web/record.aspx?UID=MEM22088-Exmoor-Royal-Forest> Accessed May 2020.
[53] H. Riley and R. Wilson-North (2001) *The Field Archaeology of Exmoor*. English Heritage, Swindon, Chapter 2.
[54] H. Riley and R. Wilson-North (2001) *The Field Archaeology of Exmoor*. English Heritage, Swindon, Chapter 4.

amount of heather moorland has been declining due to the expansion of plantation forestry and increasing land improvement, as well as other causes.[55]

Development of upland moorland in England for agriculture has been sustained, although there have been conflicts between the economic interests of agriculture and that of grouse shooting. For example in the Peak District, reclamation of moorland and the loss (specifically) of wastes and common grazing land was progressive between in the late eighteenth and early nineteenth centuries.[56] The creation of grouse-shooting estates was for the benefit of elites, yet a lack of access for ramblers onto moorland left for shooting led to the mass trespass of Kinder Scout, in Derbyshire, in 1932, an event that would lead to the access to the countryside legislation and the establishment of National Parks in England and Wales, and more recently in Scotland[57]. They may be separated by some 330km via the motorway system, but Exmoor National Park ticks many of the same boxes as the peaty moorland areas of the Peak District National Park.

Settlement and agriculture on Exmoor can be demonstrated from the Bronze Age but may go back further to the Neolithic.[58] Infield-outfield agriculture was practiced before the 19th century. In this system of upland agriculture, fields close to a farmstead were intensively managed as the' infield', whereas further away, the 'outfield' comprised rough moorland grazing and intakes of land from the pre-existing commons. This would be largely rough grazing or was only periodically cultivated and left fallow to recover its fertility. To take one example, it is likely that this system was practiced at Clogg's farm, Hawksridge on Exmoor prior to improvement.[59]

Published in 1869, at the height of Victorian romanticism, R. D. Blackmore's Lorna Doone: A Romance of Exmoor is set in the late 17th century when the Forest Laws applied. The landscape plays a prominent role in the novel, with one character perishing in a mire.[60] Legally speaking, 'Exmoor Forest' had been a royal (hunting) forest from some time earlier, probably during the reign of Henry II (1154-89) or possibly earlier still.[61] Henry was a great proponent of royal forests, and between this time and the Charter of the Forests in 1217 it is likely that areas designated under the Forest Laws reached their maximum.[62]

Ironically, Exmoor contained relatively few areas of woodland or forest in the non-legal sense, for it had been leased by the Crown to others, including the Walpole family prior to the sale.[63] The main source of income was sheep pasturage, but there was also income from turf (peat) cutting and quarrying. The dominant land cover would have been moorland.

[55] H. Armstrong (1991) Britain's Declining Heather Moorlands. *Outlook on Agriculture* 20 (2), 103-107. https://doi.org/10.1177/003072709102000208
[56] D. Hey (2014) *A History of the Peak District Moors*. Pen and Sword, Barnsley.
[57] National Parks UK (2018) History of the National Parks <https://www.nationalparks.uk/students/whatisanationalpark/history> Accessed September 2018.
[58] H. Riley and R. Wilson-North (2001) *The Field Archaeology of Exmoor*. English Heritage, Swindon, 20-21.
[59] H. Riley and R. Wilson-North (2001) *The Field Archaeology of Exmoor*. English Heritage, Swindon, 132.
[60] R.D. Blackmore (1869) *Lorna Doone*. Sampson Low, Son and Marston, London.
[61] H.L. Bourne (1968) *A little History of Exmoor*. J.M. Dent and sons, London, 56.
[62] H. Cook (2018) *New Forest: The Forging of a Landscape*. Windgather Press, Oxford, Chapter 5.
[63] R.A. Burton (1989) *The Heritage of Exmoor*. Maslande, Tiverton, 52-55.

The term 'disafforestation' refers to the removal of privileges of a forest to the state of ordinary land, making it legally like other land. What followed on Exmoor was remarkable investment in land hitherto considered marginal for agriculture, although its essential character was only partially altered. In the following century, specifically in 1954, the area was designated Exmoor National Park under the 1949 National Parks and Access to the Countryside Act.[64] At 694km² in area it is relatively small in comparison to most other Parks.[65]

Legal disafforestation occurred in 1815.[66] The forests were originally set aside for hunting, largely red, roe and fallow deer and wild boar (the latter likely became extinct by the 16th century). The Commissioners of HM Woods, Forests, and Land Revenues organised the sale in a process that had been instigated in order to pay for the Napoleonic Wars[67]. John Knight, from a family of midlands ironmasters, was able to purchase the newly disafforested land in 1818.[68] He proved to be the highest bidder and subsequently also purchased some neighbouring areas to enlarge his estate. Exmoor became a focus, not for outmoded Forest Laws or romantic novels, but for agricultural improvement.

Knight also obtained the mineral rights, but these were not regarded as very valuable.[69] His position as an ironmaster, however, likely did provide the incentive to acquire the former Royal Forest; he had planned large-scale exploitation of iron ore. Knight also attempted copper mining, without much success. Together with his son Fredrick and Robert Smith, the latter's steward, fame was ultimately earned not from mining but for land reclamation. Knight's objectives were especially ambitious because there was little existing infrastructure across Exmoor.

John Knight first set about constructing roads and enclosing fields with characteristic strong stone walls.[70] His purpose was to attempt large-scale 'demesne farming'[71]. He also set about enclosing his area with almost 30 miles (48km) of walls.[72] Knight's major improvements comprised the construction of boundary walls, construction of roads (already helped by Crown Commissioners who marked out the roads)[73] and the cutting of open drains to facilitate drainage and remove boggy areas. Some 1,000ha of Exmoor was improved for agriculture by paring off 5 to 7.5cm of turf, burning the pared-off turf, spreading the ash to release nutrients, liming (to raise the pH) and then ploughing the land. John Knight took up residence in Simonsbath House, evidently he had grand plans for its setting. Just south of Simonsbath

---

[64] A Brief history of the Exmoor National Park Authority (n.d). http://www.exmoor-nationalpark.gov.uk/?a=513519 Accessed October 2018.
[65] National Parks (2018) <http://www.nationalparks.gov.uk/students/whatisanationalpark/factsandfigures> Accessed May 2020.
[66] Exmoor National Park (2020) The Historic Monument Record for Exmoor National Park <https://www.exmoorher.co.uk/hbsmr-web/record.aspx?UID=MEM22088-Exmoor-Royal-Forest> Accessed May 2020.
[67] The Regency Redingote (2019) <https://regencyredingote.wordpress.com/2018/07/20/regency-bicentennial-the-disafforestation-of-exmoor-forest/> Accessed May 2020.
[68] R.A. Burton (1989) *The Heritage of Exmoor.* Maslande, Tiverton, 61; Exmoor National Park (2020) The Historic Monument Record for Exmoor National Park <https://www.exmoorher.co.uk/hbsmr-web/record.aspx?UID=MEM22088-Exmoor-Royal-Forest> Accessed May 2020.
[69] R.A. Burton (1989) *The Heritage of Exmoor.* Maslande, Tiverton, 60.
[70] H. Riley and R. Wilson-North (2001) *The Field Archaeology of Exmoor.* English Heritage, Swindon, 138-142.
[71] L. Curtis (1971). *Soils of Exmoor Forest* Soil Survey of England and Wales Spec. Survey no 5. Harpenden, 3
[72] Exmoor National Park (2020) The Historic Monument Record for Exmoor National Park <https://www.exmoorher.co.uk/hbsmr-web/record.aspx?UID=MEM22088-Exmoor-Royal-Forest> Accessed May 2020.
[73] H.L. Bourne (1968) *A little History of Exmoor.* J.M. Dent and sons, London, 119-121.

a deer park was created and there were beech plantations on nearby Birchcleave .[74] An industrialist had acquired many accoutrements of the English landed gentry.

John Knight died in 1850, although his son, Frederick Winn Knight MP had taken over management of the estate in 1841. Frederick's policy was to discontinue the practice of demesne farming, preferring to build new 'model farm' type farmsteads and to rent out allotments to new tenant farmers. During this period, the number of farms increased to 15, comprising about 2,000ha of improved land. This phase of land-improving tenant farmers largely gave Exmoor its present character.[75] Frederick mined iron ore later in the 19th century; iron mining was already established in the adjacent Brendon Hills. Lime, required to reduce soil acidity, was imported from South Wales or from further west where limestone was available. The remains of limekilns today can be seen both along the coast[76] as well as across central Exmoor.[77]

Figure 7.7. Exmoor enclosures with woods and moorlands in the distance, near Selworthy (the author).

---

[74] H. Riley and R. Wilson-North (2001) *The Field Archaeology of Exmoor*. English Heritage, Swindon, 138-142; R.A. Burton (1989) *The Heritage of Exmoor*. Maslande, Tiverton, 70.
[75] L. Curtis (1971) *Soils of Exmoor Forest* Soil Survey of England and Wales Spec. Survey no 5. Harpenden, 3.
[76] See Chapter 4, this volume.
[77] H. Riley and R. Wilson-North (2001) *The Field Archaeology of Exmoor*. English Heritage, Swindon, 147-152.

Figure 7.8. Exmoor enclosure wall with laid hedge (the author).

Frederick created rectangular farmsteads and enclosed large fields of some 50 acres (20 hectares) as shown in Figure 7.7. Substantial earth banks were faced with stone on each side to a height of 4 feet (1.3m) and topped with turf, making the barrier six feet (1.8m) high. A double row of beech was then planted upon the crown with wattle fencing protection on either side to make them sheep-proof, although this was later substituted by wire (Figure 7.8).[78] Initially it was hoped that cereal crops would be grown, something rather ambitious given the topography and climate. Instead, less profitable ranch farming of sheep was developed, with cottages and sheepfolds constructed for that purpose.

Robert Smith of Emmet's Grange farm (a model farm *c.* 1844) was land agent or steward to Frederick Knight from 1848. Smith (a native of Lincolnshire) was evidently impressed by the irrigated meadows, presumably dating from the time of John Knight, or maybe earlier.[79] Smith could write:

> Nothing could exceed my admiration of the 'water-meadows' in early spring, a period (in the east) when I had been wont to value a blade of grass as a rare production. To

[78] M. Werret (2011) Hedgerows: the story of our landscape. *Exmoor Magazine* February 21st 2011. <https://www.exmoormagazine.co.uk/hedgerows-the-story-of-our-landscape/> Accessed May 2020.

[79] H. Cook and T. Williamson (2007) The Later History of Water Meadows in H. Cook and T. Williamson (eds) *Water Meadows.* Windgather Press, Macclesfield, 57.

see the Exmoor Ewes, with their early lambs feeding upon the verdant meadow, and secondly, the green meadow at such an inclement season.[80]

These water meadows are of 'catchwork' design (see Chapter 4) and typically laid out downslope 6-12m producing 2-2.5 ha sections. There were some 14 locations with parallel sub-contour carriers numbering between one and five within the 8,000ha of the parish of Exmoor.[81] Catchworks were fed from springs and streams and are commonly found associated with farmsteads where waste from cowsheds was flushed out to fertilise the catchwork meadows. Irrigation events could be rather more episodic instead of continuing for several days, as with bedwork water meadows.[82]

Naturally, the soils of Exmoor are often very acid unless cultivation practice has raised the pH. Despite the efforts of the Knights and others, gleyed mineral and peat soils, podzols and shallow soils are commonplace alongside brown earths. Vegetation away from improved areas

Figure 7.9. Redundant catchwork system, Halse Farm nr Winsford (the author).

---

[80] R. Smith (1851). Some Account of the Formation of Hill-side Catch-Meadows on Exmoor. *Journal of the Royal Agricultural Society of England* 12, 139-148.
[81] L. Curtis (1971) *Soils of Exmoor Forest* Soil Survey of England and Wales Spec. Survey no 5. Harpenden, Figure 12.
[82] H. Riley and R. Wilson-North (2001) *The Field Archaeology of Exmoor*. English Heritage, Swindon, 138-142.

is characteristic of upland moorland (including *Calluna, Erica*, sedges and rushes, bilberry, cotton grass, *Agrostis-Festuca*, (acidic) grassland and bracken invasion.[83]

Irrigation of heath or moorland (Figure 7.9), here and elsewhere, was a recognised method of changing the vegetation towards grass for grazing and cutting, particularly on acidic soils.[84] Several methods were used that were aimed towards improving upland areas that were economically marginal for agriculture. Typically these are paring, liming, drainage and irrigation. One consequence has been the fragmentation of upland moorland.

The rate of loss of moorland on Exmoor has been estimated by Elizabeth Rowan from 1086ha between 1954 and 1966, to 483ha between 1966 and 1977, 224ha between 1977 and 1988 and 14ha between 1988 and 1995. Across Britain, loss has been similarly dramatic, and although habitat loss has ceased in recent times across Exmoor (probably due to agri-environmental schemes), there remain concerns around the quality of the habitat despite efforts to reverse drainage.[85] The main reasons for the loss of moorland were reclamation to agriculture, followed by afforestation on areas such as the Brendon Heaths, with an area around 500ha in total being planted with conifers in the early 20th century.[86]

7.7 Heaths, moors and their surroundings

Heathlands and moors were originally areas that were unlikely candidates for intensive agricultural development, or where it was tried with limited success. Most notorious in their destruction has been the enclosure movement, eventually sanctioned by Parliament. However, as the Hartland Moor example shows, 'improvement' continued into the 20th century, when fertiliser and other treatments were employed.

Characteristic heathland habitats were replaced with more intensive grass production to increase grazing density. Overall, the post-Second World War situation provided an opportunity for investment into hitherto economically marginal land. Alongside chalk grasslands, wetlands, deciduous forests and other semi-natural habitats many areas of heathland perished.

Even if heaths and moors were shrinking (provided they were not fully enclosed and converted to other uses), it is instructive to examine their relationship with surrounding land uses. Figure 7.10 shows the drovers roads linking the New Forest and the lower Avon valley in the 1790s. These routes were closed in the 1960s when the forest was fully enclosed, with cattle grids constructed on the roads in and out, with the objective of constraining livestock that were otherwise free to graze the forest heaths and wood pasture. Outside the Forest to the west (left side) is agricultural land.

---

[83] L. Curtis (1971) *Soils of Exmoor Forest* Soil Survey of England and Wales Spec. Survey no 5. Harpenden, Chapter 1.

[84] J. Hillman and H. Cook (2016) 'By floating and watering such land as lieth capable thereof': recovering meadow irrigation in Nottinghamshire. *Transactions of the Thoroton Society of Nottinghamshire* 120, 75-94.

[85] Rural Focus (2016) *Exmoor's Moorland: Where Next?* <https://www.exmoorsociety.com/wp-content/uploads/2016/04/Exmoor_Moorland_Report_2016.pdf> Accessed May 2020, 9.

[86] Land Use Consultants (2004) Moorlands at a Crossroads: the State of the Moorlands of Exmoor, 2004. <https://www.exmoorsociety.com/wp-content/uploads/2016/07/Moorlands-of-Exmoor-Report-2004.pdf> Accessed May 2020, 16.

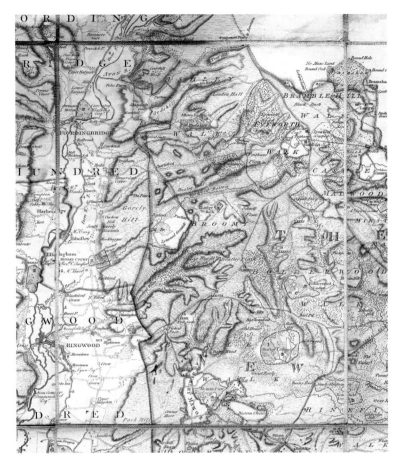

Figure 7.10. Extract from Thomas Milne's map of the New Forest (1791) (public domain).

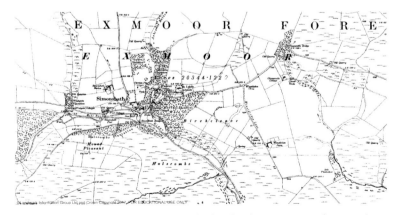

Figure 7.11. Simonsbath showing the key land uses. OS 1904/1906 scale mapping. The width of the extract is about 3.5km (© Crown Copyright and Landmark Information Group Limited 2023).

Figure 7.11 shows the relationship between the open unenclosed moor and surrounding agricultural land around Simonsbath, Somerset on Exmoor.

Here there is an example of agricultural land improvement dating largely from the 19th century that constrains the extent of the moorland rough grazing that is shown around Halscombe on the south side and in the northeast of the extract. Today around Simonsbath, the lower slopes are forested, including the Halscombe Plantation. Encroachment on the moorland through enclosure is evident at Mound Pleasant as elsewhere on the map.

7.8 Conclusions: Heaths and moors

Heath and moor are important landscape and semi-natural ecosystems, famed for their conservation value and well-represented in English literature. Geological factors that produce gravel and sandy base-poor soils, or base-poor rocks, acidic soils, heath, and moor were seen as an impediment to agriculture, and hence these landscapes have been encroached upon and fragmented by enclosure.

The *structures* of the landscapes are the result of vegetation management more than anything else. Moors and heaths are defined not only by their vegetation communities but also growth habit resulting from such activity as grazing, vegetation burning and cutting. Structures are changed by discontinuing these processes, for example where scrub has been allowed to develop, where wood pasture is maintained or where afforestation is deliberately undertaken.

The *functions* within the landscape are considerable. Historically heathlands were farmed in prehistory, an activity that clearly failed in comparison to surrounding areas that developed more intensive agriculture, industry or settlement. That failure is a function of the exhaustion of otherwise poor soils. Agriculture continued, largely as heaths and moors providing unimproved grazing, but also other products useful for fuel and animal husbandry. Other activities include gravel extraction and turf cutting for fuel, which affects not only present topography but also soil type, further depleting already nutrient poor soils. Subsequently moorland particularly became a focus for game bird hunting, suggesting appropriation by elites.

The *value* has clearly changed, from largely marginal areas supporting limited agricultural activity (direct economic value) or hunting by elite groups to areas of conservation, landscape and recreational value. Certain areas may also be targeted for afforestation, particularly for conifers, in an effort to restore economic value given sufficient time. Actually, when the New Forest and Exmoor are considered, these National Parks uphold new values of conservation and amenity supplanting older imperatives of direct economic exploitation.

The *scale* of description varies, but the fragmentation of the Dorset heaths has been dramatic, producing isolated smaller areas, while the Exmoor example of moorland encroachment and development for agriculture clearly shows activity at a more or less regional scale that shrinks surviving habitat areas.

Change applies not only to land use change due to fragmentation but also to restoration where heathland has been 'improved' for agricultural purposes or forestry. Climate change is a matter of concern for these semi-natural habitats and will no doubt determine management prescriptions in the future. Evidently the value has changed in recent decades through economic gain from former marginal land to landscape, biodiversity and habitat creation. Ironically, whatever the land use over moorland and heath areas, present, past or future, it is always determined by human intervention.

Chapter 8

Woods and forests

8.1 What are woods and forests for?

Woods and forests are integral to the Wessex landscape, and once covered all of it to some degree. Woodlands naturally have associated industries, including charcoal making (used in metal ore smelting), fuel supply and wood for construction; the present chapter will deal with woodlands in a wider rural context. Control of wooded areas was crystallised under the manorial system, including within Royal Forests, for example, the New Forest (NCA 131) and Exmoor (NCA 145).[1]

In chapter 3 the term 'forest' was described as a legal entity, where elites held sway over hunting and where management on the ground very much had this in mind. This should be contrasted with modern 'forestry', that is typically a plantation monoculture of a tree crop (typically conifers), a term that in effect replaced 'silviculture' as used in the 17th century. Another essential distinction is with 'woodland management', which requires a high skill set in managing areas including coppice, pollards and high 'standard trees'. There is good evidence that woodland management is prehistoric in origin. Indeed, many deciduous trees can be readily coppiced including alder, ash, beech birch, field maple, hazel, hornbeam, oak, willow, small-leaved lime, and sweet chestnut. Most conifers do not coppice as they will not grow back after cutting, an exception is yew.[2] This topic will be returned to in chapter 10.3.

Hedges and associated features will also be discussed. Apart from manors, institutions influencing modern forestry and woodland management have included the Crown, the Office of Woods, and the succeeding Forestry Commission. Aside from measurable economic benefits, woodland and forests are of deep cultural significance, they provide a rage of habitats for birds, plants and animals and have great habitat and amenity value. Tree growth also has the potential for carbon sequestration from the atmosphere.

Yet, understanding the role of woodland in landscape evolution is not always clear-cut with uncertainty around the influence of early humans and large herbivores, or ecological preference for trees and other plant species dictated by soil, topography and ambient climate. Humans and forests go together, and in terms of the functionality of wooded areas, we must consider both market and non-market values.[3] For present purposes, we may draw distinctions:

- Natural forests regarded as habitats for plants and animals
- Natural forests understood as a resource base for hunter-gather societies

---

[1] Natural England (2014) National Character Area Profiles: data for local decision making https://www.gov.uk/government/publications/national-character-area-profiles-data-for-local-decision-making/national-character-area-profiles> Accessed November 2018.
[2] http://www.pondheadconservation.org.uk/wp-content/uploads/2018/12/COPPICING-download.-PCT-12.181.pdf
[3] H. Cook (2018) *New Forest, the Forging of a Landscape*. Windgather Press, Oxford, Chapter 2.

- Semi-natural forests or woodland as a resource base for agricultural societies (e.g. wood products used for burning, timber for construction)
- Semi-natural forests, chases, 'parklands', or forests for hunting by elites
- Plantations and planted trees for amenity and indicating prestige
- Forests and woodland for amenity, spiritual, health or recreational value
- Plantations and (modern) forestry for production purposes

The first two categories apply to the post-glacial 'wildwood', the third, agricultural phase (largely) from the Neolithic Age. The next two concern appropriation of woodland and forest by elites, especially hunting by kings and nobility, known from Anglo-Saxon times onwards. The last theme applies to industrialising societies concerned with construction, ship building, mining, and other industrial purposes. During this later time, institutions such as the Office of Woods and later the Forestry Commission came to represent state intervention, replacing Crown interests in royal forests with their raft of officials that sat on top of manorial arrangements. Woodland management outside royal forests experienced (historically speaking) the manor as the only regulating institution.

The nature and dynamics of the wooded landscape remain a matter of considerable debate. Key questions include: How continuous was the post-glacial wildwood? Was it (in structure) more like a dense plantation or more open like a savannah? What were the mechanisms for opening-up the canopy in order to produce clearings, and to what extend did the process involve large herbivores, humans, wind, lightning strikes or fire?[4] How might the forest recover from the creation of open glades?

The transition to agriculture implies a destruction of the wildwood, although gradual by modern standards. In many societies, hunting has been shown to continue even when most of the population would be engaged in agriculture. In the New Forest area, this transition may have occurred 1500 years later than elsewhere in its environs, notably on the downs. We might assume that agricultural activity was minimal and that this area was merely penetrated by hunting bands, much as it had been since the Mesolithic. However, this must remain an assumption as there is hitherto no clear information on the Neolithic, although the elm decline, now attributed to disease, was once considered to have been selectively harvested for fodder.[5] Forests remained within the Wessex landscape as important sources of fuel, charcoal and construction as well as providing habitats in which to hunt animals.

Structurally, deciduous woodlands comprise four layers:[6]

- Canopy layer: highest trees for example oak, ash, beech, birch
- Underwood layer: small trees and shrubs growing beneath taller timber trees, for example field maple, hawthorn, hazel, holly, rowan, wild cherry, wild service and rhododendron, an invasive species (Figure 8.1)

[4] O. Rackham (2006) *Woodlands*. Collins, London, 47. While fire has a strong impact on forest ecology and management in many warmer climates, it is unlikely this was ever important in Britain as native deciduous forests seem resistant to burning. This is in contrast with heathland and moorland where burning can maintain an open landscape (see Chapter 7, this volume).

[5] O. Rackham (2006) *Woodlands*. Collins, London, 92-4.

[6] Woodland Structure (n.d) <http://www.countrysideinfo.co.uk/woodland_manage/struct.htm> Accessed November 2018.

- Field layer: ferns, grasses, sedges, herbs (including vernal flowering plants of the 'groundflora', bluebells and anemone)
- Ground layer: mosses, ivy, lichens and fungi

In general, the amount of light available will condition the diversity and density of each layer. Other factors influencing the structure of a native deciduous woodland will be water, nutrient availability, and grazing pressures (including deer). This contrasts with coniferous plantations where little or no light reaches the forest floor, permitting lower layer growth within the canopy. The rural landscape that developed since the Neolithic is more complicated as it is the outcome of human-impact as well as physical environmental and ecological considerations. It contains a mosaic of fields, meadows, woodland, and plantations (commonly comprising conifers grown on economically marginal land). Frequently separating each unit are hedges.

Hedges also link landscape units, particularly woods and forests. In age they vary from prehistoric through to the last 300 years, many being the result of enclosure of a more open landscape. Some hedges are derived from ancient woodlands and will contain indicator species as well as providing a corridor for species dispersion. For example, it may take 100 years for bluebells to colonise a secondary woodland from an adjacent area where they are

Figure 8.1. Small oak woodland with coppiced hazel understory, near Stockbridge Hampshire, that has the structure of an ancient coppice (the author)

established.[7] Whatever the case, some hedges were deliberately planted during enclosure or otherwise they define (or defined) a frontier limit to cultivation left by woodland assarting.[8]

A correlation between age and number of species has been proposed.[9] Although conventional ecological wisdom points to woody plant species within hedgerows increasing in number over time (one extra per hundred years in each 30m of hedgerow),[10] it is likely hedge planting involved any species available to workmen undertaking the enclosure. That being the case, simple correlations might be misleading.

8.2 What is 'ancient'?

Another convention that is presently under scrutiny is the concept of 'ancient woodland'. The term is attractive because it potentially links to notions of the wildwood, indeed it has been presumed that certain woods may be descended from the wildwood.[11] The convention is that to be 'ancient', a woodland existed by AD 1600 (for England and Wales) demonstrated by provable sources. The evidence is derived from documents, maps, place names, archaeology (such as wood-banks delineating the edge of a wood), woodland structure (including giant standing 'standard trees') and coppice stools. The pollen record also helps to establish local environmental history.[12]

Importantly, specific plant species are considered to survive from the wildwood.[13] Most ancient woods are affected by humans, they may have been managed for timber and other products, but they have long displayed woodland cover.

The presence of a particular species may be deemed to indicate an 'ancient wood'. Over 200 plant species are used as ancient woodland indicators with key characteristics being that they are:

- Slow at colonising a new habitat, so less likely to be found in new woodland (species with slow powers of dispersal)
- In need of conditions created by stable and continuous woodland cover
- Less likely to survive outside woodland where they could be more exposed
- Easily identifiable in survey

Other factors also influence the number of species found in woodlands. These include wood size, site management and structure. Indicators are one clue to a wood's history. Some more commonly used indicators in the UK are:[14]

7   I. Cummings (2020) pers. comm.
8   <https://historicengland.org.uk/advice/technical-advice/parks-gardens-and-landscapes/hedges/>   Accessed April 2020.
9   E. Pollard, M.D. Hooper and N.W. Moore (1974) *Hedges*. Collins, London.
10   M. Hooper (2014) The History of Hooper's Hedgerow Hypothesis. British Naturalists Association <http://www.bna-naturalists.org/hooper.pdf> Accessed November 2018.
11   O. Rackham (1986) *A History of the Countryside*, J.M. Dent, London, 67.
12   A.D. Brown (2010) Pollen analysis and planted ancient woodland restoration strategies: a case study from the Wentwood, southeast Wales, UK. *Vegetation History and Archaeobotany* 19 (2),79-90. DOI: 10.1007/s00334-009-0227-5
13   O. Rackham (2006) *Woodlands*. Collins, London, 23-4.
14   Woodland Trust (2016) https://www.woodlandtrust.org.uk/blog/2016/04/plant-ancient-woodland/ Accessed January 2024.

- Wood anemone (*Anemone nemorosa*)
- Ramsons (*Allium ursinum*)
- Wood spurge (*Euphorbia amygdaloides*)
- Bluebells (*Hyacinthoides non-scripta*)
- Wood sorrel (*Oxalis acetosella*)
- Wild service (*Sorbus torminalis*)
- Small leaved lime (*Tilia cordata*)
- Guelder rose (*Viburnum opulus*)

Any other extant woodland may be the result of secondary growth or result from deliberate plantation. 'Secondary woodland' can be scrubby to mature in structure, but it is developed on land that at some time has not been woodland. Secondary woodland tends to be species poor: the main tree species are pioneer trees (such as oak, birch, hawthorn, or ash).[15] To be even more specific, we may differentiate two kinds of ancient woodland:

Ancient semi-natural woods have developed naturally. This means they *may* have existed since the end of the last (Devensian) glaciation. Alternatively, such woodlands represent a re-colonisation of land that was previously cleared, but this occurred hundreds or thousands of years ago. Neither does it follow that 'indicator species' re-colonised with the new woodland. Ancient woodland cannot necessarily be assumed to be a relic of original post-glacial woodland, and the `ancient woodland indicator` vascular plants do occur in secondary woodland.[16] Where a woodland is derived from the immediate post-glacial woodland, it is sometimes termed **`primary woodland`**. **Plantations on ancient woodland sites** occur where woods were felled and planted with non-native trees, often conifers. Large areas of ancient woodland were replanted during the 20th century to make the UK more self-sufficient in timber. The effects of felling, drainage and replanting and the impact of dense shade (generally by closely planted conifers) threatens the survival of the fragile ancient woodland ecosystem. Restoration may enable a return of native species.

It is argued that remaining ancient woodland is irreplaceable because woods planted today will not become ancient woods in 400 years' time. This is because the soils on which they have developed have been modified by modern agriculture or industry, and the fragmentation of natural habitats hampers species migration, natural movements and interactions. In any case, characteristic species of ancient woods disperse slowly and do not easily colonise new areas.[17]

However, given time, the outcome is floristically diverse and dominated by 'vernal' (spring flowering) species, which complete their growth cycles before the coppice canopy reduces light penetration to the woodland floor. Such are bluebells (*Hyacinthoides non-scripta*) and wood anemone (*Anemone nemorosa*), originally 'woodland edge plants' that have limited tolerance to direct sunlight. Centuries – or millenia – of woodland cultivation and removal of biomass as wood will work to reduce soil nutrient status (notably of phosphorous),[18] and increase root

[15] O. Rackham (2006) *Woodlands*. Collins, London, 69-70.

[16] S.P. Day (1993) Woodland origin and `ancient woodland indicators': a case-study from Sidings Copse, Oxfordshire, UK. *Holocene* 3 (1), 45-53. https://doi.org/10.1177/095968369300300105

[17] Ancient Woodland (n.d) <www.woodlandtrust.org.uk/visiting-woods/trees-woods-and-wildlife/woodland-habitats/ancient-woodland> Accessed November 2018.

[18] O. Rackham (1975) *Hayley Wood: Its History and Ecology*. Cambridge and Isle of Ely Naturalists' Trust, Cambridge.

competition. The abundance of *Anemone nemorosa* is influenced by soil water stress, suggesting that root competition between ground-flora and trees might be improved by reintroducing coppice, for ancient woodland ground flora may produce spectacular blooms within two years of coppicing.[19] The interplay of soil nutrients, soil water and shade (i.e. light availability and heat energy factors) determines the abundance of ground vegetation.[20]

Further disambiguation is required. For example, medieval plantation of woods has been identified in several locations by Oliver Rackham. These are in the east of England and the practice was evidently encouraged by the monasteries. He cautions, however, that a general rise in timber and wood production in England from private farms occurred later. In part this may be traced to the later agricultural writer Arthur Standish (1552-1615) who proposed the temporary establishment of copses in the corners of fields, which could later be returned to tillage.[21]

Plantations on ancient woodland sites are generally another matter altogether. In the New Forest, governmental concern over a potential shortage of timber (oak and beech) for ship construction for the Royal Navy gave rise to legislation, commencing with the 1698 'Act for the Increase and Preservation of Timber in the New Forest'. This legislation may have regularised earlier modest attempts at hardwood plantation, and certainly allowed for the first phase of large-scale timber production.[22]

These plantations are generally pasture woodland where grazing by livestock helps to maintain the characteristic appearance and species diversity. They are characterised by towering groves of beech (*Fagus sylvatica*), durmast oak or pedunculate oak (*Quercus petrea*), with an under-story of holly (*Ilex* sp.) with an admixture of birch (*Betula* sp.), thorny species and yew (*Taxus baccata*). The woods have a great structural diversity, ranging from trees of closed high canopy forest to open stands with heathy or grassy lawns and glades. However, many have been incorporated into inclosures[23] and replaced by plantations, although some old growth beech and oak stands remain as 'pre-inclosed woodland' within inclosures. They have high biodiversity value.[24] A similar fate in the New Forest may be ascribed to former coppice enclosures that have become incorporated in inclosures, many dating from the 17th and 18th centuries[25]. The 1698 Act furthermore legislated to prevent pollarding to promote the growth of tall timber. It had some success, suggesting that large pollards of beech or oak would predate the Act.[26]

The change to conifer plantation is, however, much more drastic and has occasioned considerable controversy nationwide. Within 'coniferised' forestry plots, it has been shown

---

[19] I.P.F. Cummings and H.F. Cook (1992) Soil water relations in an ancient coppice woodland in G.P. Buckley (ed.) *Ecology and Management in Coppice Woodland*. Chapman Hall, London, Chapter 4; I. Cummings (2020) pers. comm.

[20] P.L. Mitchell (1982) Growth stages and microclimate in coppice and high forest, in G.P. Buckley (ed.) *Ecology and Management in Coppice Woodland*. Chapman Hall, London, Chapter 3.

[21] O. Rackham (2006) *Woodlands*. Collins, London, Chapter 18.

[22] O. Rackham (2006) *Woodlands*. Collins, London, Chapter 18.

[23] The spelling 'inclosure' is merely a term that has come to mean enclosures for growing timber in the Forest.

[24] Forestry Commission (n.d) The Ancient and Ornamental Woodlands <https://www.forestry.gov.uk/pdf/eng-new-forest-b4-a-and-o-plan.pdf/$FILE/eng-new-forest-b4-a-and-o-plan.pdf> Accessed November 2018.

[25] H. Cook (2018) *New Forest, the Forging of a Landscape*. Windgather Press, Oxford, Chapter 6.

[26] <http://www.newforestexplorersguide.co.uk/heritage/history-in-the-landscape/pollard-and-coppice-trees.html> Accessed January 2024.

that the seed bank of the broadleaved woodland ground flora that was replaced may partly survive, although its role in restoration would be limited.[27] For example, where coppicing has ceased or where a broadleaved woodland has been replanted with conifers, the seed bank from the previous management may largely remain viable for about 50 years.[28] Probably most successful for broadleaved woodland restoration, is the translocation of some shade-tolerant elements of the ground flora from a donor wood to a new site, including over 70% of original ancient woodland indicator species.[29] Since 1985 restocking woodland using natural regeneration has been actively encouraged through grant aid for broadleaved woodlands alongside replanting.[30]

8.3 Clearing wood and wastes

To the modern mind, removing established deciduous woodland constitutes sacrilege. This was not always the case. Known from the middle ages, 'assarts' cause characteristic curvilinear boundaries resulting from uneven cuts into a wood.[31] In prehistory, clearance would also have been to produce monuments. Within ancient parks and woodland are grassy compartments, often including pollards, which were open for deer at all times. These are termed 'launds'.[32] They occur in the New Forest and were once common in the Lyndhurst area.[33] A laund is not the same as an assart.

Fritham in the New Forest with its quintessentially Anglo-Saxon place name may have its origins at that time as an assart. It displays the characteristics of this encroachment on original woodland to produce a curvilinear boundary including pollards and coppice trees. It also has characteristics of a medieval vaccary ('cow farm') but cannot be identified in Domesday (1086). The internal sub-divisions of the enclosure at Fritham include a funnel-like feature on its eastern side that may have been for the herding of cattle.[34] In any case, Fritham, being situated in the heads of two valleys, is sheltered and hence more suitable for settlement and agriculture. Most human activity at this time was concentrated in valley settlements centred on arable production; the open forest was probably used for grazing.[35] Assarting of established woodland is likely.

---

[27] R. Ferris and W.E. Simmons (2000) Plant Communities and Soil Seedbanks in Broadleaved-Conifer Mixtures on Ancient Woodland Sites in Lowland Britain Forestry Commission <www.forestry.gov.uk/PDF/FCIN032.pdf/$FILE/FCIN032.pdf> Accessed November 2018.

[28] A.H.F. Brown and S.J. Warr (1992) The effects of changing management on seed banks in ancient coppices in G.P. Buckley (ed.) *Ecology and Management in Coppice Woodland*. Chapman Hall, London, Chapter 8.

[29] P. Buckley, D.R. Helliwell, S. Milne and R. Howell (2017) Twenty-five years on – vegetation succession on a translocated ancient woodland soil at Biggins Wood, Kent, UK. *Forestry* 90 (4), 561–572. https://doi.org/10.1093/forestry/cpx015

[30] B.G. Hibbard (1991) *Forestry Practice*. Forestry Commission Handbook no. 6 <https://www.forestry.gov.uk/pdf/FCHB006.pdf/$file/FCHB006.pdf> Accessed December 2018, 7, 59; <https://www.forestry.gov.uk/ewgs> Accessed December 2018.

[31] Assarting is the clearance of waste (rough grazing) scrub or woodland for more productive agricultural activities.

[32] O. Rackham (1986) *A History of the Countryside*, J.M. Dent, London, 126-127.

[33] W. Page (1911) Lyndhurst in W. Page (ed.) *A History of the County of Hampshire, Volume 4*. Victoria County History, London, 630-634.

[34] S. Davies, L. Walker and L. Coleman (1998) *The New Forest Historical Landscape*. Wessex Archaeology, Salisbury; N. Smith (1999) The Earthwork Remains of Enclosure in the New Forest. *Proceedings of the Hampshire Field Club and Archaeological Society* 54, 21.

[35] Forestry Commission (n.d) The Ancient and Ornamental Woodlands <https://www.forestry.gov.uk/pdf/eng-new-forest-b4-a-and-o-plan.pdf/$FILE/eng-new-forest-b4-a-and-o-plan.pdf> Accessed November 2018.

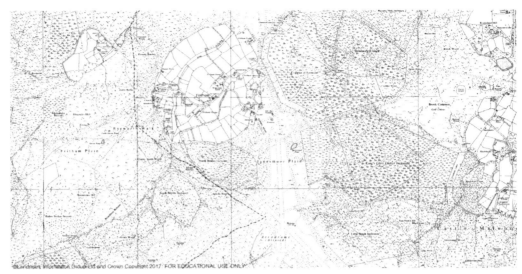

Figure 8.2. Fritham area in the 1960s (© Digimap).

In south-east England, coppice compartments may be termed 'cants'. Thus, lines of pollards that do not necessarily grow in woodland banks are termed 'cant-markers' and were sought out when coppiced woods were to be auctioned.[36] In Epping Forest, pollarding was the manorial court's responsibility and it is argued that this was a hangover from before the imposition of forest law,[37] much as commoners' rights would pre-date the Conquest. However, in the New Forest, pollarding has not always been a common or popular practice. There is no clear evidence for it being the responsibility of adjacent manors at any point, perhaps partly because it is often difficult to disentangle the origin of a wood as being from pollards, coppices, or other forms of trees.[38]

The antiquity of coppices outside Fritham is suggested by their irregular boundaries. North and South Bentley Inclosure boundaries follow the line of old coppice enclosure south of Fritham (Figure 8.2). These had been recorded as coppiced until the early 1700s. After that they were wholly or partly cleared and planted with acorns of sessile oak (*Q. petrea*) because timber was required for the Navy. However, although a holly shrub layer may be commonplace, overall, the biodiversity of such plantations may be limited due to the slow recolonisation of many species of plants following planting.[39]

Assarts could be made for settlement or created to define coppice woodland. The Bentley Inclosures may contain some of the oldest surviving planted oaks in western Europe. By 1700, coppice areas were being planted with acorns alongside the timber following the 1698 Act.[40]

---

[36] O. Rackham (2006) *Woodlands*. Collins, London, 197-8.
[37] M.W. Hanson (1992) The Pollard Trees, in R.W. Hanson (ed.) *Epping Forest - Through the eye of the naturalist*. Essex Field Club, 18.
[38] R. Reeves, P. Roberts (2014) pers. comm. The term *robora* is a medieval Latin terms also used for pollard trees <http://info.sjc.ox.ac.uk/forests/glossary.htm>
[39] C.R. Tubbs (2001) *New Forest, the history, ecology and conservation*. New Forest Ninth Centenary Trust, Lyndhurst, 216.
[40] H. Cook (2018) *New Forest, the Forging of a Landscape*. Windgather Press, Oxford, xxii, Chapter 7.

This Act reintroduced the regulation to remove stock from the forest during the midsummer 'fence month' and during 'winter heyning' when winter forage was low. In the New Forest Winter heyning occurred from 22nd November to 4th May, a period when commonable stock could not be 'depastured', and thus did not compete with deer for the browse. Domestic animals should have been removed when deer were calving in June and to preserve winter fodder for the deer.

8.4 Coppicing and pollarding

The word 'coppice' (sometimes 'copse') likely shares a common Latin origin with the French *couper* 'to cut'. The term refers to a small wood consisting of the understory, small trees grown for periodic cutting close to ground level, or as a verb, to cutting back understory trees to their base so they will shoot again.

After cutting, what remains to shoot again are low stumps termed 'coppice stools.' In most cases, it is essential to exclude grazing animals that would eat the new shoots. Pollarding refers to cutting higher up, the word being related to 'head' (Figure 8.3). Cutting for a pollard will depend in part on the height desired to protect from grazing animals. For example, the browse line at the bottom of the tree canopy is approximately 1.3m for fallow deer, 2.1m for

Figure 8.3. Willow pollards at the Harnham Water Meadows, Salisbury with llama (the author).

cattle, 2.7m - 3.0m for horses.[41] This means a minimum height of 2m is typical; the point of cutting should be readily accessible to a person standing on a cart or using a ladder.

Both operations are best carried out between October and March before the sap rises in the trees. Overall, around 2% of the area of the UK is considered ancient woodland.[42] Coppicing potentially allows this natural deciduous woodland to survive in modified form because of its exploitation for fuel, building wood and other purposes. Periodic cutting greatly extends the life of most deciduous trees (and yew, which is a conifer), so that coppiced stools may be many hundreds of years old. Coppicing remained the most widespread method of woodland management until the mid-19th century; woodsmen used simple hand-tools.

Coppice (or pollards where animals are present) produces wood or small timber products that are easier to handle than are large timber trees. However, it is unlikely that the 'underwood' was selected for particular tree species although in some places, coppices were 'improved'

Figure 8.4. Recently cut coppice with standard trees (the author).

---

[41] H. Read (2000) Veteran trees: A guide to good management. <http://ancienttreeforum.co.uk/wp-content/uploads/2015/03/Veteran-Trees-A-Guide-to-Good-Management-almost-complete.pdf> Accessed January 2024.
[42] Ancient woods (n.d) <https://www.woodlandtrust.org.uk/visiting-woods/trees-woods-and-wildlife/woodland-habitats/ancient-woodland/> Accessed January 2024.

through encouraging selected species by layering, planting and natural regeneration, filling in gaps where old stools died. Unwanted species could be removed.[43]

Across Britain, 'pure coppice' includes only one species, such as sweet chestnut, essentially in a monoculture, whereas mixed coppice may contain hazel, birch, willow, ash, hawthorn and alder among others. Also coppiced are hornbeam, oak, small-leaved lime, beech and field maple.[44]

The typical coppice in lowland is a compact wood 5-50ha, set in farmland, with a sharp boundary, often defined by a ditch.[45] Cutting is generally on a cycle of 8-25 years depending on the required length and thickness of the poles created or the intended industrial purpose. A well-managed coppice wood is organised into compartments (termed 'coupes, or 'cants'), often fenced off to prevent browsing by livestock or deer, each coupe having a different stage of growth. The stools (the bases of individual trees) provide a continuous supply of wood products. As a woodland management technique, repeatedly cutting trees at the base (or 'stool'), allowing the trees to regrow, provides a sustainable supply of timber that does not require replanting after felling. Coppice products from hazel and oak have been identified from the British Mesolithic.[46] Coppicing was likely systematically practiced in the Neolithic landscape when wooden trackways across the Somerset Moors were constructed using coppiced timber.[47]

Standard trees are those left to grow to full height and will eventually be felled for timber. They are typically oak, ash, sycamore and sweet chestnut. The term 'coppice-with-standards' refers to a coppice woodland (or 'copse') including some trees that are allowed to grow to full height, providing timber across more than one coppice rotation, as in the New Forest. Figure 8.4 shows coppice in Dorset with standard trees that remain after felling. In practice, most coppice woodlands incorporate standard trees, and this woodland structure is recorded from the Middle Ages. The wood is therefore in effect a means of manipulating the natural woodland structure so that the canopy layer is thinned to isolated standards and the underwood is available for coppicing. It is therefore significant that both oak and hazel coppice can flourish as the light penetration in improved. Furthermore, regular coppicing increases light penetration to the woodland floor for the ground flora.

8.5 Products and producers

Palaeoenvironmental evidence is consistent with woodland clearance, perhaps for fuel and increased grazing activities, into the period of Roman occupation.[48] During the Roman period, woodlands were doubtless exploited, perhaps as coppice, a move that would have sustained

[43] The Conservation Volunteers (2018) <https://www.conservationhandbooks.com/woodlands/a-brief-history-of-woodlands-in-britain/coppicing/> Accessed November 2018.

[44] Small woodland owners group (2018). What is coppicing? <http://www.coppice.co.uk/> Accessed November 2018.

[45] G.F Peterken (1992) Introduction, in G.P. Buckley (ed.) *Ecology and Management in Coppice Woodland*. Chapman Hall, London, Chapter 1.

[46] R.R. Bishop, M.J. Church and P.A. Rowley-Conwy (2015) Firewood, food and niche construction: the potential role of Mesolithic hunter-gatherers in actively structuring Scotland's woodlands. *Quaternary Science Reviews* 108, 51-75.

[47] The Somerset Historic Environment Record (2018). 10739: Sweet Track, Shapwick Heath. <https://www.somersetheritage.org.uk/record/10739> Accessed November 2018.

[48] M.J. Grant (2005) The palaeoecology of human impact in the New Forest. Unpublished Ph.D. Thesis, University of Southampton, 330-331.

resources. It is likely that the Romans introduced the planting of sweet chestnut for coppicing. It probably took around 23ha of coppiced woodland to supply the hypocaust for a Roman villa.[49] There is no evidence for timber exploitation (as opposed to coppice) in the Roman period, but this does not prove that it did not occur.

Coppiced woods provided poles cut from the understory, timber obtained from the standard trees and firewood often sold in bundles as faggots. Other products included wattle (for the 'wattle and daub' of medieval houses), basketry (mainly from osier or 'withy' beds), thatching spars, hurdles, tool handles, charcoal, oak bark for tanning, besom brooms and hop poles. Recently pulpwood and chestnut paling has been harvested and the system of coppice may be employed to grow energy crops. There is still demand for firewood, and small markets remain for hazel coppice cut for such things as thatching spars, hurdles (used as fences), bean poles and hedging stakes. The field (lowest) layer provided occasional grazing for domestic animals, who were only sometimes allowed into the wood compartment, and only after four to seven years growth, to prevent them eating young coppice shoots. Otherwise, animals were not usually allowed into the coppice. Animal exclusion, always a challenge in woodland management, is today greatly assisted using modern fencing.

Coppices have been maintained by woodsmen; their skills included cutting and stacking or preparing timber products for the market and many have been skilled in making objects from wood. Today there is a problem maintaining demand for coppice products. Coppice is, however, popular as a conservation technique for the benefits it offers to wildlife and to perennial spring flowering flora, such as anemone and bluebells, that grow beneath the tree canopy, for over a coppice cycle, a range of shading affects the woodland floor.

The work of a woodsman sounds romantic, but like any land-based industry in the past, it was a hard life. Its importance is undeniable though, as before the Industrial Revolution, woods were the single most significant source of raw materials. Nearly every nearly every trade and craft relied on a supply of wood, and neither were the woods located conveniently close to settlements.

A woodsman would typically rise by moonlight for an early start at the copse - hiding their heavy tools as these might be stolen. Wages would have been similar to agricultural rates, but with fewer weekend working hours, and there was free wood for domestic purposes, as there was for those who dwelt in the New Forest. Woodsmen could be employed (typically by an estate), or self-employed and individuals might earn more, more depending on how hard they worked. In large woodlands men worked in teams, otherwise it was an isolated life, and there were serious risks of injury and even death through accidents.

Most woodsmen started young (8 or 9 years old), usually working with their father when schooling finished, and the training would have been long and slow, starting with the most basic of skills. Speed and accuracy were essential with each job. Then there was the matter of demand for wood and timber. Depending on both local demand (for example when construction was ongoing), and also national demand (such as during times of war), demand for wood products varied from year to year. In bad years, a shortage of work for woodsmen

---

[49] T. Rook (2002) *Roman Baths in Britain* Shire Books; O. Rackham (2006) *Woodlands.* Collins, London, 447.

led to them undercutting each other. Unlike agriculture (for example with the Tolpuddle Martyrs) it is thought that a lack of organisation within the woodland industry was a key factor in their demise.[50]

8.6 Hedgelaying

Hedgelaying is an ancient art of enclosing land by utilising the natural properties of trees and shrubs to form a living boundary that defines ownership and (importantly) constrains livestock. Non-living boundaries have included wattles and stone and brick walls; in the industrial age various forms of wire fencing have been developed. Hedgelaying has been commonplace in the British Isles and regional practices vary, giving rise to hedges of different character. Figure 7.8 shows an Exmoor enclosure wall with laid hedge. The practice of hedgelaying can be complicated, and its practitioners are skilled.

The craft of hedgelaying reached its zenith during the times of enclosure, although it was also practiced in ancient countryside, the hedges being a product of assarts. The practice declined after the Second World War, and by the 1960s, hedges were declining at an alarming rate. The National Hedgelaying Society dates from the 1970s.[51] Without attention, hedges became tall and gappy, with sparse growth at the base, and will tend to grow into a line of trees instead of a solid boundary that can constrain livestock. To add insult to injury, some hedges were grubbed out to make larger fields accommodating larger machinery in the drive for food production in the post-War period.[52]

Traditional tools included billhooks, axes and saws. Modern hedgelayers may also use chainsaws. All methods employed involve bending and partially cutting (or 'pleaching') through the stems of a line of planted shrubs or small trees near to ground level. The skill is not to actually break through the pleacher (it must remain alive), rather to arch the stems at an angle above the partial cut (they are known as 'liggers' once laid) along the hedge boundary so they become intertwined along the direction of the hedge-line.[53]

Upright stakes are set at regular intervals along the line of the hedge to give the finished barrier more strength. In time, the pleaches may die but will be replaced by shoots growing from ground level, much as with coppicing. The process may need to be repeated every eight to fifteen years unless the hedge is regularly trimmed. Small shoots that branch off the pleachers (known as 'brash') may be partly removed or partly woven to add further solidity to the hedge. Commonly there may be hazel binders inter-woven along the top of a laid hedge, also designed to give stability and strength.

Many species of trees and shrubs are used in hedgelaying; blackthorn, hawthorn, willow, and birch are common. Many other plants, both woody and herbs, are found, for example ash,

---

[50] Dorset County Council (2010) <http://dorsetcoppicegroup.co.uk/Dorset_Coppice_Group_History.pdf> Accessed June 2010.
[51] The National Hedgelaying Society (n.d) <https://www.hedgelaying.org.uk> Accessed November 2018.
[52] RSPB (n.d.) A history of hedgerows. <https://www.rspb.org.uk/our-work/conservation/conservation-and-sustainability/advice/conservation-land-management-advice/farm-hedges/history-of-hedgerows> Accessed November 2018.
[53] A. Warner (n.d.) All you ever wanted to know about hedge laying. National Trust. <https://www.nationaltrust.org.uk/buttermere-valley/features/all-you-ever-wanted-to-know-about-hedge-laying> Accessed November 2018.

field maple, beech, plum, crab apple, holly, sweet chestnut, beech, hornbeam, whitebeam, wild privet, hazel, spindle, ivy, wild rose, bramble, convolvulus, primrose, campions, cranesbills and vetches.[54] Flowering and berry-bearing bushes encourage biodiversity among invertebrate and bird species. It must be borne in mind that the initial planting may have involved several species gathered from the surrounding area, including common land.

8.7 Timber!

The term 'timber' is different from 'wood', for it has an economic meaning, referring to valuable products of woods and forests as used in construction. Ancient woods, especially coppice-with-standards, produce both. Wood can be anything including offcuts used for burning. It is furthermore unlikely that ships timbers were recycled for domestic or any other building. Former cuts, channels, joints and the like on timbers in old buildings relate either to shaping the timber or reflect earlier building phases. Indeed, to presume that carpenters were in the habit of taking former ships timbers is both impractical and does a disservice to the 'chippies' of old who would set off into the forest with a woodsman in order to select timbers best fitted for the desired construction.[55] In ancient woodlands, 'standard trees' provided timber for felling. These were typically, but not exclusively, oak, ash being another example.[56]

In the New Forest, the transition from coppicing and the attempt to ban pollarding to encourage large oak and beech for shipbuilding has been noted. In all, four surveys of resources were undertaken between 1608 and 1783. Trees considered as growing timber fit for the Navy stood at 123,927 in 1608, falling to 12,476 by 1707 (the survey being prompted by a storm). Subsequently, an increase to 19,836 by 1764, then a fall once more to 12,447 by 1783 failed to impress the commissioners writing the 'Fifth Report'.[57] There was a crisis in production, and inquisitions into New Forest governance and management during this period cried out for reform; the consequences reached far into governmental and Crown interests. If, in the early 17th century, the New Forest contained almost 124,000 timber trees; 24,000 could be spared. There were 118,000 loads of 'Fyrewood and decayed Trees'. Coppice then amounted to (only) 1,304 acres (520ha) of which there were some 96 acres (39ha) at Aldermoor. Between about 1670 and 1673 some 100 acres (40ha) of coppice was enclosed at Holmhill, and a further 300 acres (121ha) at Aldridgehill and Holiday's Hill for 'nursery and supply of timber'.[58]

Inclosure areas with trees of sufficient size were opened for grazing, and elsewhere further areas were subsequently planted. This was 'rolling afforestation' and inevitably it would lead to friction between commoners and the Crown. There was also evidence of neglect. After 1760, management was in decline, and the resulting record levels of grazing seemingly prevented natural regeneration, allowing the Forest to tend towards becoming a mature canopy of trees. This situation was to continue until 1851 when it was noted the understory dominated by holly developed.[59] Two further enabling Acts for 'statutory inclosures' followed

---

[54] K. Stearne (2018) Pers comm.

[55] T. Tatton-Brown (2018) Pers. comm.

[56] R.J. Fuller and M.S. Warren (1993) Coppiced woodlands: their management for wildlife. JNCC <http://jncc.defra. gov.uk/pdf/pubs93_Coppicedwoodlands.pdf> Accessed November 2019.

[57] Fifth Report of the commissioners appointed to enquire into the state and condition of the woods, forests and land revenues of the Crown. 22nd July 1789, 22.

[58] C.R. Tubbs (2001) New Forest, the history, ecology and conservation. New Forest Ninth Centenary Trust, Lyndhurst, [85-6].

[59] G.F. Peterken (1981) Woodland Conservation and Management. Chapman and Hall, London, 17.

in 1808 and the Deer Removal Act in 1851.[60] Thus the total area under statutory inclosure (Table 7.1, see footnote 61 below) amounted to 7023 acres (2844ha) in the mid-19th century and was still rising! While these figures are indicative of land use change, some enclosures were thrown open for grazing and subsequently wholly or partly re-enclosed.[61] The greater obstacle to intensifying broadleaved timber production in the New Forest was the advent of iron battleships in the 19th century.

One response was to move towards coniferisation. The planting of conifers at a significant scale in the New Forest dates from the 1850s. Following the creation of the Forestry Commission, this planting increased. Between 1940 and 1970, coniferisation mostly involved planting of Douglas fir and larch, Scots pine, Corsican pine and Norway spruce.[62]

Elsewhere in Wessex, plantation of both broadleaved trees and conifers has occurred on a large scale. The shortage of timber for ships originally stimulated demand for hardwoods but rapidly dwindled in the 19th century as iron hulled ships took over. The creation of the Forestry Commission (from the Office of Woods) in 1919 was a response to a need to rebuild the strategic timber reserve following the First World War. They had, under the 1919 the Forestry Act, a good deal of freedom to acquire and plant land.[63] There was widespread coniferisation throughout Britain, often causing the demise of broadleaved woods, particularly coppice.[64]

Today, the South West region (including Devon and Cornwall) of the Forestry Commission contains over 212,000ha of woodland in parcels 0.1ha and over, or 8.9% of the land area, of which 17% is owned or managed by the Forestry Commission. In this region, broadleaved woodland is the dominant forest type (56.7% of all woodland), while Conifer woodland represents 22.8%, Mixed woodland 14.2% and Open Space within woodlands 4.8%.[65] Coniferisation is now well in reverse, for it is seen as ecologically and archaeologically destructive in creating a bland non eco-diverse monoculture of trees.[66]

8.8 Charcoal making

Charcoal making is ancient, and it has many uses. Archaeological evidence for its use goes back about 30,000 years.[67] The most common use historically has been for smelting; charcoal has been used by smiths because it may be heated with a metal ore in reduction processes to produce metals such as copper, tin or iron. Other uses may include cooking, art, medicine, cosmetics and gunpowder production. Today charcoal from organic sources is much used in industrial processes.

---

[60] New Forest Association op cit
[61] P. Roberts (2014) Pers. comm; H. Cook New Forest (2018) op. cit. Chapter 7
[62] C.R. Tubbs (2001) *New Forest, the history, ecology and conservation*. New Forest Ninth Centenary Trust, Lyndhurst, 222
[63] Forestry Commission (2017) History of the Forestry Commission <https://www.forestry.gov.uk/forestry/cmon-4uum6r> Accessed November 2019.
[64] S. Atkinson and M. Townsend (2011) The State of the UK's Forests, Woods and Trees: Perspectives from the sector. Woodland Trust. <https://www.woodlandtrust.org.uk/mediafile/100229275/stake-of-uk-forest-report.pdf> Accessed November 2018.
[65] Forestry Commission (2020) <https://www.forestresearch.gov.uk/documents/3061/nisouthwest.pdf> Accessed April 2021.
[66] S. Bell (1999) Plantation Management for Landscapes in Britain. *The International Forestry Review* 1 (3), 177–181.
[67] P. Harris (n.d) On Charcoal <http://www.personal.rdg.ac.uk/~scsharip/Charcoal.htm> Accessed April 2021.

The charcoal industry in Britain was important in the middle ages, however the use of coal replaced it in many industrial applications. Historically, the trade of 'collier' changed in meaning from charcoal burner to coalminer. Collier is an English surname. Elsewhere, on the European mainland, charcoal making has been blamed as a significant cause of deforestation, in the absence of coal.

Charcoal is a lightweight black carbon (and ash) residue produced by removing water and volatile constituents from animal and vegetation substances, but generally wood products are used as they are most commonplace. Charcoal is made by heating wood while excluding air and hence most oxygen.[68] As a product for combustion, charcoal has the advantage of burning at a higher temperature than the wood from which it is derived. It produces little smoke as its combustion produces mainly carbon dioxide. It is argued that re-growth of the wood takes in carbon from the air, so charcoal production is more sustainable than use of fossil fuels such as coal and oil.[69]

Historically charcoal burning took place inside earthen clamps, however today the burning takes place in large steel or even concrete silos. Figure 8.5 shows modern charcoal burning equipment. An advantage of charcoal production is that it uses offcuts from wood, and also wood of poor quality that has no other uses other than burning for heating, so it need not

Figure 8.5. Charcoal burning in modern times (the author).

[68] N. Wilde (2000) Charcoal Burning in the Wyre Forest. Wyre Forest Study Group
[69] M. Broadmeadow and R. Matthews (2003) Forests, Carbon and Climate Change: the UK Contribution Forestry Commission Information note June 2003.

interfere with the production of poles and timber destined for other purposes. Coppicing has been the usual way to provide wood for producing charcoal as part of a cycle of tree cutting and re-growth, so it can be a by-product of timber production.

The process is stopped before all the wood turns to ash, leaving black lumps and powder, about 25% of the original weight. Charcoal produces more energy per unit mass than the original wood, and the advent of copper, bronze, iron and even the industrial revolution may be attributed to charcoal, although in more modern times coal was mined and used instead in many countries, including Britain.[70]

8.9 Case study: Coppicing and pollarding in the New Forest

Today the New Forest is hardly synonymous with coppicing, although there are a few areas where it is still practiced, for example near to Hale on the western border of the Forest, and it has even been noted that the practice died out. Trees such as hazel, ash and underwood, once cut, suffered from animals grazing and hence inhibiting re-growth. Clearly only modest coppicing occurred by the late medieval period. However, there was need to legislate for wood production as coppice, and small though it was, this was a portent of what was to follow, albeit largely in areas outside the forest boundary (called the 'perambulation'). Yet general 'underwood exploitation' seems to have been ancient and required some form of regularisation.

'Coppicing' was first mentioned in the New Forest in 1389 but the records of wood sales are unclear about earlier times. Under Richard III (reigned 1483-1485), the Encoppicement Act of 1483[71] encouraged the enclosure of private forest 'coppices' for several coppice cycles to provide for large timber.[72] To keep grazing animals excluded, enclosures thus created had a ditch and bank topped with (unsharp) palings.[73] Thorns could then be planted on the inside of the enclosure to deter grazing. Concern was not so much for the tall trees, but the underwood – that is in effect the coppice. Coppices were leased to individuals by the Crown, although the interest for these would have been largely the underwood, standards being harvested to the benefit of the Treasury, with the Surveyor of Woods presenting annual accounts to the Treasury.

These changes should not be seen in terms of systematic wood production, rather the aim was to preserve the underwood for domestic fuel and charcoal, both significant aspects of the forest economy for its inhabitants. Although private coppices were protected by the 1483 Act, parallel activities were likely applied subsequently where Crown coppices were leased to individuals.[74] Coppices remaining in Crown land were actually not affected by the legislation of 1483. However, there must have been a developing state interest in large timber production and the exploitation of royal forest, specifically for the navy; earlier, in 1416/17, the New Forest had supplied timber for Henry V's flagship, *Grace Dieu*.[75]

---

[70] A. Shevchenko (2014) How to make wood charcoal. <http://ukrfuel.com/news-how-to-make-wood-charcoal-19.html> Accessed November 2018.
[71] H. Cook (2018) *The New Forest: The Forging of a Landscape*. Windgather Press, Oxford, 89.
[72] C.R. Tubbs (1964) Early encoppicements in the New Forest. *Forestry* 37, 95-105.
[73] H. Cook (2018) *The New Forest: The Forging of a Landscape*. Windgather Press, Oxford, 92.
[74] C.R. Tubbs (1964) Early encoppicements in the New Forest. *Forestry* 37, 95-105.
[75] R. Reeves (2008) *To enquire and conspire, New Forest Documents 1533-1615*. New Forest Record Series Vol. V. New Forest

Better known among foresters today is the 'Statute of Woods' (1544).[76] This was effectively a working plan for coppice management and perhaps the first recorded 'regularisation' of silviculture as opposed to canopy protection merely as protection of the vert (vegetation) for hunting. Royal Forests were moving away from a role that was a blend of aristocratic hunting and politics by the reign of Elizabeth I (1558-1603); the New Forest was poised to become a supplier of commercial timber. Coppice management also developed into this reign when coppice clearance was extended to nine years (see below). The Regarders of the Forest reported that markets existed for charcoal, fuel, fence stakes and house repairs. The demand for large timber trees was not apparent – or at least there may not yet have been a supply problem.

The 'Statute of Woods' provided for 12 standels (standard trees) to remain per acre through several coppice cycles to provide for large timber. A later Act of 1588 extended the length of enclosure after felling to nine years. In 1595 coppice was present at Castle Hill, Broadstone and South Bentley Coppice.[77] Preservators were officials introduced in Elizabethan times charged with protecting timber assets of the Government as distinct from Crown officials of the Royal Forests.[78] Preservators' returns show the coppices were leased to groups of individual 'wood seller tenants' at between £3 and £4 per acre, although one William Hobbes let five acres near to Harmansgrove.[79]

These tenants could cut the hazel underwood and lop pollarded oak for fuel, charcoal, fence stakes, hurdles, house walls, baskets, barrel hoops, frameworks for bee skeps or hives and general house repairs. Fuel wood and charcoal were taken from remnants not used for any other purpose. However, sapling oaks had to be left to mature for use by the navy; coppicing is not helpful when large ship's timbers are called for. Demand for underwood products was slackening nationally, although demand seems to have held in the New Forest. The coppice system on Crown lands ended with the *Act for the Increase and Preservation of Timber in the New Forest* (1698).[80] By this time there was only two surviving coppices, North and South Bentley, and these were cleared, re-fenced and replanted.[81]

To provide larger timbers, pollarding was banned in the New Forest. Royal forests had supplied hardwood to local markets for centuries: oak for fittings and furnishings, also beech for furniture.[82] The deliberate planting of oaks and (to a lesser extent) beech to supply the navy was a priority. The policy under William III, set out in the 1698 Act was to plant 2,000 acres (810ha) of oak, increasing by 200 acres (81ha) per year for 20 years, until an area of 6000 acres (2430ha) was achieved.[83] In the event, only slightly over 1,000 acres (414ha) was planted

Ninth Centenary Trust, Lyndhurst, xxiv.

[76] C.R. Tubbs (2001) *New Forest, the history, ecology and conservation.* New Forest Ninth Centenary Trust, Lyndhurst, 84-85.

[77] R. Reeves (2008) *To enquire and conspire, New Forest Documents 1533-1615.* New Forest Record Series Vol. V. New Forest Ninth Centenary Trust, Lyndhurst, xxxv.

[78] P. Roberts (2001) Elizabethan conservators of the New forest. *Proceedings of the Hampshire Field Club and Archaeological Society* 56, 246-253 .

[79] P. Roberts (2002) *Minstead: life in a New Forest community.* Nova Foresta Publishing, Lyndhurst.

[80] C.R. Tubbs (2001) *New Forest, the history, ecology and conservation.* New Forest Ninth Centenary Trust, Lyndhurst, 86.

[81] C.R. Tubbs (2001) *New Forest, the history, ecology and conservation.* New Forest Ninth Centenary Trust, Lyndhurst, 86.

[82] J. Bond (1994) Forests, chases, warrens and parks in medieval Wessex, in M. Aston and C. Lewis (eds) *The Medieval Landscape of Wessex.* Oxbow monograph 6, Oxbow Books, Oxford, 130.

[83] D. Cobb (1974) Forest trees and ships. *Proceedings of the Hampshire Field Club and*

in 15 years.[84] With the practice of pollarding made illegal,[85] presumably full tree growth was preferred. This Act also gave Verderers powers to fine miscreants £5 for breaking inclosure fences, burning heather and fern, destroying covert (apparently a catch-all offence) and stealing wood.[86] Venison offences could be dealt with by the (apparently harsher) common law.

The 1698 Act made it illegal to start new pollarding, specifically on oak trees within the New Forest, because this would not have produced timber suitable for shipbuilding. Trees that had already been pollarded were permitted to continue until the Deer Removal Act of 1851 when it was considered no longer necessary to provide winter fodder for the deer, ponies and cattle. Today, holly pollarding still occurs each winter to provide additional food for ponies.

8.10 Case Study: Garston Woods, Dorset

RSPB Garston Woods nature reserve is near to Sixpenny Handley, Dorset,[87] and constitutes a part of NCA 134: Dorset Downs and Cranbourne Chase, viz:

> 'This strongly rural and agricultural NCA is characterised by large, open fields of pasture and arable, punctuated by blocks of woodland all draped over the undulating chalk topography.'[88]

Comprising 33ha in area, there are large areas of hazel (coppiced every 10-12 years) and maple. Other habitats include oak woodland, scrub, and mixed plantation, with important features such as glades, rides, and scrub management. Work on rides and glade will enhance parts of the woods for orchids and other flowering plants such as wood spurge (food plant of the rare drab looper moth). This plant is also important for silver washed fritillary and white admiral.

The soils are predominantly Carstens association (well drained plateau drift and Clay-with-Flints, Figure 8.6), fine silty and clayey often very flinty (and suited to tree cultivation) or Andover 1 association, shallow well-drained silty soils developed on 'Seaford chalk', a formation of the former Upper Chalk.

Figure 8.7 shows a map extract from the 1960s, including the present Garston Woods Nature Reserve. Adjacent wooded areas are marked as Hoe Coppice, Pribdean Woods, Captain's Wood and Mistleberry (or more accurately Mistlebury) Wood where there is a 'camp' marked at National Grid Reference ST 99587 19486. These names alone suggest complexity in structure of this woodland, something not uncommon within ancient woods.

Adjacent woods are bounded on the east side by a relatively straight road that climbs from south to north along the dry valley bottom. The north side is a presumably very ancient

Archaeological Society 13, 28-29.

[84] H. Cook (2018) *The New Forest: The Forging of a Landscape*. Windgather Press, Oxford,132

[85] G.F. Peterkin (1991) Woodland conservation and management. Chapman Hall, London

[86] C.R. Tubbs (2001) *New Forest, the history, ecology and conservation*. New Forest Ninth Centenary Trust, Lyndhurst, 82.

[87] RSPB (n.d) Garston Wood <www.rspb.org.uk/reserves-and-events/reserves-a-z/garston-wood/>

[88] Natural England (2014) National Character Area Profiles: data for local decision making https://www.gcv.uk/government/publications/national-character-area-profiles-data-for-local-decision-making/national-character-area-profiles> Accessed November 2018.

Figure 8.6. Carstens association 'clay-with-flint' soil in root plate of uprooted tree (the author).

boundary. It is marked on old maps as an unmade roadway referred to as the 'Shire Rack' (a hollow path associated with a boundary); there are sections of boundary ditch and good evidence for hedgelaying. Shire Rack is also the Dorset-Wiltshire boundary; the wood is located on the south side, just within Dorset.

There was extensive late Iron Age and Romano-British occupation, with a range of settlement types, across Cranborne Chase, which was later to form a royal hunting ground. Mistlebury Camp is one of a range of archaeological survivals, for as a hunting ground the area was relatively protected. This Iron Age defended settlement lies within the 'neck' of woodland named as Mistleberry Wood, situated on the relatively gentle south facing slopes of a ridge overlooking the dry valley today marked as 'Deanland'. An oval enclosure is defined by a rampart bank standing up to 7m wide and 0.6m high with an outer ditch with an entrance to the south east, although it may be unfinished as there is a considerable missing section to the south west. The enclosure has been linked with an Anglo-Saxon burh mentioned in a charter of AD 956 as 'Mealeburg'.[89]

[89] Historic England (2018) Camp in Mistlebury Wood <https://historicengland.org.uk/listing/the-list/list-entry/1002455> Accessed November 2018.

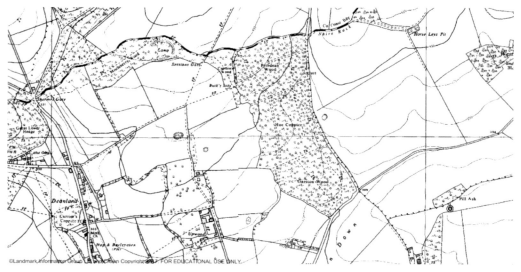

Figure 8.7. Garston and adjacent woods in the 1960s. OS mapping (© Digimap).

Figure 8.8. Former layered hedge north side of Garston woods, Dorset (the author).

Woodbanks on the north side of the wooded area are present, but indistinct, suggesting boundary definition along the Shire Rack. Impressive evidence in the form of certain trees gives clues as to management.

There is little evidence throughout the woods of pollarding, while coppicing is the dominant form of tree management that causes deviation from natural growth habits for the trees. Figure 8.8 shows a tree developed from a former layered ash hedge with the main horizontal bough set parallel to the path. Note also the wood anemone in the bank, an indicator species for ancient woodland.

The implication is that the major means of defining the boundaries of Garston woods was hedge laying.

To the south and west, the farmsteads are relatively modern. Upwood Farm was established *c.*1790 although the present buildings are modern. Deanland and Newtown (to the west of the wood) are 18th and 19th century encroachments on the woodland. Here the open fields were finally enclosed in 1797. The irregular shape of the western side of Garston Wood

Figure 8.9. Curvilinear field boundary, grassed field margin and multi-stemmed boundary tree at the western edge of Garston Woods (the author).

Figure 8.10. Standard hedgerow trees (RHS) abutting assarted edge, western side of Garston Woods (the author).

and Hoe Coppice is the result of assarting, but is unlikely to be medieval.[90] In Figure 8.9 the enclosures on the west side clearly show encroachment into the woodland.

The boundary features on this side of Garston wood are indistinct and there are ground features that are clearly attributable to ploughing to the edge. Most trees show evidence of coppicing, and there are several standard trees along the boundary with little evidence of either pollarding or hedge-laying.[91]

The field boundaries on the western side of the wood are historic, for they include some standard hedge trees (Figure 8.10), and these boundaries likely developed from 18th and 19th century encroachment on the wood. Woodland plants, including bluebells (*Hyacinthoides non-scripta*), may be observed as forming part of the flora of these hedges. Today the wood is protected by deer fences located within the boundaries.

[90] Royal Commission on Historic Monuments (1075) Sixpenny Handley, in *An Inventory of the Historical Monuments in Dorset, Volume 5, East*. HMSO, London, 64-72.
[91] The Author acknowledges Dr Ian Cummings for discussion of these topics.

Figure 8.11. Ramsons or wild garlic (*Allium ursinum*) with white flowers dominating the bank and ditch of a presumed prehistoric enclosure in Mistleberry Wood, part of the Garston wood complex (the author).

Figure 8.12. Wood pasture adjacent to Shermal Gate, Garston wood complex (the author).

Figure 8.13. Beech plantation LHS of Shire Rack at edge of wood (the author).

Figure 8.11 shows areas of bluebells and wild garlic associated with the prehistoric enclosure. The bank is a monoculture of wild garlic, while the bluebells form a carpet in the middle distance on the left and right of the bank.

Bluebells may germinate under shady conditions and lack of management disturbance may lead to homogeneity. On the other hand, the wild garlic grows largely on the bank.[92]

Figure 8.12 shows a small area of wood pasture south of the Shire Rack, located east of Shermal Gate (Figure 8.7) and west of the enclosed Mistleberry Wood. This is located between the arable areas and defined woodland. On the other side of the same path is an area of beech plantation. The line of the presumed boundary of the ancient wood is clear along the pathway. On the left-hand side of the photograph, a younger, even beech plantation may be observed (Figure 8.13). The continuation of the ground flora cover into this plantation suggests this is re-planted ancient woodland.

RSPB Garston Wood presents several observations. On the one hand it corresponds to what is normally understood as 'ancient woodland' and its past and present management bear witness to this. However, archaeological evidence points to there being an open area in the

---

[92] With thanks to Ian Cummings and also Graeme Bathe.

Iron Age where a defensible settlement was located overlooking a valley. Today this is within woodland, yet there are modern fields both to the north and south of this monument. On the south side, apparently woodland clearance only occurred from the 18th century. It is likely there have been several cycles of clearance and re-establishment of woodland since prehistory, but the modern wooded area would have likely been included in a large woodland area in AD 1600.

8.11 Conclusions: Woods and forests

Ancient woodlands are complicated and are managed, so they constitute semi-natural ecosystems. Like later commercial forestry, they often occupy poor or problematic soils and hence are a typical land cover for economically marginal areas within an agricultural landscape. Hence, they show all the signs of being encroached upon during times of agricultural expansion. The extent to which ancient woods constitute fragments of the 'wildwood' remains uncertain. Against this idea, increasingly archaeological features are suggesting many areas were likely open landscapes at various points in the past. In Britain, furthermore, the pollen record gives an indication of the history of woodlands including periods of clearance.[93]

Like heath and moorlands, the structures of the landscapes result from vegetation management, although ancient woodlands are unlikely to have been affected by fire in any significant way. Managed ancient woods remain structured, constituting ground, field, underwood and canopy layers. Management is largely conducive to this, including the common instances when 'standard' trees are retained for timber production. This all contrasts with the monoculture of plantation trees, be they hardwood or conifer.

The *functions* of the landscape are considerable. In purely economic terms, woods and forests provided for fuel, construction materials and other forest products including food and medicinal herbs as well as small mammals and larger herbivores, such as deer, that provided meat. Today they also function as areas of plant and animal conservation.

Indirect economic *values* include ecosystem services for habitat creation and wood products, and carbon sequestration. Non-market values include habitats and aesthetic value. *Scale* depends largely on the area of individual woodland and how contiguous these areas are. The care often taken to name quite small areas reflects past economic value and ownership. Finally, *change* has been ever-present, not only through seasonal or year-by-year management but also through removal and efforts towards coniferisation of woodlands. Sadly, in recent decades, native tree populations in Wessex have been affected by Dutch elm disease and ash dieback. It is a reflection of the values in general that we attach to woods and forests that there is a drive to conserve, even re-establish broadleaved trees.

---

[93] A.E. Caseldine (2018) Humans and Landscape, *Internet Archaeology* 48. https://doi.org/10.11141/ia.48.4

Chapter 9

Between two seas

9.1 Wessex coasts

Scrutiny of Figure 1.1 (Chapter 1) shows Wessex to be approximately and irregularly hexagonal in shape. Four of the region's sides are land boundaries: north of the upper Thames and southern Cotswolds, between Oxford and Windsor, Windsor to Portsmouth and in the west the boundary with Devon. The remaining two are the English Channel Coast between Portsmouth and Lyme Regis, and the south eastern shore of the Bristol Channel from Exmoor to approximately Thornbury.

Geologically speaking, the southern coast is relatively straightforward. The Lower Lias group around Lyme Regis comprises predominantly grey, well bedded, marine calcareous mudstone and silty mudstone that are late Triassic (specifically Rhaetian age mudstones and shales of the Blue Lias) and early Jurassic in age. The bulk of the coast (travelling east) comprises Jurassic age rocks (the internationally famous 'Jurassic Coast', see Chapter 10),[1] but in general becomes younger, with chalk outcropping, where the Dorset Downs meet the coast, west of Lulworth Cove and at Swanage in the Old Harry Rocks (see Chapter 1). The solid geology of the area displays a range of soft sedimentary strata of Tertiary Age (Figure 1.3).

The Chalk crops out once more on the mainland coast eastwards at Bognor Regis. The Isle of Wight may, for present purposes, be divided into a northern area comprising the same range of Tertiary strata and the southern area comprising Cretaceous deposits ranging from Wealden through lower Greensand and Gault to the Chalk that forms an east-west ridge across the central part of the Island and caps the hills above Ventnor. The Isle, because of the varied geology, includes some impressive and unstable cliffs facing the English Channel.

Few areas allow such direct insights on geology, human intervention and the humans themselves than this stretch of the Dorset coast. In terms of National Character Area profiles, most are determined by their landward characters. Two notable possible exceptions are NCA 136: South Purbeck and NCA 137: Isle of Portland. NCA 142 which borders the Bristol Channel (Somerset Levels and Moors) is discussed in Chapter 4 and Exmoor (NCA 145), which has a most rugged and impressive coastline, is discussed in Chapter 7. The Isle of Wight comprises its own NCA.

To provide context, some coastal sites that were natural sheltered havens had their origins in the Mesolithic with a potential to became early ports through which both objects and raw materials could be imported. Between the early Neolithic and Early Bronze Age, the regions

---

[1] T. Badman, D. Brunsden, R. Edmonds, S. King, C. Pamplin and M. Turnbull (2003) *The Official Guide to the Jurassic Coast.* Coastal Publishing, Wareham.

Figure 9.1. Durdle Door, Dorset (the author).

of Britain were connected by sea.[2] Christchurch harbour is a case in point, and archaeological research has established that Poole Harbour has been an important port since the Iron Age.[3]

9.2 The English Channel Coast: Lyme Regis to Portland

The physical basis for the Coasts of Wessex is introduced in Chapter 1, which presents an idealised 'coastal cell' to describe key coastal processes. While in Dorset there are some dramatic features, including the stunning arch of Durdle Door (Figure 9.1) developed in the Portland Limestone and located to the west of Lulworth Cove. Elsewhere (typically around Poole Harbour) there are softer coastal features, including sand dunes (for example) at Studland. The idea of coastal cells is applied here to the English Channel coast.

In Figure 9.2, a total of nine such coastal sediment transport cells are identified between Lyme Regis and Brighton in East Sussex. These cells are discrete in that they are more-or-less fixed boundaries to sediment movement (dotted lines). The dominance of south-westerly winds should enable longshore drift (see Chapter 1). This process creates shingle-bar features along

---

[2] R. Bradley, A. Rogers, F. Sturt, A. Watson, D. Coles, J. Gardiner and R. Scott (2016) Maritime Havens in Earlier Prehistoric Britain, *Proceedings of the Prehistoric Society* 82, 125-159.
[3] M. Markey, E. Wilkes and T. Darvil (2004) Poole Harbour: An Iron age port <http://www.wessexportal.co.uk/wp-content/uploads/2014/07/Poole-Harbour-Iron-Age-Port.pdf> Accessed August 2021.

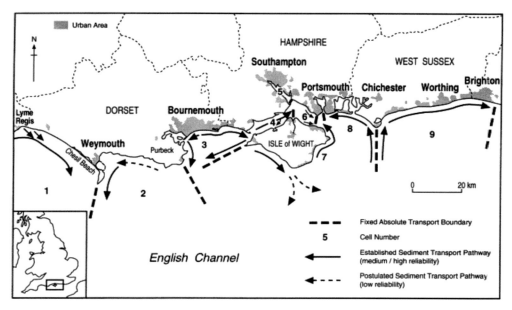

Figure 9.2. Coastal sediment transfer cells (J.M. Hooke, 1999, Decades of change: contributions of geomorphology to fluvial and coastal engineering and management. *Geomorphology* 31, 1-4, 373-389).

the south coast of England, such as Slapton Ley, Chesil Beach (cell 1 in Figure 9.2), the shingle beaches along the Sussex coast east of Chichester and the Dungeness foreland in Kent.[4] Shingle bar features have attracted attention due not only for curiosity, but also on account of their ecological potential for primary vegetation succession, soil formation and possibility of water supply.[5]

East of Chesil Beach and around the Isle of Wight this pattern is less simple. East of the tombolo of Portland the situation changes (cell 2), and there are similarities in pattern to that of Purbeck (cell 3), while the complicated estuarine and tidal situation and around Wight (four tides per day)[6] impacts on sediment transport as far east as Selsey Bill (cell 8). On the mainland side of the Isle of Wight and east of Hurst Point (cells 4,5 and 6) are appreciable saltmarsh and grazing marshes. Only cell 9 restores the dominant west to east longshore transportation of sediment.

Returning to the western section of the English Channel Wessex Coast, Figure 9.3 shows in more detail the situation for sediment transport between Lyme Bay and Portland Bill. Key here is not only the longshore drift but human intervention in the form of the harbour or

[4] J.D. Orford, S.C. Jennings and D.L. Forbes (2001) Origin, development, reworking and breakdown of gravel-dominated Coastal Barriers in Atlantic Canada: Future scenarios for the British Coast. in J.R. Packham, R.E. Randall, R.S.K. Barnes and A. Neal (eds) *Ecology & Geomorphology of Coastal Shingle*. Westbury, Otley.

[5] C.P. Burnham and H.F. Cook (2001) Hydrology and soils of coastal shingle with specific reference to Dungeness in J.R. Packham, R.E. Randall, R.S.K. Barnes and A. Neal (eds) *Ecology & Geomorphology of Coastal Shingle*. Westbury, Otley, Chapter 5.

[6] SCOPAC (2004) North East Isle of Wight (East Cowes to Culver Cliff) <http://www.scopac.org.uk/scopac_sedimentdb/neiow/neiow.htm> Accessed May 2020.

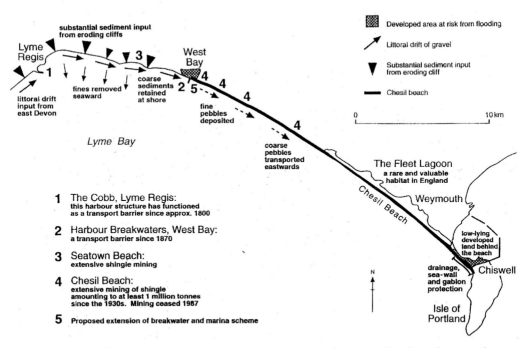

Figure 9.3. Longshore sediment movement Lyme Bay (D. Brunsden, 1999, Chesil Beach - Two Ideas. In R.J. Allison (ed.) *Dorset Revisited: Position Papers and Research Statements.* West Dorset Coastal Research Group).

'Cobb' at Lyme (1). Since its construction in its present form in 1754, it has functioned as a large terminal groyne and has promoted the updrift (west) side accretion of Monmouth Beach (Figure 9.4).

Longshore sediment supply from updrift has thus steadily diminished, although the Cobb does not present an absolute boundary to littoral movement because there is accumulation of gravel in Lyme Regis harbour under high energy wave conditions.[7] The impact of the Cobb has been to block finer sediment movement eastward; a further impediment to longshore movement exists in the form of the breakwaters at West Bay (2). In the past there was extensive shingle mining at Seatown Beach (3) and to the southwest extensive mining along Chesil Beach (Figure 9.6).

This interference in the longshore movement of sediment has made the cliffs Between Lyme Regis town and Charmouth (such as Black Ven, the largest active coastal landslide system in Europe) especially unstable, with mudslides commonplace. Scrutiny of large-scale maps and satellite images furthermore shows the aggressive nature of coastal erosion at this point. The river Lim gives its name to Lyme Regis, and only 2km to the east, the river Char does the same for Charmouth. The confluence of these two rivers would have been on the site of Lyme

[7]   SCOPAC (2012) Beer Head to Lyme Regis <http://www.scopac.org.uk/sts/bh-lr-literature-review.html> Accessed May 2020.

Figure 9.4. The Cobb at Lyme Regis looking beyond towards Monmouth Beach (the author).

Bay and the cutting across of field systems by this actively eroding clastline is evident from mapping.[8]

The story of the coast at Lyme Regis has interesting cultural and scientific ramifications. The Georgian novel *Persuasion* by Jane Austen was published posthumously in 1818. In it, Louisa Musgrove twice jumps down from some steps on the Cobb, as she enjoys being caught by the honourable Captain Wentworth. However, on the second attempt she falls and has to be carried away unconscious. *The French Lieutenant's Woman* (the film was released in 1981 and was inspired by the 1969 novel of the same name by John Fowles) features a romance within its Victorian period drama between a gentleman palaeontologist called Charles Smithson, and the complex and troubled Sarah Woodruff, known as 'the French Lieutenant's Woman'.

'Fowles's postmodern take on narrative convention and literary sensibility produced a metafiction that comments on reader expectations in both the Victorian and the contemporary period and explores the social hypocrisies of the earlier era through an excoriating depiction of class, gender, science, and economics.'[9]

---

[8] A. Dykes (2020) pers. comm., after D. Brunsden pers. comm. (1999).
[9] L. Bolton (2015) The French Lieutenant's Woman: A Room of Her Own. <https://www.criterion.com/current/posts/3655-the-french-lieutenant-s-woman-a-room-of-her-own> Accessed May 2020.

Anyway, the Cobb features during the filming as well as the book. It should first also be noted that (by strange coincidence) the American 'Wentworth Scale' is used by sedimentary geologists to define size ranges, giving class limits with names such as pebbles, sand, silt, or clay (with further subdivisions) in a logarithmic scale. Captain Wentworth was a true gentleman (but probably not a sedimentologist); Sarah's seducer in the story was a palaeontologist. But what of reality?

The Cobb at Lyme Regis, by reducing longshore drift, caused accelerated erosion of the undercliff to the east over two and one-half centuries, a process accelerated by landslips. While this has produced massive structural problems for the slope behind the town of Lyme Regis, it greatly increased the possibilities of fossil collection from the repeated landslips.

Mary Anning (1799-1847) was a young fossil collector and palaeontologist to whom only in recent years has due acknowledgement been given.[10] Anning's finds and reconstructions of such extinct marine reptiles as the *Ichthyosaur* and *Plesiosaurus* make her a pioneer. And she has fired the imagination of scientists and children alike. A Westcountry girl of limited education, who was forced to help support her family from the age of 11 when her father died, she stood little chance of fame in her own day – in any conventional sense.

Anning suffered from the attitudes of her contemporary world, as did William Smith (1769-1839). He was another geological genius, also largely self-educated, and was the surveyor who produced the first geological map of Great Britain in 1815. Also grandly titled 'The Father of English Geology', Smith surveyed for the Kennett and Avon Canal, worked for the Somerset Coal Canal Society, and helped interpret the structure of the Somerset Coalfield (Figure 9.5). Earning the nickname 'Strata Smith', he established the branch of geology known as stratigraphy, Earth's history told in rocks and in fossils.[11] His humble birth led to rejection by the scientific establishment of his day, and it was not until late in life that his achievements were acknowledged.

Recognition for William Smith finally came in 1831 when the Geological Society of London (GSL) awarded him the inaugural Wollaston Medal, and in 1832, the King awarded him a pension.[12] Like Mary Anning, Smith's ideas were plagiarised, they both suffered financial hardships and had to sell their collections. Anning had it worse, for she was totally shunned by the GSL despite some supporting her and recognising her work. She was, after all, female.[13] Full recognition of her great scientific insights remains long overdue.

If Smith can be claimed as its father, English geology also has a mother in Etheldred Benett (1775-1845), only six years his junior. That Mary Anning had a female precursor comes as further confirmation that not all pioneers gain recognition when outside the establishment.[14]

---

[10] T. Badman, D. Brunsden, R. Edmonds, S. King, C. Pamplin and M. Turnbull (2003) *The Official Guide to the Jurassic Coast*. Coastal Publishing, Wareham.

[11] BGS (2020) William Smith — a man who changed the world. <https://www.bgs.ac.uk/discoveringGeology/geologyOfBritain/archives/williamsmith/home.html> Accessed May 2020.

[12] York Museums Trust (2020) The Map that Changed the World <https://www.yorkmuseumstrust.org.uk/news-media/latest-news/the-map-that-changed-the-world/> Accessed May 2020.

[13] T. Begum (2019) Natural History Museum news <https://www.nhm.ac.uk/discover/news/2019/july/a-series-of-mary-anning-films-are-about-to-hit-the-cinemas.html> Accessed May 2020.

[14] The Geological Society (2012) Etheldred Benett (1775-1845). https://www.geolsoc.org.uk/Library-and-Information-Services/Exhibitions/Women-and-Geology/Etheldred-Benett> Accessed May 2020; Steve Hannath pers. comm.

Figure 9.5. Inscription honouring William Smith at Tucking Mill Cottage near Bath, Somerset. Whether he actually lived here is disputed (the author).

Miss Benett was a woman of independent means who in 1815 (the year of Smith's map) produced the first annotated bed-by-bed section of the Chicksgrove Quarry near Tisbury, where stone from the Purbeck and Portland beds in the Vale of Wardour is still extracted today.

From here a unique range of fossils, particularly plant and reptile remains, are to be found. Despite her having connections, no learned society recognised Benett and her work. However, Gideon Mantell (Figure 10.4) was so impressed with the work of Miss Benett that he named a Cretaceous sponge after her – *Doryderma benetti*. She was generous in sharing her palaeontological specimens and, like William Smith, was a pioneer of the new earth science sub-discipline of stratigraphy.

At the other end of longshore sediment cell 1 (Figure 9.3), where the coarsest pebbles are deposited from the high energy environment facing the full impact of the English Channel, lies the Isle of Portland (Figure 9.6). In an extreme manifestation of 'We don't like strangers round here', Thomas Hardy described Portland as 'The Isle of Slingers'; apparently the inhospitable 'Portlanders' threw stones to keep strangers away. Here there is no shortage of stones.

Chesil Beach meets the Isle of Portland to form an amazing tombolo feature. NCA 137 is that of Portland Bill (see Chapter 5); the fragility of Chesil beach is mentioned in Chapter 1. It connects the Isle with the 'Weymouth Lowlands' and despite risks of overtopping and breaching by the sea, provides a barrier sheltering the land to the east from storm waves coming from the west. Behind Chesil Beach, the original coastline includes low cliffs and coves. Between the Beach and the shoreline the East and West Fleet runs for 29km. Chesil and the Fleet SAC constitutes marine areas, sea inlets (35.5%), tidal rivers, estuaries, mud flats, sand flats, lagoons (including saltwork basins) (30.5%), salt marshes, salt pastures, salt steppes (2%) and shingle, sea cliffs and islets (32%).[15]

The Isle of Portland (Dorset, England) formed part of a cross-channel trading system during the first century BC.[16] Portland stone from the Island has been widely used across the country, most notably in London including St Paul's Cathedral. The Portland Limestone formation dips

Figure 9.6. Chesil Beach from the Isle of Portland (the author)

---

[15] JNCC (n.d.) Chesil and the Fleet <https://sac.jncc.gov.uk/site/UK0017076> Accessed May 2020.
[16] J. Taylor 'The Isle of Portland: An Iron Age Port-of-trade', *Oxford Journal of Archaeology* 20, 2001

Figure 9.7. a. Left after the quarrying of Portland Freestone at Portland Bill.
b. the 'Red Crane' now preserved and once used to load blocks of Portland Stone on to ships (the author)

southwards by between 1.5 and 2.5° (degrees) typically, and up to 6° (degrees) locally, reaching the sea at Portland Bill (Figure 9.7), where there is a raised beach from the Pleistocene above the area of quarrying around the Old Lower Lighthouse. Above the limestone are Purbeck beds and below Portland sand underlain by Kimmeridge clay, and thus the northern part of the island reaches around 150m above OD at the Verne, a former military fortification, today a prison. However, this geology also imparts incredibly unstable cliffs.[17] It furthermore gives rise to calcareous grassland which supports important populations of butterflies and moths, notably silver-studded blue, small blue, Adonis blue, chalkhill blue, Lulworth skipper, dingy skipper, grayling and the chalk carpet moth. A butterfly sanctuary is to be found on the Isle at Broadcroft Quarry.[18] The Bill hosts an important observatory for migratory birds.[19]

Quarries, some still active, are a major part of the scenery. Figure 9.7 shows a hoist, the 'Red Crane' (Figure 9.7 inset). The southern part of the island allowed for easy extraction and loading of stone blocks on to ships. From here, weather permitting, there would be relatively easy export including London via the south coast and Thames estuary. In the 19thC Portland building stone was exported from the port of Weymouth.[20]

A third aspect of the Portland landscape (after habitats and quarrying) are strip lynchets. The open field system commenced during the Anglo-Saxon period and flourished after the Norman Conquest. Distinctive strip fields were produced by areas of unploughed land being left between allotments and their shape was thus determined by ploughing. The size of the strips was roughly an acre (0.405ha). An acre represented a day's work with a plough, its length determined by the distance an ox team could plough before needing a rest, i.e. a furlong (201.2m). Part of a medieval open field system to the east and northeast of the Old Higher Lighthouse survives well around Lawnsheds, close to Portland Bill.[21]

9.3 Purbeck, the Isle of Wight and into Hampshire

East of Portland is the classic, rugged and internationally famous stretch between Weymouth and Swanage, a generally concordant coastline of the Isle of Purbeck introduced in Chapter 1. Like the Isle of Portland, stone was quarried from such places as Dancing Ledge, St Aldhelm's Head and Tilly Whim Caves. Turning the corner at Durlston Head near to Swanage, concordant coastline turns to discordant as the geological section cuts the upper Jurassic Portland and Purbeck beds, Wealden, Upper Greensand, the Chalk at Old Harry's rocks and, north (and west) of Studland, Tertiary beds where heathland may be found such as at Hartland Moor (see Chapter 7).

South Purbeck National Character Area (NCA) is a 'compact but highly diverse landscape'. The coast displays a wide range of Jurassic strata, including Kimmeridge clays, limestone and

---

[17] I. M. West (2019) The Isle of Portland <http://www.southampton.ac.uk/~imw/Portland-Isle-Geological-Introduction.htm> Accessed May 2020

[18] Natural England (2015) NCAP 137 Isle of Portland <https://publications.naturalengland.org.uk/publication/3495352> Accessed May 2020.

[19] Portland Bill Observatory and Field Centre (n.d.) <http://www.portlandbirdobs.com/> Accessed May 2020.

[20] J.H. Bettey (1986) *Wessex from AD 1000*. Longman, London, 250; I.M. West (2019) Portland Bill and Sandholes <http://www.southampton.ac.uk/~imw/Portland-Bill-2017.htm> Accessed May 2020.

[21] Historic England (1952) Portland Open Fields < https://historicengland.org.uk/listing/the-list/list-entry/1002729> Accessed May 2020.

chalk downland. Some 42km of coast reveals exposures of strata creating cliffs, bays, stacks, arches and coves.[22] The area between Swanage and Poole (including the harbour) is at the edge of the Dorset Area of Outstanding Natural Beauty (urban areas excluded),[23] and the Dorset and East Devon AONBs may (together with the East Devon AONB) form the core of a new Dorset National Park, including (significantly) the World Heritage status Jurassic Coast.[24] From the Old Harry Rocks at the end of the Purbeck hills there is the sandy beach seaward of Studland Heath and a ferry that provided a link with Sandbanks spit which partially closes the mouth of Poole Harbour. In Figure 9.2, cell 3 shows the longshore currents that caused this feature. Behind is Brownsea Island, owned by the National Trust and managed by the Dorset Wildlife Trust. It is a rare English habitat for the Red Squirrel.[25]

Poole harbour is large in area (c.36km^2) but shallow. In the late and post-medieval period there was a ceramic export industry. In the early 17th century, the local economy was strengthened by links with the Newfoundland fishing grounds from whence fish were sold throughout southern England. Poole exported cloth, clothing, farming equipment, nets, ropes and more to the new colony; the port remained relatively prosperous into the 18th and 19th centuries and also exported potter's clay and corn from the chalklands.[26] However, there was a general decline after the Napoleonic period with only coastal and limited foreign trade remaining.[27] Today, draft for some shipping is maintained by dredging and there is oil extraction (Chapter 2).

Still in cell 3 (Figure 9.2) and eastwards from Poole is the commercial and resort town of Bournemouth. It grew from virtually nothing in the 19th century due to its impressive sandy beaches but today (in addition to tourism) has important functions in finance, insurance, digital and education.[28] East of here is Hengistbury Head, which displays defensive features on the headland. This was likely the main port of the area in prehistory and the Head shelters Christchurch Harbour behind.

The Dorset Stour allowed access into the interior of the Dorset Downs and on to the Jurassic Ridge and Parrett, potentially linking the two shores of the region. The Salisbury Avon similarly linked the coast to the chalk Salisbury Plain, thence on to the Kennet and Thames valleys. Both rivers meet in Christchurch Harbour, enabling trade with the continent and access to the Solent. This situation pertained until the Roman occupation when trade via both the Atlantic and English Channel declined. In the Bronze Age, there were clear links with the continent and by the Iron age (for example) Kimmeridge shale jewelry items were exported

---

[22] Natural England (2012) NCA Profile:136 South Purbeck (NE370). <http://publications.naturalengland.org.uk/publication/3504906> Accessed May 2020.

[23] T. Badman, D. Brunsden, R. Edmonds, S. King, C. Pamplin and M. Turnbull (2003) *The Official Guide to the Jurassic Coast*. Coastal Publishing, Wareham.

[24] Dorset National Park (n.d.) The Case for a National Park <https://www.dorsetnationalpark.com/the-case> Accessed May 2020.

[25] Dorset Wildlife Trust (n.d.) Brownsea Island <https://www.dorsetwildlifetrust.org.uk/brownsea-island> Accessed May 2020.

[26] J.H. Bettey (1986) *Wessex from AD 1000*. Longman, London, 117, 145-6, 208, 250.

[27] Royal Commission on Historic Monuments (1970) Poole, in *An Inventory of the Historical Monuments in Dorset, Volume 2, South east* HMSO, London, 189-240.

[28] Bournemouth Tourism (2020) History of Bournemouth <https://www.bournemouth.co.uk/explore/history-of-bournemouth> Accessed May 2020.

via Hengistbury. By the Roman arrival, amphorae from western Italy were imported as was pottery from Brittany.[29]

East of Christchurch Harbour is the highly problematic eroding coastline at Barton on Sea, situated northwest of the entrance to the Solent. Instability arises from both coastal erosion and ground water movements; the main hazard along this stretch of coastline is instability of the cliff face and cliff top. Small artificial 'headlands' have been built to try to nudge this stretch into naturally creating stable mini-bays and beaches.[30] The dominant current through the Solent was west to east (Figure 9.2). At the western end, there are saltmarshes behind Hurst Castle (built to protect the coast by Henry VIII). Hurst Spit is a key feature at the western entry to the Solent[31] and represents the eastward prolongation of a pronounced drift of sand and shingle derived from the soft Tertiary cliffs of Christchurch and Barton Bays.[32] Figure 9.8 shows the characteristic 'cuspate foreland' shape of Hurst Spit. Such features are mobile over time and will tend to move with the prevailing currents, in this case coming from the southwest.

The entire Solent, from Hurst Spit in the west to Lee-on-Solent (just beyond Tichfield Haven) in the east as well as all the estuaries and river valleys in the north of the Isle of Wight and much of the northeast facing coast, is a Special Protection Area (SPA).[33] SPA is a designation under the European Union Directive on the Conservation of Wild Birds whereby member states have a duty to safeguard the habitats of migratory birds and certain particularly threatened birds (Chapter 10).

There is an SSSI defined between Hurst Castle and Lymington River Estuary. Its significance lies in its remarkable assemblage of salt, brackish and freshwater habitats which support a great variety of wildlife, in particular birds and certain rare invertebrates and plant species, many in numbers of international significance. On the lee side of the spit developed saltmarshes extend eastwards along the coast (Figure 9.8). However, erosion has affected the spit and shingle replenishment is necessary. The saltmarshes are also under attack, a process that is being hastened by a dieback of the *Spartina anglica* marsh grass. The seaward edge of the marsh is eroding at about 3m per year.[34]

The importance of saltmarshes in history is varied, but they represent a rich environment for humans and may experience varying degrees of change through modification (Chapter 4), ranging from natural, primary wetlands to intensive arable production. Depending on management, they may be exploited for food gathering, wildfowling, fish, salt making, grazing and intensive agriculture.

---

[29] B. Cunliffe (1993) *Wessex to AD 1000*. Longman, London, 3,189,201, 224.
[30] A.P. Dykes (2020) Pers. comm.
[31] I. M. West (2020) Hurst Point, Hampshire: Historic record of storm events <http://www.southampton.ac.uk/~imw/Hurst-Spit-Historic-Coastal-Events.htm> Accessed May 2020.
[32] R.L. Collin (1996) Digital Mapping of Hurst Castle Spit, Hampshire. *Earth surface process and landforms* 21(11), 1049-1054. <https://doi.org/10.1002/(SICI)1096-9837(199611)21:11<1049::AID-ESP705>3.0.CO;2-I> Accessed May 2020.
[33] Life Project (n.d.) Solent <http://roseatetern.org/solent.html> Accessed May 2020
[34] Poole and Christchurch Coastal Group (2011) Hurst Spit <http://www.twobays.net/hurst_spit.htm> Accessed May 2020.

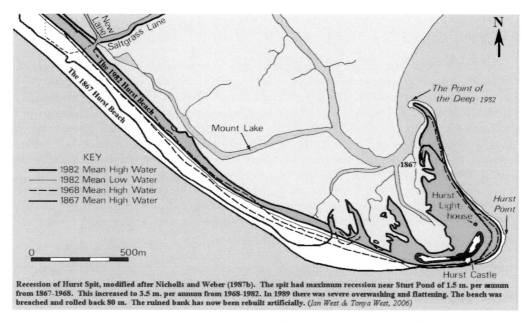

Recession of Hurst Spit, modified after Nicholls and Weber (1987b). The spit had maximum recession near Sturt Pond of 1.5 m. per annum from 1867-1968. This increased to 3.5 m. per annum from 1968-1982. In 1989 there was severe overwashing and flattening. The beach was breached and rolled back 80 m. The ruined bank has now been rebuilt artificially. (*Ian West & Tonya West, 2006*)

Figure 9.8. Recession of Hurst Spit (© Ian West and Tonya West)

Opposite Hurst point, the Isle of Wight covers an area of 380km² and boasts 92km of coastline. The Isle comprises NCA 127.[35] The Chalk ends at the western tip in a series of three chalk stacks called the Needles. Wight exhibits the key characteristics of much of lowland England, including arable coastal plains, pastures, and woodland, and from steep chalk downs, diverse estuarine areas and dramatic sea cliffs and stacks. There are a range of internationally, nationally, and locally important nature conservation sites recognised for their important habitats and species. These include maritime cliffs and slopes, coastal and flood plain grazing marsh, lowland heathland, saline lagoons, intertidal mudflats, coastal sand dunes, intertidal flats and seagrass beds, and coastal vegetated shingle. The intervening Solent and Southampton Water is designated as a Ramsar site and as a Special Protection Area, as it supports wintering waterfowl and rare invertebrates and plants.

Wight (from the Roman 'Vectis') inevitably has a close relationship with the sea, forming its cultural heritage since prehistoric times. The complex landscape (outlined above) is the outcome of its geological history and is dominated by a central chalk east-west ridge. The northern part is characterised by (low-lying) Tertiary clays sometimes overlain by gravel-capped ridges. To the south there is both chalk and Upper Greensand overlying Gault, Lower Greensand older Wealden Beds. The outcome is varied including downs, a dissected plateau and soft, unstable cliffs and steep-sided valleys called 'chines'. There is a southern coastal plain including arable farmland with large fields and few trees, giving an open aspect. There

---

[35] Natural England (2014) NCA Profile:127 Isle of Wight (NE561). <http://publications.naturalengland.org.uk/publication/6225459138265088> Accessed Aug 2021.

Figure 9.9. a. View of Lymington Reedbeds and; b. Nature Reserve sign (the author).

are three main ports: Yarmouth, Cowes and Fishbourne and many towns have a Victorian aspect including redbrick and also local building materials.[36]

Returning to the mainland, the Lymington and Keyhaven Marshes Local Nature Reserve belongs to Hampshire County Council. It comprises coastal marshes, lagoons and former salt marshes, particularly important for wintering birds. The saline lagoons are bodies of salt or brackish water that are partially connected to the sea through narrow openings or permeable barriers. They lie just inside the sea wall, connected to the sea by sluices and vary in width from less that 2m to over 200m. Generally, they are relatively narrow ditches about 0.5m deep, with muddy bottoms. They also receive fresh water as rainwater, runoff from adjacent land and from nearby streams which dilutes the sea water, reducing the salinity. There are a range of associated specialist lagoon species and birds including little egret, teal, and little tern feed in the lagoons. Waders such as curlew, sandpiper, dunlin, and little stint stop off on their long migratory journeys.

At certain times of year, high evaporation has the opposite effect of dilution: it will concentrate the salt. Salt making in the area goes back at least 2000 years; it developed particularly around the coast in the Iron Age.[37] Salt was obtained by impounding sea water in shallow lagoons, known as salterns, formed behind sea walls. Mentioned in Domesday Book (1086), salt making remained the principle economic asset of the area into the 17th century, reaching a peak early in the 18th century when 163 saltpans were in use at Lymington, with much of the salt exported. The industry declined with the arrival of railways bringing salt from the Cheshire salt mines.

Today, the reserve features the best-preserved example of medieval and later salt workings in southern England including 'salt docks' used for the importation of coal for the boiling houses and the export of salt on barges. Salt marshes have long been used for grazing and this has continued to the present. The development of grazing proved an incentive to reclaim saltmarshes from the sea and much of the reserve consists of rough grazing marsh. When flooded, these areas become dominated by rush. They are important for wintering birds and support populations of breeding waders in the summer months. On higher ground scrub has developed, and this too supports many specialist breeding birds, including linnet and Dartford warbler.[38] East of the Lymington and Keyhaven Marshes Local Nature Reserve is the Lymington Reedbeds Nature Reserve (Figure 9.9), which supports sea birds, warblers, migratory birds, wading birds, waterfowl, otters and water vole.[39]

Southampton Water is a result of the drowned confluence of the Test and Itchen rivers due to post-Pleistocene sea level rise. There are saltmarshes, and around the natural harbour at Southampton developed a range of economic activities including trade, fishing, and there was salt production around the local coast. The modern city and port developed from Anglo-Saxon *Hamwic* (see Chapter 2). From the later 12th century the port grew dramatically through

---

[36]   Natural England (2014) NCA Profile:127 Isle of Wight (NE561). <http://publications.naturalengland.org.uk/publication/6225459138265088> Accessed Aug 2021.
[37]   J.L. Kinory (2012) *Salt Production, Distribution and Use in the British Iron Age.* BAR BS559. Archaeopress, Oxford
[38]   Hampshire County Council (n.d.) Lymington and Keyhaven Marshes Local Nature Reserve <https://www.hants.gov.uk/thingstodo/countryside/finder/lymingtonkeyhaven> Accessed May 2020.
[39]   Hampshire and Isle of Wight Wildlife Trust (n.d.) <https://www.hiwwt.org.uk/nature-reserves/lymington-reedbeds-nature-reserve> Accessed May 2020.

successful commerce and defenses and merchants houses were constructed. [40] Southampton Water remains a major commercial harbour, with interesting environmental issues of its own.

The saltmarshes of Southampton water near Hythe have long been studied, and in botanical circles they are known for the hybridisation between the Old World species *Spartina maritima* and the North American *Spartina alterniflora* to produce the fertile common cord-grass (*Spartina anglica*). It is thought that *Spartina alterniflora* was originally introduced in ships' ballast water. *S. anglica* is widespread around the east of Britain and is still colonising in the west. [41] It is now regarded as an invasive species on mudflats, but there is an overall reduction of *Spartina* marsh, especially in the second half of the 20th century.

Retreat of *Spartina* marsh and consequent release of fine sediments has been widely reported for the western shore of Southampton Water. This erosion commonly takes two forms, either a frontal retreat resulting in the formation of low cliffs 0.5m - 1.5m high separating marshes and mudflats or internal dissection of the marsh by erosion of channel margins and development of 'erosion pans'. One cause appears to be dieback of *S. anglica* around inter-creek pans, which reduces the stability of accumulated saltmarsh sediments, making them susceptible to wave and/or tidal abrasion. The systematic loss of saltmarsh in the area during the last three decades of the last century looks set to continue. Mapped comparisons have revealed the retreat of the low water mark at a significantly more rapid rate than the high water mark, so that the intertidal zone has narrowed and steepened during the period 1870-1965; this was recorded for all parts of Southampton Water except the reclaimed docks between the Itchen mouth and Redbridge.

The causes of marsh and mudflat erosion are uncertain, but several, probably inter-related, possibilities include an absolute shortage of sediment, continuing relative sea level rise (see Chapter 10), land reclamation, coast defences, channel dredging and the failure of some intertidal marshes and mudflats to store sediment released by adjacent erosion. Dredging may also have played a significant role. Some localised losses have specific causes, such as oil spillage on Fawley marshes causing marsh plant mortality. Littoral drift towards the entrance of Southampton Water from both the west and east Solent is indicated by the presence of Calshot and Hook Spits respectively although no quantified study has been undertaken. [42]

Figure 9.2 shows the dominant transport at the eastern end of the Wessex coast towards the northwest along the eastern coastline of the Isle of Wight (cell 7) and from Portsmouth past Leigh-on-Solent towards the Hamble estuary (cell 6). The overall picture for sediment movement in the Solent and Southampton Water is extremely complicated.

Like the Southampton area, Portsmouth and Gosport are heavily urbanised and have long been linked with the sea, something remembered through the Portsmouth Historic Dockyard, the current home of Nelson's flagship Victory. Portsmouth was founded in 1180. The main exports

---

[40]   J.H. Bettey (1986) *Wessex from AD 1000*. Longman, London, 26, 51.
[41]   A.J. Gray, D.F. Marshall and A.F. Raybould (1991) A Century of Evolution in *Spartina anglica*. *Advances in Ecological research 21*, 1-62. https://doi.org/10.1016/S0065-2504(08)60096-3; JNCC (2006) *Spartina anglica*. <http://archive.jncc.gov.uk/default.aspx?page=1680> Accessed May 2020.
[42]   SCOPAC (2004) Southampton Water <http://www.scopac.org.uk/scopac_sedimentdb/soton/soton.htm> Accessed May 2020.

from medieval Portsmouth were wool and grain,[43] while the main imports were wine, woad for dyeing, wax for candles, and iron. Being defensible behind the Solent, it held significance for the Navy in the middle ages, however being unprotected by walls, Portsmouth suffered being burned by the French in 1338, but recovered.[44]

Henry VII strengthened the town's fortifications by building the square tower and constructed a dockyard in 1495. Here royal warships could be built or repaired so Portsmouth became a naval port. Henry VIII enlarged Portsmouth dockyard and further militarised the town. From here, in 1545 Henry watched as his warship *Mary Rose* sink in the Solent. In the late 16th century Portsmouth declined in importance as other dockyards were opened on the Thames; from this point ships were repaired at Portsmouth, but no more were built there. Thereafter the dockyard's fortunes varied.[45] Today Portsmouth combines its role as a naval base with that of a commercial port (largely retail, shipping, ferries) and tourism (for the Historic Dockyard).[46] Dredging for the new aircraft carrier Queen Elizabeth II revealed some 20,000 items 'from shoes to mines', a treasure trove for marine archaeologists.[47] Despite being developed and industrialised, Portsmouth Harbour is designated an SPA, composed of extensive intertidal mudflats and sandflats with seagrass beds, areas of saltmarsh, shallow coastal waters, coastal lagoons, and coastal grazing marsh.[48]

Along the coast, fishing was always important from a range of ports and fishing villages, although a large deep-sea fleet is today absent. Fishing was important in the early modern period[49], and Weymouth, Poole, Bridport and Lyme Regis provided Newfoundland fish across southern England.[50] While sea angling, like freshwater angling, is commonplace and important to particular local tourist economies or inland landowners, commercial fisheries remain important but are in decline. Typically, inshore fisheries (in water up to 30m deep) are fished with small and simple vessels when compared with deep sea trawlers. Along the Dorset coast (for example) there remain around 300 commercial fishing vessels. Main catches include finfish such as cod, shellfish such as scallops and other species like crab and lobster.[51]

The coast between Swanage and Selsey Bill contains Poole Harbour, the Solent, Southampton Water, Portsmouth and Langstone Harbours. These offer relatively sheltered fishing. Small boats catch such high-value species as sole, oysters, bass, and lobsters. Further west the coast is exposed (to the prevailing south-westerlies) and only Portland Bill offers protection. There

[43] T. Lambert (2019) A Brief History of Portsmouth, England <http://www.localhistories.org/portsmouth.html> Accessed May 2020.
[44] J.H. Bettey (1986) *Wessex from AD 1000*. Longman, London, 113.
[45] T. Lambert (2019) A Brief History of Portsmouth, England <http://www.localhistories.org/portsmouth.html> Accessed May 2020.
[46] Portsmouth City Council (2020) <https://www.portsmouth.gov.uk/ext/business/relocate-and-invest/local-economy> Accessed May 2020.
[47] Royal Navy (2017) Shoes to Mines <https://www.royalnavy.mod.uk/news-and-latest-activity/news/2017/august/01/170801-20-000-items-recovered-during-dredging> Accessed May 2020.
[48] Natural England (n.d.) Natural England Conservation Advice for Marine Protected Areas Portsmouth Harbour SPA Natural England guidance. <https://designatedsites.naturalengland.org.uk/Marine/MarineSiteDetail.aspx?SiteCode=UK9011051&SiteName=portsmouth&countyCode=&responsiblePerson=&SeaArea=&IFCAArea=&HasCA=1&NumMarineSeasonality=4&SiteNameDisplay=Portsmouth%20Harbour%20SPA> Accessed May 2020.
[49] J. H. Bettey (1986) *Wessex from AD 1000*. Longman, London, 142.
[50] J. H. Bettey (1986) *Wessex from AD 1000*. Longman, London, 146.
[51] B. Waycott (2018) A Fishing Industry Boost for Dorset and east Devon, DorsetCoastal Forum <https://main-hookandnetmag-hookandnet.content.pugpig.com/2018/03/20/2018-03flag/pugpig_index.html> Accessed May 2020.

are few vessels along this coast which are not day-boats. There has also been an increase in charter (and casual) angling vessels; this sector is year-round and of considerable importance to the local economy. Crab and lobster provide half of the value of all landings in the district with other shellfish making up a further 25% and finfish the balance. The welfare of the shellfish stocks is of major concern.

A large proportion of the shellfish catch is exported directly to the continent, where better prices are often offered than in Britain. Most of the registered fishing boats working from the Isle of Wight (from Bembridge, Cowes and Yarmouth) use static gear. Eels, mullet, flounders and bass are netted in the tidal reaches of the River Medina. Yarmouth has 13 boats. Poole remains important, as do Weymouth and Portland itself, but relatively few boats now operate in and around Lyme Regis.[52]

9.4 South coast resorts

The above discussion reminds us of the importance of the settlements of the south coast of Wessex, providing for fishing, port and military functions that are central to the history of England. The rise of Lyme Regis as a tourist destination bridges all these considerations.

Land was gifted by Cynewulf, King of Wessex, to the Abbot of Sherborne, who established a salt works. The 'Regis' part of the name results from a charter granted by Edward I, illustrating its significance in the medieval period from which its harbour dates and when shipbuilding became an important industry. In 1644, the town was besieged by Royalist forces who were defeated but with many casualties. In 1685, the Duke of Monmouth landed at Lyme Regis and embarked on his ill-fated rebellion to oust the King. By the mid-18th century, Lyme Regis had become a popular seaside resort, visited by writers and painters including Henry Fielding, William Turner and James Abbott McNeill Whistler, while Jane Austen wrote 'Persuasion' during her time there between 1803 and 1804. Mary Anning was born in Lyme Regis in 1799. The railway both arrived and was decommissioned during the 20th century.[53]

Other significant towns that became resorts include Bournemouth, which was uninhabited heath until 1810, but rapidly grew attracting wealthy visitors. Swanage was a small port and fishing village, its development stimulated by the demand for Portland stone, later becoming a holiday resort. Melcome Regis is a part of modern Weymouth with its mid-19th century naval fortress, the 'Nothe Fort',[54] and has the unfortunate distinction of being where the Black Death entered the British Isles (see Chapter 6). The Port of Weymouth supplied ships and mariners in 1347 for the siege of Calais which had begun the previous year, and in 1588 six of the English ships sailed from Weymouth for the fight against the Spanish Armada.

[52] S.A. Walmsley and M.G. Pawson (2007) *The coastal fisheries of England and Wales, Part V: a review of their status 2005–6* Ceefas Tech Rpt. no. 140, Chapter 8. <https://www.cefas.co.uk/publications/techrep/tech140.pdf > Accessed May 2020.

[53] World Guides (2019) Lyme Regis History Facts and Timeline <http://www.world-guides.com/europe/england/dorset/lyme-regis/lyme_regis_history.html> Accessed May 2020.

[54] J.H. Bettey (1986) *Wessex from AD 1000*. Longman, London, 243-248.

Weymouth played an important part in both World Wars .[55] During the 18th century it became a popular sea resort, favoured by George III, who was advised to take the waters for his health.[56] A later monarch, George V, notorious for his heavy smoking, was according to some stories, far from complimentary about Bognor Regis when advised to visit for his health.[57] Bognor too, had a previous existence through smuggling, fishing and farming but it lies outside Wessex as defined here.[58]

The Isle of Wight has several coastal resort towns at Cowes (famous for yachting), Yarmouth, Ryde, Shanklin and Ventnor (famous for its Botanic Gardens). Made fashionable by Queen Victoria (who died at Osborne House on the island in 1901) several former fishing villages enjoyed a certain prosperity brought by tourism in the 19th century[59].

The progression of coastal settlements from port, fishing harbour, industrial or military location towards commerce and tourism during the past 250 years is hardly surprising given that basic infrastructure was already present. The next sections will elaborate on this development.

9.5 The Bristol Channel and Severn Estuary Coast

The other coast of Wessex is the coast of Somerset that runs from the inner shores of the Bristol Channel at Exmoor to the Severn Estuary north of Bristol (Figure 1.1). The discordant nature of the coast around Weston-super-Mare has been noted in Chapter 1, yet northeast of here the 10km section from Clevedon to just beyond Portishead is actually concordant. Along this stretch, the upper Devonian Portishead Formation outcrops along the coast and the Dolomitic Conglomerate rests unconformably on the Devonian (see Chapter 1).[60]

Until Tudor Times the 'Bristol Channel' was known as the Severn Sea. It was considered to reach a long-way to the west and also into the tidal mouths of many rivers, defining a hinterland for the Port of Bristol, and running in effect into the west English Midlands. In modern Welsh, this sea is referred to as *Môr Hafren* and in Cornish as *Mor Havren*.[61] There are several features of particular interest here and which contrast with the southern coast:

- There is one seaway into the region (the Bristol Channel)
- The age range of geological formations is greater, being Devonian to Recent; the former dominates the Exmoor Coast, the latter the Somerset Levels coastline. Triassic and lower Jurassic 'Liassic' rocks commonly occur along the coast from east of Minehead

[55] Weymouth Harbour (n.d.) Weymouth Harbour History <https://www.weymouth-harbour.co.uk/history/> Accessed May 2020.
[56] Weymouth-Dorset.co.uk local History (n.d.) King George III and Weymouth <https://www.weymouth-dorset.co.uk/georgeIII.html> Accessed May 2020.
[57] Famous Last Words (n.d) <https://www.phrases.org.uk/famous-last-words/king-george-v.html> Accessed May 2020.
[58] T. Lambert (n.d.) A brief history of Bognor regis, Sussex <http://www.localhistories.org/bognor.html> Accessed May 2020.
[59] English Heritage (n.d.) History of Osborne <https://www.english-heritage.org.uk/visit/places/osborne/history-and-stories/history/> Accessed May 2020.
[60] C. M. Barton, P.J. Strange, K.R. Royse and A.R. Farrant (2002) *Geology of the Bristol District and accompanying 1:50.000 sheet 264 Bristol.* BGS, Nottingham.
[61] E.T. Jones (2019) The Severn Seas in the fifteenth-sixteenth centuries. Paper delivered to the Conference 'Recent work in the Welsh Landscape History' 19th December 2019.

to the Parrett estuary. Carboniferous and Devonian age formations occur between Weston-super-Mare and Avonmouth.

- While the coast arguably displays as much variation in *appearance* as does the coast of Dorset and Hampshire, it is more straightforward to characterise in terms of stretches of coastal type when classified as concordant, discordant or recent depositional in nature.
- There is a strong historical economic interplay between ports on the English side (Bridgwater, Bristol and Avonmouth).
- The Bristol Channel experiences one of the highest tidal ranges in the World, 12.3m mean spring range at Avonmouth.[62] This means the estuary is considered for the development of tidal power.
- Fine sediment is supplied into the Bristol Channel from the Severn, the Bristol Avon, Wye, Usk, Ebbw, Rhymney, Parrett and Axe, although impoundment at the mouths of the Ely and Taff prevents sediment from these rivers. Sand grade sediment originates in the Celtic Sea south of Ireland.
- Because significant rivers such as the Taff, Parrett and Axe enter the upper reaches of the Severn Estuary, they present difficulties for flood defense, especially the Somerset Levels (Chapter 4), and they are also of great conservation interest.
- The estuarine ecology is varied and valuable, which presents issues for development.
- The effect of tides flowing *into* this *estuary* and up the *river* gives rise *to* the famous *Severn* Bore.

The grade of sediment, as might be expected when considering the varied geology and the active hydrology and coastal processes of the region, varies from clay to gravel in grade. The tidal complexity means there is interaction of fresh water from estuaries with saline water and sediment tending to move into the river mouths. The large tidal range of the estuary leads to very strong currents throughout its main body, but while the Bristol Channel is governed by both tidal currents and Atlantic swell waves (mainly produced from the prevailing south-westerly winds), the Severn Estuary's northeast-southwest orientation gives considerable protection from incoming waves. This is useful to shipping but the Estuary remains dominated by tides, while the funnel-shape of the estuary channel and shallow water friction effects causes 'tidal asymmetry' with the rising (flood) tide dominating over the ebb tide, although the duration of the ebb tide is longer.[63]

Currents in the Bristol channel are complex and depend greatly on tidal cycles. Shipping may, depending on the skill of the seamen involved, readily move inwards towards Bristol and beyond on the rising tide, whereas currents in the falling tide make westward travel to the Atlantic relatively easy.[64]

---

[62] Associated British Ports (n.d.) <http://www.southwalesports.co.uk/Marine_Information/Marine_Information/Bristol_Channel_Tides/> Accessed May. 2020. The tidal ranges in second only in magnitude to the Bay of Fundy in Nova Scotia, Canada.

[63] P. Cannard (2016) The Sediment Regime of the Severn Estuary Literature Review Bristol City Council <https://severnestuarycoastalgroup.org.uk/wp-content/uploads/sites/4/2016/02/The-Sediment-Regime-of-the-Severn-Estuary-Literature-Review.pdf> Accessed April 2021.

[64] E.T. Jones (2019) The Severn Seas in the fifteenth-sixteenth centuries. Paper delivered to the Conference 'Recent work in the Welsh Landscape History' 19th December 2019.

The main source of sand originates from deposits in the 'Celtic Sea' (the Atlantic to the south of Ireland); these were produced by glacial rivers during the Pleistocene. This sediment is transported upwards into the estuary by strong tidal currents. The rising (flood) tide dominates on the north side, so sand is moved along that (Welsh) side to settle on sandbanks within the Estuary. The Ebb tide tends to carry sand along its central axis, and these tides are sufficiently strong not to cause deposition. The substrate here tends therefore to be compacted sand and gravel. Fine sediment is transported by both wind-driven waves and tidal currents with strong interplay of sediments between cliffs, beaches, mudflats, and saltmarsh. Spring tides have higher current velocities which mobilise fine sediment from the estuarine bed, while neap tides have lower current velocities that cannot maintain sediment in suspension, and will tend to deposit sediment once more, creating an immobile layer on the estuary bed. Suspended sediment concentrations are higher around the eastern, English side.[65] Science can now explain why the Author, as a 10-year old boy, was sorely disappointed by the muddy beach at Weston-super-Mare.

Plans for the 16km long Severn Barrage between Lavernock Point, west of Cardiff, to near Brean Down in Somerset (seen on in Figure 1.12 as an example of discordant coastline) would impound an area of 480 km² of the Estuary. There would be lock gates to allow ships access the port at Bristol and Avonmouth and other docks on the River Severn. The proposed Barrage would generate about 5.4 % of current electricity for England and Wales and hence proportionately reduce carbon emissions from electricity generation.

Alternatively, a series of tidal lagoons could be constructed which are discrete and less likely to impact on esturine ecology, would permit the passage of shipping and produce a comparable amount of electrical power close to points of consumption such as Cardiff and Bristol.[66] Progress is somewhat stop-and-go, and the debate between experts continues, considering the national interest, engineering and logistical issues,[67] and in consideration of nature conservation in the Severn Estuary. For example, Simon Brenman of the South West Wildlife Trusts can comment:

'The Wildlife Trusts recognise the huge potential that the Severn estuary has as a source of energy. However, it also provides vital habitat for a vast number of birds, supports juvenile fish species, and is the fourth largest expanse of mud and sand flats in the UK. A barrage has the potential to destroy a large proportion of the estuary's internationally recognised wildlife and habitats, which is why the Wildlife Trusts are opposed to the full barrage schemes.'[68]

It is probably an understatement to say the jury is still out.

---

[65] P. Cannard (2016) The Sediment Regime of the Severn Estuary Literature Review Bristol City Council <https://severnestuarycoastalgroup.org.uk/wp-content/uploads/sites/4/2016/02/The-Sediment-Regime-of-the-Severn-Estuary-Literature-Review.pdf> Accessed April 2021.

[66] Friends of the Earth (n.d.) Severn Barrage Lagoons <https://friendsoftheearth.uk/sites/default/files/downloads/severn_barrage_lagoons.pdf> Accessed January 2020.

[67] J. Excell (2013) Your questions answered: building a Severn barrage. The Engineer 11th February 2013. <https://www.theengineer.co.uk/your-questions-answered-building-a-severn-barrage/> Accessed May 2020.

[68] P. Hain (2012) Should a tidal barrage be built across the Severn estuary? The Guardian 15th May 2020. <https://www.theguardian.com/environment/blog/2012/may/15/severn-barrage-peter-hain-energy> Accessed May 2020.

Fisheries both commercial and for leisure are of economic value in the Severn Estuary. Over 110 fish species have been identified in the Estuary, making it one of the UK's most diverse fish assemblies. Three rare species, designated 'Annex II' features under the Severn Estuary SAC, are river lamprey (*Lampetra fluviatilis),* sea lamprey (*Petromyzon marinus)* and twaite shad (*Alosa fallax).* Additionally, salmon (*Salmo salar*), sea trout (*Salmo trutta*), eel (*Anguilla Anguilla*) and allis shad (*Alosa alosa*) are designated features of the Ramsar Site, as is the wider assemblage of fish species (which includes the migratory species, estuarine specialists and the marine and freshwater species) also a component of the estuary feature of the SAC.[69]

Birdlife, by the same token is internationally important in the Severn Estuary, both wading birds and wildfowl are features for which the Severn Estuary is designated as a SSSI, SAC, SPA and Ramsar. Migratory wintering and passage populations number over 70,000 birds in winter; the entire assemblage of waterfowl (wildfowl and waders) is a feature of the SPA and Ramsar site and is also a notable sub-feature of the 'Estuaries' feature of the SAC.[70]

Further southwest along the coast, west of Bridgwater Bay, the landward grazing marsh and decoy behind the shingle barrier of Sparkhayes Marsh at Porlock, west Somerset, was introduced in Chapter 4. Figure 9.10 shows the changes to the barrier bar after two serious storm events in October 1996 and February 1997 followed by rebuilding the following year. Figure 9.10A shows its location within the Severn Estuary/ Bristol Channel.

This largely west to east drift-aligned gravel barrier (Figure 9.10B) is vulnerable to breaching, and since the 1950s has received occasional recharge of sediment dredged from the harbour entrance at Porlock Weir. This stabilised the crest of the ridge (designated P4.PD in Figure 9.10D), but over-steepened the section of the shingle barrier (Figure 9.10C for May 1994). This part of the crest failed in a storm of October 1996 and retreated to a new position *c.*40m away that was more stable. Re-building occurred September 1998, but the illustration demonstrates that essentially the barrier remained unstable despite efforts to the contrary. There is a complicated relationship over time between major storm events, sea level rise and sediment supply.[71] The gravel bar at Porlock has not been repaired, making Porlock Bay an area of 'managed retreat' where saltmarsh and shingle habitats can be created.[72]

Figure 9.11 shows the breach in the ridge. The build-up of gravel as storm surges deposit fresh pebbles over the former grazing marsh is evident. Close inspection on the ground shows immature saltmarsh vegetation becoming established on former pasture areas. This represents a moving landward of both coastal physical features and their associated habitats. Porlock and Sparkhayes Marsh are fascinating, for they are microcosms of coastal change at a national, or even continental scale.

---

[69] ASERA (n.d.) Fish of the Severn Estuary European Marine Site.<https://www.asera.org.uk/features/fish/> Accessed May 2020.
[70] ASERA (n.d.) Birds. <https://www.asera.org.uk/features/birds/> Accessed May 2020.
[71] J.R. Packham, R.E. Randall, R.S.K. Barnes and A. Neal (eds) *Ecology & Geomorphology of Coastal Shingle*. Westbury, Otley, 41-46.
[72] Plymouth Coastal Observatory (2016) Porlock Bay <https://southwest.coastalmonitoring.org/wp-content/uploads/Learning_Materials/Case-Study-Porlock-Bay.pdf > Accessed May 2020.

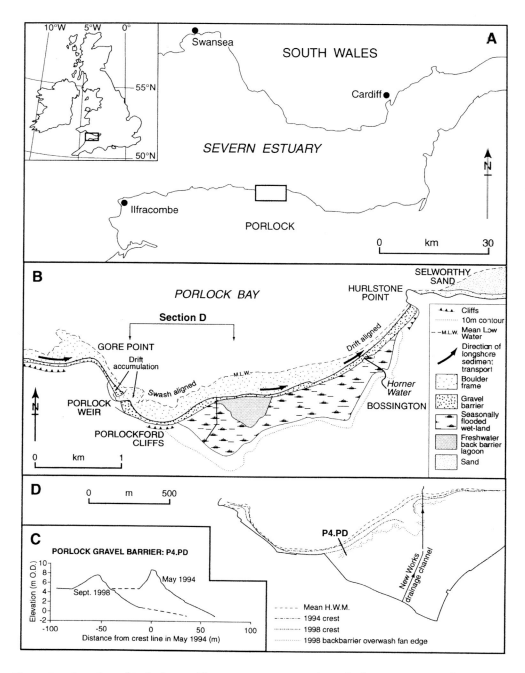

Figure 9.10. Location of Porlock gravel barrier, west Somerset (D. Orford, S.C. Jennings and D.L. Forbes (2001). Origin, development, reworking and breakdown of gravel-dominated coastal barriers in Atlantic Canada: future scenarios for the British coast. in *Ecology & Geomorphology of Coastal Shingle*, J.R. Packham, R.E. Randall, R.S.K. Barnes and A.N. Westbury (eds), Otley.

Figure 9.11. Shingle lobes encroaching on grazing marsh and saltmarsh from the gravel breach at Porlock Weir in 2016 (the author).

9.6 Ports and Coastal Towns of Somerset

Bristol, a city of over a million people, is situated between Somerset and Gloucestershire at a crossing point on the tidal River Avon called *Brig Stow* ('meeting place by the bridge' in Old-English). From humble beginnings Bristol was probably a burgh or fortified settlement in the 10th century and soon became a significant port, competing with Minehead, Bridgwater (especially) and London. Bristol was destined to became one of the country's largest ports and an economically and culturally important city.[73]

In medieval Bristol, wool was woven and dyed, and leather was made. Finished wool products were exported, as was rope, sailcloth, and lead. To assist navigation, between 1239 and 1247 the course of the Frome (a tributary of the Avon) was diverted. The new channel was more than 700m long.[74] Its maritime economy is well known from customs records dating from around 1490.[75] Bristol became a centre of exploration, most notably in 1497 when John Cabot

---

[73] T. Lambert (n.d.) A brief history of Bristol, England <http://www.localhistories.org/bristol.html> Accessed May 2020.
[74] T. Lambert (n.d.) A brief history of Bristol, England <http://www.localhistories.org/bristol.html> Accessed May 2020.
[75] E.T. Jones (2019) The Severn Seas in the fifteenth-sixteenth centuries. Paper delivered to the Conference 'Recent work in the Welsh Landscape History' 19th December 2019.

(a Genoese merchant) sailed across the Atlantic. He is said to be the first European to set foot on the American mainland since the Vikings. Following this voyage, fishermen from Bristol were to settle Newfoundland.

In Tudor times trade was dominated by cloth and wine and involved Ireland, France, Spain, and Portugal .[76] The origins of the slave trade date from this time. Renewed growth came with the 17th century rise of England's American colonies despite two significant sieges during the English Civil War. The 18th century saw the rapid expansion of Bristol's part in the "triangular trade" when African people were taken and exchanged for goods. Trade grew with sugar, cotton, cocoa, wine, tobacco, iron and timber.[77] Around 2,200 voyages from Bristol took Africans as slaves to the Americas between the late 17th century and abolition in 1807. Bristol was the third slaving port after Liverpool and London.[78]

By the 1760s the Bristol docks were losing trade to Liverpool, which had better capacity, a lower tidal range and better access to and from the sea. Bristol's ships had to be of stronger build to cope with being stranded when the tide went out, and the tidal conditions caused delays and congestion, acting as a brake on the growth of trade. The disruption of maritime commerce through war with France (from 1793) and the abolition of the slave trade (1807) further contributed to Bristol's failure to keep pace.

The late 18th and early 19th centuries saw the construction of a 'floating harbour', an impounded harbour using lock gates. Opening in 1809, it allowed ships to remain afloat at all stages of the tide, giving the city new growth, but was not a permanent solution.[79] Advances in shipbuilding and further industrialisation with the growth of the glass, paper, soap, and chemical industries aided the establishment of Bristol as the terminus of the Great Western Railway. In 1841, the broad-gauge Bristol and Exeter Railway also began to run its services from London into Bristol Temple Meads.[80]

Bristol was heavily bombed during the Second World War because it was in the forefront of aircraft manufacture, but it recovered well and today the Bristolian economy is mostly high-end. The city become an important financial centre and high technology hub by the beginning of the 21st century. It also boasts two successful universities. However, as the volume of trade grew, shipping became larger and containerisation became established, the main commercial port function was relocated to Avonmouth in the 20th century.

Avonmouth Dock opened in 1877 at the mouth of the river. Although the entrepreneurs behind its construction struggled at first due to competition with the City Docks, in 1884 the Port of Bristol Authority took over Avonmouth Docks too. The same organisation was now

[76] Women and Property: Early Tudor Bristol (n.d.) University of the West of England. <http://humanities.uwe.ac.uk/bhr/Main/women_prop/2_early_tudor.htm> Accessed May 2020.
[77] W.E. Minchinton (ed.) (1957) *The Trade of Bristol in the Eighteenth Century.* Bristol records Society Publications XX <https://www.bristol.ac.uk/Depts/History/bristolrecordsociety/publications/brs20.pdf> Accessed May 2020.
[78] A. Tibbles (2000) 'The main European ports involved in the slave trade.' TextPorts Conference April 2000. Paper given at Liverpool Hope University College, April 2000. <https://www.liverpoolmuseums.org.uk/ism/resources/slave_trade_ports.aspx> Accessed May 2020
[79] The Bristol Port Co. (2020) Our History <https://www.bristolport.co.uk/about-us/our-history> Accessed May 2020.
[80] A. Bristlin (2014) City of Bristol: History and Information. City History March 5th. <http://www.bsten.co.uk/health/item/302-city-of-bristol-history-and-information> Accessed May 2020.

running the City and Avonmouth Docks under one authority and they started to prosper. In 1908 the Docks were extended, and the Royal Edward Dock was opened by the King.

Royal Portbury Dock was opened in 1977 following the adoption of containerisation by the late 1960s and a steady increase in vessel size that had been progressive since the Industrial revolution. Bristol City Council decided to build the new dock across the river. The Port of Bristol now had a facility large enough to handle containers, but this late arrival on the scene did not prosper. Privatisation in 1991 led to increased investment, and Royal Portbury Dock is now one of the country's major ports.[81]

On the English side of the Severn Sea, massive investment at different times up to the present enabled accommodation of volume of trade, ship size and containerisation against a less than ideal physical configuration of coast and river mouth. The City of Bristol Docks today combine heritage, water transport and pleasure boating where once a major part of English trade passed.

Bridgwater was a Saxon foundation and in 1086 it probably had a population of about 160. The town gained its charter in 1200, including the right to hold a market. Bridgwater had a fair and weekly market, a situation that provided considerable economic stimulus. Prior to that it was largely agricultural. The construction of Bridgwater Bridge c.1200 and pontage (tolls for crossing, used to maintain the structure) levied on its users indicate the importance of the east-west river crossing, and routes both east and south of the town over low-lying marshes had to be causewayed. From the 12th century, it is likely there were quaysides on both sides of the river Parrett.[82] Medieval Bridgwater was also a significant port with a market, but until 1402 was officially part of the port of Bristol. Imports included wine, cloth, grain, beans, peas, and hides.[83]

The town's manufactures also grew lucratively during the Middle Ages. Wool was woven, fulled, and dyed – the town was famous for its drapers and tailors – and a corn market was established. There was also some pottery manufacture.[84] To protect the town, medieval Bridgwater was surrounded by a ditch with an earth rampart topped by a wooden stockade. There were four stone gates where tolls were charged on goods entering the town. Early in the 13th century a castle was built overlooking Bridgwater. Tudor Bridgwater was still a busy port importing fish and millstones from the Forest of Dean, but the cloth trade declined in the 16th century.

Fortunately, shipbuilding developed and by the end of the 17th century production of tiles, terracotta plaques and brickmaking also became prominent, remaining important local industries into the 20th century.[85] Tile drains for agriculture were also manufactured

---

[81] The Bristol Port Co. (2020) Our History <https://www.bristolport.co.uk/about-us/our-history> Accessed May 2020.

[82] A.P. Baggs and M.C. Siraut (1992) Bridgwater, in R.W. Dunning and C.R. Elrington (eds) *A History of the County of Somerset: v6, Andersfield, Cannington, and North Petherton Hundreds (Bridgwater and Neighbouring Parishes)*. Victoria County History, London, 192-206.

[83] T. Lambert (n.d.) A Brief History of Bridgwater, Somerset. <http://www.localhistories.org/bridgwater.html> Accessed May 2020.

[84] J.H. Bettey (1986) *Wessex from AD 1000*. Longman, London, 117.

[85] South West Heritage Trust (2020) Brick and tile Museum <https://swheritage.org.uk/our-sites/brick-and-tile-museum/> Accessed May 2020

in Bridgwater in the 19th century as were agricultural edge-tools. A glass industry was established in the 18th century and along with bricks and tiles its products were exported through the port.[86] The Bristol-Exeter railway was opened as far as Bridgwater in 1841. In the 20th century, the port of Bridgwater went into decline and the docks closed in 1971. Warehouses were converted to flats, while shipbuilding had already ceased in the early 20th century. From the 1930s there was a preserves industry and a cellophane manufacturer. In the late 20th century, surviving industry included heavy engineering, manufacture of electrical equipment, brick making, tile making and brewing.[87]

In the west of our area, the place-name Minehead is not redolent of deep mining, rather it is a compound of Welsh Mynydd (mountain) and English 'head' (or hill), for it presents, especially when approached from the east, an impressive promontory of North Hill. Minehead is an agglomeration of several small hamlets and, by the 14th century, was a small port that traded with Ireland and France as well as along the coast. Reference to a town, port and fair at Minehead occurred in 1380, before the powerful Luttrell family of nearby Dunster Castle gained full control; historically Minehead's market was overshadowed by that of nearby Dunster. During the 15th century Minehead became, thanks to the influence of the Lutrell family, both a successful fishing and trading port, for the family's connections with France encouraged continental trade. Minehead was also a departure point for pilgrims to Santiago de Compostela in northern Spain. By the early fifteenth century, Dunster was losing its port function to Minehead, probably because the haven was becoming too shallow and the shingle bank was rising creating with changes to the coastline and to tides.[88]

Trade developed further in particular with South Wales and the colonies in the West Indies and Virginia, but it was never easy. The town experienced turbulent times and by the 18th century the harbour was in poor repair and charged high duties. The local woollen industry on which much of Minehead's trade depended fell into decline as did its fisheries, particularly herring. The town contracted, despite attempts to reverse its fortunes which included the setting up of a turnpike trust. Two serious fires in the 1790s destroyed much of the lower town. Plans for re-building were slow to emerge and the Luttrells were blamed for their agent's failure to relieve the suffering of the townspeople, occasioning some unrest.

Coastal shipping along the Bristol Channel remained a mainstay of the port; West Somerset timber, grain and other produce were exported, returning with coal, limestone, slate, and groceries. The railway terminus arrived at Minehead in 1874 but was located about 1km away from the harbour, suggesting it was conceived more for commerce and visitors than commodity transport. The expanding South Wales coalfields created a demand for food as well as timber for pit props.[89] Pit props from Exmoor were exported to South Wales and passengers on the return trip helped to develop tourism,[90] which had started to attract visitors in the 18th

[86] J. H. Bettey (1986) *Wessex from AD 1000*. Longman, London, 205, 208, 258.
[87] T. Lambert (n.d.) A Brief History of Bridgwater, Somerset <http://www.localhistories.org/bridgwater.html> Accessed May 2020.
[88] Dunster Haven (n.d.) <www.EnglandsPastForEveryone.org.uk/Explore> Accessed January 2020; D. Taylor (2006) The Overseas Trade of Mid-Sixteenth Century Bridgwater unpub. MA Thesis, University of Bristol, Chapter 4. <http://www.bris.ac.uk/Depts/History/Maritime/Sources/2006taylor.pdf> Accessed May 2020.
[89] VCH (2020) Watchet Harbour <https://www.victoriacountyhistory.ac.uk/explore/items/watchet-harbour> Accessed May 2020.
[90] Minehead Bay Co. (n.d.) History of Minehead <https://www.mineheadbay.co.uk/history-of-minehead> Accessed May 2020.

century. Evidently the tourist aspect of the town would serve to give a new economic impetus. However, the pier was demolished in 1940 as part of the military coastal defense preparations and was not rebuilt, preventing larger boats from visiting Minehead. In 1951, the harbour was given to the Urban District Council by the Luttrells and repaired such that pleasure boats returned. The holiday industry revived in 1962 when Butlins opened in Minehead.[91]

Weston-super-Mare was a small coastal settlement in 1800 that would develop fast in the 19th century; the first hotel opened in 1810. The population of the small fishing hamlet in 1801 was 487, 4,594 in 1851 and 19,448 in 1901.[92] Unlike Minehead it was never originally economically diverse in its activity, rather it was longterm a fishing port.[93] The railway arrived in 1841 and a market was established in the next year. As at Minehead, day-trippers arrived by paddle steamer from South Wales, especially miners. As the town developed and grew into a handsome town during the 19th century, middle-class 'villas' were built. Tourism developed further during the inter-war period (as did the population) including the construction of a pier, Winter Gardens Pavilion and an open-air swimming pool.

Until about 1820, **Clevedon** was an agricultural village, although it was granted a market under a charter of 1346. In the first half of the 19th century, poor quality land for agriculture on the hillside overlooking the sea was developed for housing which afforded good views.[94] Urban development was underway, and shops and hotels likewise grew in number. The railway arrived in 1847 (a branch of the Bristol to Exeter railway) and the middle classes settled. Private schools proliferated, educating the children of the upper middle classes employed abroad in the diplomatic service, by the East India Company, in the Navy and the Army. Similarly, medical doctors, solicitors and other professional people prospered. Clevedon Pier (1869) was constructed from eight spans made of rails left over from a previous failed project of Brunel's (Figure 9.12).[95] This enabled steamers to land their passengers from South Wales and Devon. Opened in 1912, the Curzon cinema is the oldest continually operated cinema in the country.[96]

9.7 Conclusions: Wessex coasts

The two stretches of coast described are of great intrinsic interest and come to define Wessex in many ways. Any notions of an isolated rural hinterland would be erroneous. Due to connectivity with the sea, inland areas have been highly responsive to social and economic change since prehistory, the result not only of available natural resources, but also the essential point that coasts provide for contact beyond the British Isles and Europe via the Atlantic seaways. Hence, Wessex played an unquantifiable role (for better and worse) in the

---

[91] C. Gathercole (2003) *An Archaeological Assessment of Minehead*. Somerset County Council, 4-7, 27. <https://www.somersetheritage.org.uk/downloads/eus/Somerset_EUS_Minehead.pdf > Accessed January 2020.
[92] [96]J.H. Bettey (1986) *Wessex from AD 1000*. Longman, London, Chapter 7.
[93] GB Historical GIS / University of Portsmouth (n.d.) History of Weston super Mare in North Somerset | Map and description. A Vision of Britain through Time. <URL: http://www.visionofbritain.org.uk/place/703> Accessed May 2020.
[94] Clevedon Civic Society (n.d.) The History of Clevedon <http://clevedon-civic-society.org.uk/histbuilders.html> Accessed April 2021.
[95] BBC (2005) Clevedon Pier <http://www.bbc.co.uk/bristol/content/articles/2005/06/20/pier_feature.shtml> Accessed May 2020.
[96] Clevedon Civic Society (n.d.) The History of Clevedon <http://clevedon-civic-society.org.uk/histbuilders.html> Accessed April 2021.

Figure 9.12. Clevedon Pier (the author).

colonial development of the British Empire, and its coastal towns and cities grew to national importance. They remain so today.

Physically speaking these coasts display great diversity not only between but also within each stretch. There are classic, even globally recognised, physical features associated with the variety and forms of coasts that today attract visitors and hence are of economic significance themselves. While the large tidal range for the Somerset coast is a mixed blessing, the nature of the inlets and the shelter provided by the Isle of Wight played a great role in both mercantile and military seafaring, something that dates from prehistory but is demonstrably important since the middle ages. The wealth of Wessex owes much to this relationship, be it through honest exchange of goods and services, or through economic exploitation, including the worse form of all – slavery.

As will be explored in Chapter 10, the Wessex Coast contains considerable sites of conservation *interest* as well as conservation *conflict*. The temptation with the funnel-shaped Bristol Channel is to develop tidal power. It is argued that, while the investment requirement is considerable, the gains in terms of 'green energy' are worth it. The losses are in terms of habitat and maybe convenience to shipping; both coasts display a wide range of habitats. Low coastal areas present inter-tidal zone variation, mudflats, saltmarsh, shingle and sand spits and reclaimed marshland. There is an enormous variation in cliff height, geology, and form. Human activity has altered the very appearance of cliffs by changing sediment supply conditions or by directly quarrying for building.

Bristol was, and through the development of Avonmouth remains, a major port. Its relationship with its hinterland is complicated. A similar thing can be said for Bridgwater in the shadow of Bristol, but its industries over time strongly reflect local economic activity, both agricultural and extractive. Minehead started as a small port (itself in the shadow of Bridgwater), based in fishing and local agricultural produce and timber (as was Watchet). Clevedon and Weston-super-Mare's development as coastal towns was later and occurred less because of their importance to fishing or commodities, but more because of that modern phenomenon, tourism, as well as a residential function from the 19th century. While sea bathing started in the 1700s, real transformation occurred in the next century as first railways, then leisure time, disposable income and eventually the motor car became available to most people. Since the 1960s, it is well-know that holidays abroad have hit British resort towns in the pocket, but day-trippers, retirement and 'dormitory town' functions continue.

The *structure* of the coastal belts of Wessex reflects not just the different rock formations but their disposition with respect to the sea. The outcome is not just the discordant and concordant stretches, but also opportunities for different coastal landforms and habitats.

The *functions* of these coastal landscapes include provision for shipping whilst their ability to provide construction materials is rich as is their heritage and conservation role. Their *value* consequently may be non-market and commercial alike. On the one hand are areas valued for physical, heritage and ecological aspects including geology, archaeology and biodiversity, on the other are primary industry products such as building materials, coastal protection, fisheries and aspects of agricultural production, typically timber and grazing for animals. Up the value chain are the service industries including transport, recreation, tourism and more.

The *scale* involved ranges from minute consideration of physical processes such as sediment particle transport or erosion of cliffs up to whole-coast processes such as longshore drift or the interruption thereof by construction of harbours or potentially tidal energy developments. Meanwhile *change* is omnipresent on the Wessex coastlines. The relentless erosion and transport of geological materials is natural, yet human activity and now climate change, affecting the severity of storms and seal level rise, are ever present. Human intervention in these areas has been ongoing since prehistory, but discussion around how the coast will be managed is ever-present.

Chapter 10

Landscape, value and change

'….The ship was cheered, the harbour cleared,
Merrily did we drop
Below the kirk, below the hill,
Below the lighthouse top.

The Sun came up upon the left,
Out of the sea came he!
And he shone bright, and on the right
Went down into the sea.….'

The Rime of the Ancient Mariner
by Samuel Taylor Coleridge

10.1 Bad times for environmental action?

The leaving of the UK from the European Union underpins a worsening domestic environmental situation. For the first time in world history a supra-national organisation had centralised the 'best practice' of so many developed countries, made a good attempt at codifying what is good, then invited each to work with the others in the common purpose of improving the environment for more than 500 million people.[1] If the blow to environmental goals which 'Brexit' represents were not enough, outside the Wessex region governmental plans to create an upgraded railway network throughout England looks set to damage in excess of 100 conservation sites, including many ancient woodlands, as 'HS2' tears through the counties between London and Birmingham.

An example of success lies in improving water quality through river basin management. At the point of accession to the European Economic Community in 1973, the UK was branded the 'dirty man of Europe'. By the second decade of the 21st century, implementation of the EU Water Framework Directive 2000 led to widespread improvement in the quality of Europe's water bodies.[2] Dangers that water quality gains are being reversed have been highlighted for several years and recently manifest through the UK's ongoing 'Sewage Crisis' whereby several water companies are discharging raw sewage into rivers and the sea. Arguably this is able to persist due to a lack of EU oversight.[3]

Long-standing concern over climate change has led, around the world, to national (including the EU *en bloc*) and local governmental declarations of a 'Climate Crisis'.[4] The declaration of

1   H. F. Cook (2017) *The Protection and Conservation of Water Resources*. Wiley-Blackwell, Chichester, Chapter 4.
2   H. F. Cook (2017) *The Protection and Conservation of Water Resources*. Wiley-Blackwell, Chichester, Chapter 10.
3   Guardian (2016) Brexit would return Britain to being 'dirty man of Europe' 03.02.2016. <https://www.theguardian.com/environment/2016/feb/03/brexit-would-return-britain-to-being-dirty-man-of-europe> Accessed May 2020; https://www.bbc.co.uk/news/explainers-62631320
4   Climate Emergency declaration (2021) <https://climateemergencydeclaration.org/climate-emergency-

a Climate Emergency has been supported by the devolved administrations and many local authorities within the UK, although not by the Westminster Government. Following the intention to withdraw from the 'ground-breaking' Paris Climate Accord of 2015 by former US President Donald Trump, his successor President Joe Biden has resolutely re-affirmed US support for this important agreement.[5] Climate change concern is likewise matched by concerns over losses of biodiversity and habitat loss in the Anthropocene. The outcome has included direct action by the 'Extinction Rebellion' on the grounds that:

> We are facing an unprecedented global emergency. Life on Earth is in crisis: scientists agree we have entered a period of abrupt climate breakdown, and we are in the midst of a mass extinction of our own making.[6]

While the UK's recent past membership of the EU may be seen positively in terms of environmental improvement, events since 1973 do not stand in historic isolation. Even if sometimes ineffective, there has been a long history of campaigning and of dedicated organisations (both public and otherwise) that led to the present regulatory environment.[7]

Wessex is especially famous for its semi-natural landscapes but has certainly not been immune from negative environmental impacts caused by humans. For example, Dorset has been noted for loss of plant species, although the rate is relatively low by national standards. Rates of plant species loss has been greatest since the 1960s, and in Wessex overall the Corncrake vanished progressively throughout the 20th century.[8] While serious efforts to conserve (semi-natural) habitats and prevent large-scale species loss have been noted, one celebrated re-introduction has been the Great Bustard to Salisbury Plain.[9]

The drive for agricultural production is ancient. It incorporates drainage and other forms of 'improvement' leading to monocultures that favour economic gain. Most modern criticism of damage to the structure of the countryside and to natural and semi-natural habitats is aimed at agriculture since 1940. At that time, much land went under the plough to achieve wartime food security. In concert, land drainage at both regional and field-scale became supported by public money and the administrative procedures to that end streamlined.[10]

It is the purpose of this Chapter to review the organisational mechanisms of conservation to the present, and to outline the challenges. The apparent spontaneity of environmental protests in recent years emphasises the value of non-governmental action. Less sober in style

---

declarations-cover-15-million-citizens/> Accessed April 2021.

[5]   The White House (2021) Paris Climate Agreement <https://www.whitehouse.gov/briefing-room/statements-releases/2021/01/20/paris-climate-agreement/> Accessed April 2021.

[6]   Extinction Rebellion (n.d.) Alone Together <https://rebellion.earth/> Accessed April 2021.

[7]   H. Cook and A. Inman (2012) The voluntary sector and conservation for England: achievements, expanding roles and uncertain future. *Journal of Environmental Management* 112, 170-177. http://dx.doi.org/10.1016/j.jenvman.2012.07.013

[8]   Natural England (2010) England's lost and threatened species. <http://www.biodiversitysouthwest.org.uk/docs/EnglandsLostAndThreatenedSpecies2010.pdf> Accessed May 2020.

[9]   RSPB (n.d.) Reintroducing the great bustard to southern England. <https://www.rspb.org.uk/our-work/conservation/projects/reintroducing-the-great-bustard-to-southern-england/> Accessed May 2020.

[10]   H.F. Cook (2010) Floodplain agricultural systems: functionality, heritage and conservation. *Journal of Flood Risk Management* 3, 1-9. DOI:10.1111/j.1753-318X.2010.01069.x; H. F. Cook (2010) Boom, slump and intervention: changing agricultural landscapes 1790 to1990, in M.P. Waller, E. Edwards and L. Barber (eds) *Romney Marsh: Persistence and Change in a Coastal Lowland*. Romney Marsh Research Trust, Sevenoaks, 155-183.

that its Victorian and early 20th century antecedents, contemporary radicalism seems to have its origins in the post-Second World War milieu that saw not only the birth of many protest movements, but also recognised new threats to humanity and the environment, starting with the Campaign for Nuclear Disarmament (CND).

A good starting point is that of non-governmental action. It is, however, not entirely appropriate to ascribe clearly differentiated roles to 'official' or to 'Non-Governmental Organisations' (NGOs) for the interplay of the two is something of a strength in UK conservation history. Sections 10.2 and 10.3 are largely based upon a published review of the topic.[11]

10.2 Voluntary conservation: a background

In his perceptive account of environmental history in the 20th century, John Sheail identifies the emergence of a rural economic 'third force' additional to farming and forestry. This provides for 'conscious stewardship of rural landscapes for their amenity and wildlife'; it is linked with opportunity for city and town dwellers to visit the countryside. With its origins in 19th century state concerns over human welfare, the third force reflects a 'philanthropic and voluntary response' from those in positions of responsibility and influence.[12] Alan Rogers can sum it up:

> as presenting a long and honourable history that promotes response to needs and inducing change.[13]

UK Conservation was born of campaigning as well as an instrumental need merely to conserve, be it habitats, species, buildings, archaeology, or geological sites. Small charities such as the Harnham Water Meadows Trust in Salisbury tend to be locally focussed, while larger bodies, capable of redressing the balance between   production, agriculture and conservation, should fulfil a campaigning role. The National Trust was formed in 1895 with an Act of Parliament enabling it to own property passed in 1907. Majorly influential, the Trust is clear about its political and economic independence. The Royal Society for the Protection of Birds (RSPB) was formed in the late 19th century and today manages many nature reserves. Other nation-wide organisations include the Wild Fowl and Wetlands Trust (1946) and the Woodland Trust (1972).[14]

During the inter-war period, concern over unplanned urbanisation led first to an appeal to the counter-industrial 'English rural idyll' followed by the politicisation of countryside conservation forged in the formation of the Campaign to Protect Rural England (the modern name for the 'CPRE'). This NGO dates from 1926. Its founders, the Earl of Crawford and Balcarres (politician and art historian), Sir Guy Dawber (architect) and the pioneer town planner, Sir Patrick Abercrombie, represented not only a cultural elite but also reflected a hostility, at the

---

[11] H. Cook and A. Inman (2012) The voluntary sector and conservation for England: achievements, expanding roles and uncertain future. *Journal of Environmental Management* 112, 170-177.
[12] J. Sheail (2002) *An Environmental History of the Twentieth Century*. Palgrave London, 10, 16.
[13] A.W. Rogers (1987). Voluntarism, self-help and rural communities development: some current approaches. *Journal of Rural Studies* 3, 353-360.
[14] H. Cook and A. Inman (2012) The voluntary sector and conservation for England: achievements, expanding roles and uncertain future. *Journal of Environmental Management* 112, 170-177.

Figure 10.1. Kingcombe Meadows Nature Reserve (Ben Heaney).

top level of the British establishment, to large commercial and urban centres. Abercrombie could state that 'the thing that is England, is the countryside'.[15]

County Wildlife Trusts also began to be set up between the two world wars and form a significant basis for voluntary sector countryside management. Founded in 1926, the Norfolk Wildlife Trust was the first, the Wiltshire Wildlife Trust was formed as late as 1962; the expansion of country trusts represented concern at the extent of post-Second World War agricultural expansion.[16]

---

[15] P. Lowe, G. Cox, M. MacEwen, T. O'Riordan and M. Winter (1986) *Countryside Conflicts*. Gower/ Maurice Temple Smith, London, 12; J. Sheail (2002) *An Environmental History of the Twentieth Century*. Palgrave London, 106.
[16] J. Sheail (2002) *An Environmental History of the Twentieth Century*. Palgrave London, 130; H. Cook and A. Inman (2012) The voluntary sector and conservation for England: achievements, expanding roles and uncertain future. *Journal of Environmental Management* 112, 170-177.

The outcome is a plethora of *environmental non-governmental organisations* (ENGOs) today found at work across the UK. For instance, there are (following amalgamations) some 46 county-based Wildlife Trusts, the majority being in England, and they are organised within the Royal Society of Wildlife Trusts. The Wiltshire Wildlife Trust works with landowners and managers to help them to manage their land in sympathy with the needs of wildlife. The Dorset Wildlife Trust (1961),[17] manages Brownsea Island Nature Reserve in partnership with the National Trust, and in 1987 purchased Kingcombe Meadows Nature Reserve (area 185ha) at Toller Porcorum, near Dorchester, in order to protect a part of the Dorset countryside that had been free from agrochemicals. It also contains several features of archaeological interest.

The wildlife interest of Kingcombe includes:

- Typical wildflowers in woodland and marshy areas
- Grassy meadows supporting many species, rarer examples including lady's mantle, corky-fruited water dropwort, pepper saxifrage, devil's-bit scabious and knapweed
- Butterflies including varieties of skipper & fritillaries
- Birds abound with visiting summer warblers and woodland birds[18]

Figure 10.1 shows the estate at Kingcombe Meadows. The impact of enclosure and the prevalence of woodlands and riverside meadows is clear. Within the estate boundary East of Higher Kingcombe, and reaching to the modern A356 is largely enclosed downland. The River Hooke flows through several meads (riverside meadows and watermeadow). South of the river is predominantly enclosed common. The enclosure is likely early 19th century.[19]

Before considering the statutory side of planning, it is worth stating that in international terms, the UK is rare in both its established tradition and modern *actualité* of voluntary sector action in environmental management and conservation. It has been noted that ENGOs vary in scale from the very local, with a specific remit to manage a defined area deemed worthy of conserving, to county level or catchment-wide remits (embracing multi-functionality) up to national bodies (such as the Woodland Trust, Wildfowl and Wetland Trust, or National Trust) whose remits vary from nation-wide thematic concern for specific habitats to wide-ranging functions including agricultural land, protected sites and buildings. Exceptionally large ENGOs are internationally organised and are concerned with planetary-scale habitat conservation, climate change and pollution. In this category are the World Wildlife Fund and Greenpeace.

In both continental Europe and the USA, ENGOs function at a large scale or engage in thematic roles such as combating climate change. The brilliantly named Ducks Unlimited (1937) manages some 14 million acres (5.7 million hectares) of wetland conservation across the USA. It is volunteer based and raises revenue from evets such as shooting, fishing and golf.[20] Additional to wildlife conservation, wetlands trap sediment, nutrients and toxins while increasing the wetland and lake area in a watershed (catchment) dramatically reduces flood

---

[17] Dorset wildlife trust (n.d.) Our vision and mission <https://www.dorsetwildlifetrust.org.uk/what-we-do/about-us/our-vision-and-mission> Accessed May 2020.

[18] Dorset wildlife trust (n.d.) Kingcombe Meadows <https://www.kingcombe.org/kingcombe-meadows-nature-reserve> Accessed May 2020.

[19] Legislation.gov.uk (n.d.) <https://www.legislation.gov.uk/changes/chron-tables/private/24> Accessed May 2020.

[20] Ducks Unlimited (n.d.) About Ducks Unlimited <https://www.ducks.org/About-DU?po=footer-m> Accessed May 2020.

peaks. Such ecosystem services make wetland re-instatement or establishment an attractive and cost-effective solution.

The USA has also pioneered citizen-led initiatives, among them coalitions enabling river basin management. For example, the Upper Susquehanna Coalition (USC) has responsibility for some 444 miles (710km) from its upper watershed at Otsego Lake to Chesapeake Bay, where there are water pollution problems. The Chesapeake Bay watershed program engages professionals to develop strategies, and in partnership with the USC network, provides upstream support to county members, local watershed organisations, town and county public works, planning officials, farmers and businesses. 'Citizens for Catatonk Creek' is one USC committee established following serious flooding during the 1990s. It amasses technical and land use information and is resident led. Funding is procured from US federal, state, and local sources.[21]

In western continental Europe, there are plenty of operational ENGOs, many of which have familiar names as international players. By the last decade of the twentieth century, four large organisations, the German Nature Protection League (NABU), the Worldwide Fund for Nature (WWF), the German League for Environment and Nature Protection (BUND) and Greenpeace, each with over 250,000 supporters, came to be recognised as the most important German environmental organisations.[22] The situation in France seems not materially different at a national level,[23] and the same pattern can be seen in many European countries.[24] However, in France, the second half of the last century saw a rise of natural area conservation societies that are independent and operate under statute.[25]

In former Communist Bloc countries, a lack of funding caused ENGOs to adopt a role focused on lobbying and monitoring of governmental institutions.[26] Today in south-east Europe they play an important role in environmental education, debating ecological issues, working on environmental projects, and increasing public awareness, and in promoting to work of indigenous conservation organisations. Some organisations, however, 'appeared to compromise themselves by becoming political parties'.[27]

While a comprehensive analysis of state *vs.* private *vs.* voluntary sector management of natural areas around the world is by no means possible here, the more direct role of national and local government is stronger in managing areas of environmental vulnerability across the European continent. To take two examples, groundwater protection is led by the city of Aalborg in Denmark, while in the Netherlands 'spatial planning' plays a wider role in the

---

[21] H. F. Cook (2017) *The Protection and Conservation of Water Resources*. Wiley-Blackwell, Chichester, Chapter 11.
[22] W.T. Markham (2008) *Environmental Organizations in Modern Germany: Hardy Survivors in the Twentieth Century and Beyond*. Berghahn Books, Oxford.
[23] M. Herry (2020) Environmental Organizations in France: a Structural Analysis. European Consortium for Political Research <https://ecpr.eu/Events/PaperDetails.aspx?PaperID=42128&EventID=115> Accessed May 2020.
[24] The Green 10 (2020) <https://green10.org> Accessed May 2020.
[25] A. Guignier and M. Prieur (2010) Legal Framework for protected areas: France <https://www.iucn.org/downloads/france_en.pdf> Accessed May 2020.
[26] T. Borzel and A. Buzogany (2011) Environmental organisations and the Feuropeanisation of public policy in central and eastern Europe: the case of biodiversity governance in A. Fagan and J Carmin (eds) *Green Activism in Post-Socialist Europe and the Former Soviet Union*. Routledge, Abingdon, Chapter 2.
[27] D. Turnock (2004) The Role of NGOs in Environmental Education in South-eastern Europe. *International Research in Geographical and Environmental Education* 13 (1), 103-109. https://doi.org/10.1080/10382040408668800

protection of water resources and in flood protection.[28] Elsewhere in Europe it seems public sector responsibilities are generally met directly, while in the UK the tradition of conservation by voluntary and NGO action is much more to the fore in delivering environmental objectives.

10.3 The rise of the statutory planning process and nature conservation

The modern planning process is largely post-Second World War and reflects concerns typified by Sir Patrick Abercrombie who, along with his associates, re-designed London after the Blitz.[29] There had been some pre-War planning legislation, for example the Restriction of Ribbon Development Act (1935). However, major legislation in the Town and Country Planning Act (1947) established planning permission as a requirement for development affecting rural areas.[30]

Yet despite this, planning permission affecting cropping, forestry and farm building development was scarcely affected, and in the latter case there remain major exemptions from the planning process. The impact has inevitably affected biodiversity and conservation. Recent plans for HS2 across England, somehow seem an echo of this.

The 1945 Labour Government had nonetheless recognised the role of the third or voluntary sector in establishing the welfare state and it was enshrined within welfarism before 1945. This multiple approach was seen integral to the 'body politic', with the third sector often working with the same principles as government in providing social services, while establishing its own (separate) sphere of action to the state. There emerged an enduring post-Second World War precedent - even a consensus - that such voluntary sector organisations should take on responsible roles. Environmental conservation was similarly engaged.[31]

The National Parks and Access to the Countryside Act (1949) also established the statutory Nature Conservancy ('Council' was added to the full title when re-constituted in 1973), a forerunner of Natural England (formed in 2006). Nature reserves became a matter of direct national intervention; hence mid-20th century solutions were still seen in planning terms although the dialogue had opened to include a range of heritage interests (landscape, heritage farming and architectural features) as well as providing for open-air recreation.

The definition of National Nature Reserves (NNRs) and Sites of Special Scientific Interest (SSSI) notified on private land became possible. Pluralist action was enabled by statutory agency and by voluntary sector bodies. Yet this was not the end of the matter, for environmental pressure groups once more became manifest through the activities of Friends of the Earth from the 1970s, a dramatic expansion of the RSPB later in the last century and the rise and achievements of the UK-wide Farming and Wildlife Advisory Groups (FWAG).

---

[28] H.F. Cook (2017) *The Protection and Conservation of Water Resources*. Wiley-Blackwell, Chichester, Chapter 11.
[29] The Encyclopaedia Britannica (2020) *Sir Patrick Abercrombie* <https://www.britannica.com/biography/Patrick-Abercrombie> Accessed May 2020.
[30] P. Brandon (1998). *The South Downs*. Phillimore, Chichester, Chapter 14.
[31] H. Cook and A. Inman (2012) The voluntary sector and conservation for England: achievements, expanding roles and uncertain future. *Journal of Environmental Management* 112, 170-177.

The more direct action-based international organisation, Greenpeace, dates from 1971.[32] Founded in an effort to influence the more conventional political process, the Green Party of England and Wales also has its origins in the 1970s. This period constitutes a second campaigning phase.

These developments successfully bridge two worlds – one of production and the other of conservation. FWAG may be a means of redressing the damage to biodiversity arising from omissions in the planning process. Formed in 1969, it is an organisation to which farmers may turn for advice in matters of wildlife conservation.[33] There was a consequent shift from anthropocentric-motivated conservation (perhaps manifest in notions of the romantic and of environmental stewardship) towards eco-centric motivations, with 'purer green' objectives.[34]

A switch has been proposed in local government from dominance by landowner and farming interests in the 1970s towards a more diverse range of 'counter-urbanised and service classes' of rural dwellers, generally with 'middle-class values.' These were concerned with local government and NGO function in environmental action alongside a will from the 1980s to 'reduce central state intervention and bureaucracies.'[35] Rural interests revolve around a number of axes: production in the form of agriculture and forestry, heritage and biological conservation and rural social issues, and delivery is made through each of the public, voluntary and private sectors. Armed with a level of legal recognition, responsibilities and a remit for conservation management, environmental charities became enmeshed with the array of national government statutory bodies and local government roles, including the management of nature reserves.

Others, including the RSPB, also operate as campaigning organisations. Lacking both governmental and democratic remits, mass membership confers legitimacy and they are regularly consulted as a part of the statutory process. Larger NGOs, by dint of mission, may display limited remits and questions may be asked in respect of smaller charities regarding their ability to embrace a wider vision, or more problematically they might appear from the outside as 'single-issue' groups. And legally, voluntary sector groups properly constituted with boards of trustees etc., can own and manage land. Voluntary sector bodies are thus increasingly deriving legitimacy from central government and from its agencies (as well as from public support and membership), achieving recognition for their contribution.

This clear shift in power sharing is manifest in the formation of Natural England (2006). This agency was formed in a reorganisation of English Nature, and prior to that its key functions were performed through the Nature Conservancy Council until 1990. A statement of widening roles in governance are manifest through the Rural Communities Act 2006:

> Natural England should champion England's natural environment, and must have the authority, resources and capacity to deliver its general purpose, while working

---

[32] Greenpeace (n.d.) About Greenpeace <https://www.greenpeace.org.uk/about-greenpeace/> Accessed May 2020.
[33] J. Sheail (2002) *An Environmental History of the Twentieth Century*. Palgrave London, Chapter 6.
[34] C.V. Burek (2008) The role of the voluntary sector in the evolving geoconservation movement, in C.V. Burek and C.D. Prosser (eds) *The History of Geoconservation*. Geological Society, London, 61-89.
[35] G.E. Cherry and A. Rogers (1996) *Rural Change and Planning*. E. and F.N. Spon, London, 172-4.

alongside farmers, landowners and NGOs. Successive reductions to its budget, however, have limited its ability to perform key functions, and reduced its wider influence.[36]

The phrase 'supra-national legislation' may be anathema to euro-sceptics, but it has a technical meaning implying that sovereign governments are in some way subject to higher legislative authority. This applies as much to United Nations backed international law and conventions as it does to anything from the European Union. For example, the United Nations Environment Programme acts as a catalyst, advocate, educator, and facilitator to promote the wise use and sustainable development of the global environment.[37] International law is concerned with protection of marine environments, fisheries, atmospheric pollution, water, sanitation and more. The World Meteorological Organisation (a United Nations body) has expressed grave concern about the continuing rise of carbon dioxide levels in the atmosphere.[38]

European Directives are more immediately visible and hence, like most EU pronouncements are prone to criticism from certain quarters of the UK's political landscape. These legal acts are created reflecting best practice internationally and involving member states. The expression 'Framework Directive' invites these member states to implement these Directives in the light of their domestic legislation. European Directives set the bounds within which common environmental objectives are sought. They are distinct from a 'regulation' that is a legally binding act.[39]

The future of EU environmental protection legislation is uncertain for the UK.[40] Key EU environmental Directives still apply to the UK's environment and are included because they define key areas of environmental protection.[41] For example, the Water Framework Directive (WFD) 2000/60/EC and the Habitats Directive (Council Directive 92/43/EEC) remain in force. WFD requires Member States to use their River Basin Management Plans (RBMPs) and Programmes of Measures (PoMs) to protect and, where necessary, restore water bodies in order to reach 'good' status and to prevent deterioration. 'Good' status indicates both good chemical and good ecological status.[42] Post-Brexit, the immediate challenge for waters in and around UK, including Wessex, is raw sewage discharges into rivers, lakes and the seas.[43]

The Habitats Directive (that remains in force in the UK) is highly significant for Wessex. It is concerned with the conservation of natural habitats and of wild fauna and flora aims to promote the maintenance of biodiversity, taking account of economic, social, cultural, and regional requirements. The Habitats Directive forms the cornerstone of Europe's nature conservation policy and works together with the Birds Directive (which aims to protect all

---

[36] House of Lords (2018) The Countryside at a crossroads HL paper 99. <https://publications.parliament.uk/pa/ld201719/ldselect/ldnerc/99/99.pdf> Accessed May 2020.

[37] United Nations (n.d.) United Nations and the Rule of Law <https://www.un.org/ruleoflaw/un-and-the-rule-of-law/united-nations-environment-programme/> Accessed May 2020.

[38] WMO (2019) Greenhouse gas concentrations in atmosphere reach yet another high. <https://public.wmo.int/en/media/press-release/greenhouse-gas-concentrations-atmosphere-reach-yet-another-high> Accessed May 2020.

[39] European Union (2019) Regulations, Directives and other acts <https://europa.eu/european-union/eu-law/legal-acts_en> Accessed May 2020.

[40] https://www.clientearth.org/latest/latest-updates/news/why-the-uk-environment-bill-matters/

[41] H.F. Cook (2017) The Protection and Conservation of Water Resources. Wiley-Blackwell, Chichester, Chapter 8.

[42] European Commission (2000) Water Framework Directive. <https://environment.ec.europa.eu/topics/water/water-framework-directive_en> Accessed May 2020.

[43] Guardian (2021) <https://www.theguardian.com/environment/2021/mar/31/water-firms-discharged-raw-sewage-into-english-waters-400000-times-last-year> Accessed January 2023.

the 500 wild bird species naturally occurring in the EU). The Habitats Directive establishes the EU wide Natura 2000 ecological network of protected areas, safeguarded against potentially damaging developments.[44]

Special Areas of Conservation (SAC) are designated under the Habitats Directive. This requires the UK Government and devolved administrations to establish a network of important high-quality conservation sites that will make a significant contribution to conserving the identified habitats and species. Special Protection Areas (SPA) are classified in accordance with the Birds Directive to protect birds identified as vulnerable including migratory species.[45]

In Wessex, examples of SACs (which constitute a range of vulnerable habitats) are Pewsey Downs and Fyfield Downs; these are also National Nature Reserves under older UK legislation. At Salisbury, the Avon is joined by three of its major tributaries – the Rivers Bourne, Nadder and Wylye – and a short distance downstream by the River Ebble. They support habitats and species of national and international importance, reflected in their designation as Sites of Special Scientific Interest (under older UK legislation), SAC and SPA.[46] Ramsar sites are wetland sites designated under much older international legislation and include the Solent and Southampton Water site.

10.4 Forestry and woods

Regarding forestry specifically, the establishment of the Forestry Commission (1919), replacing the Office of Woods, reflected a strategic concern over security of timber production, and in times of agricultural recession, timber, coppice, and forestry provided employment in rural areas. The Commission, like post-1940 agriculture, was set up to meet national imperatives, which would bring it into conflict with conservation and landscape considerations.[47] The New Forest provides one such example of conservation conflict, following the transfer in 1924 of the Royal Forests (New Forest and the Forest of Dean) to the Forestry Commission (FC).

It may be difficult to comprehend the role of the FC in the New Forest, which had clearly been valued as a 'national gem' pre-1900. The 20th century clashes – between timber production and ecological and landscape conservation imperatives – may have been fuelled by 20th century optimism outside the perambulation of the Royal Forest, but pessimism ruled within. Friction would prove especially manifest after the Second World War and was resolved first by the curtailment of FC activities (generally around the extensive planting of conifers rather than broadleaved trees) and finally by the establishment of the New Forest National Park in 2005.

There was considerable military activity in the Forest during the Second World war, including bomb testing and the imposition of airfields. Additionally, there was limited agricultural

---

[44] European Commission (2019) The Habitats Directive. <https://ec.europa.eu/environment/nature/legislation/habitatsdirective/index_en.htm> Accessed May 2020.

[45] JNCC (2020) Special Areas of Conservation: An overview <https://jncc.gov.uk/our-work/special-areas-of-conservation-overview/> Accessed May 2020.

[46] P. Davison and S. Green (2019) River Avon SAC -Phosphate neutral development. Wood Environmental and Infrastructure Solutions UK Ltd. <https://cms.wiltshire.gov.uk/documents/s157886/HRA0501RiverAvonSACPhosphateIDPMainReport.pdf> Accessed May 2020.

[47] H. Cook (2018) New Forest. The Forging of a Landscape. Windgather Press, Oxford, Chapter 8.

cultivation. The Attlee Government from 1945 was to produce statutory responses to onslaughts on the countryside by both agriculture and urban development. The publication of a report by John Dower first laid out a need for 'relatively wild country' to be preserved for open air enjoyment, while conserving historic buildings and wildlife and maintaining farming.[48] The report included the New Forest in the 'other amenity areas not suggested as national parks' category, although it should be noted this was primarily because Dower felt the New Forest was already adequately dealt with by the FC, rather than it not meeting the criteria. The Hobhouse Report (1947) identified the New Forest as one of 12 proposed national parks,[49] yet it took until the present century for the Forest to be formally designated as such, viz: '...it should have great natural beauty, a high value for open-air recreation and substantial continuous extent'.

The overall legislative response is surprising; one incredibly positive aspect was the National Parks and Access to the Countryside Act (1949):

> An Act to make provision for National Parks and the establishment of a National Parks Commission; to confer on the Nature Conservancy and local authorities powers for the establishment and maintenance of nature reserves; to make further provision for the recording, creation, maintenance and improvement of public paths and for securing access to open country, and to amend the law relating to rights of way; to confer further powers for preserving and enhancing natural beauty; and for matters connected with the purposes aforesaid. [16 December 1949].

Within this Act was also provision for SSSIs and NNRs which are today designated by Natural England (formerly the Nature Conservancy) as key places for wildlife and natural features. These can also reside in public ownership. During the 1950s, there were ten national parks in England and Wales. These are the protected landscapes of the Peak District, Lake District, Snowdonia, Dartmoor, Pembrokeshire Coast, the North York Moors, Yorkshire Dales, Exmoor, Northumberland and Brecon Beacons. The National Parks and Access to the Countryside Act 1949 would have significant influence outside the New Forest perambulation, at least in 'upland' Britain.

The New Forest had lost out. This was perhaps an unexpected victory for the 'pro-timber lobby'.[50] The response to the needs within the New Forest was the New Forest Act of 1949. There was, however, acknowledgement of 'due regard to the existing rights and interests', and enigmatically, 'provision for adjusting the forest to modern requirements'. While consultation with the verderers by the FC was a requirement in setting management prescriptions, and there were (for example) measures to protect the 'ancient and ornamental woodland', the presumption was otherwise towards timber production. Many of these concerns were addressed in the New Forest Act of 1964 including 'due regard to the interests of amenity', and

---

[48] Ministry of Town and Country Planning (1945) Report on National Parks in England and Wales by Mr John Dower: consideration of draft and subsequent publicity and policy. National Archives Ref HLG 92/49 <https://discovery.nationalarchives.gov.uk/details/r/C2147250> Accessed May 2020

[49] Department for Environment, Food and Rural Affairs (2005) Explanatory memorandum to the New Forest National Park Authority (Establishment) Order 2005 no. 421. <http://www.legislation.gov.uk/uksi/2005/421/pdfs/uksiem_20050421_en.pdf> Accessed May 2020.

[50] H. Cook (2018) *New Forest. The Forging of a Landscape*. Windgather Press, Oxford, Chapter 8.

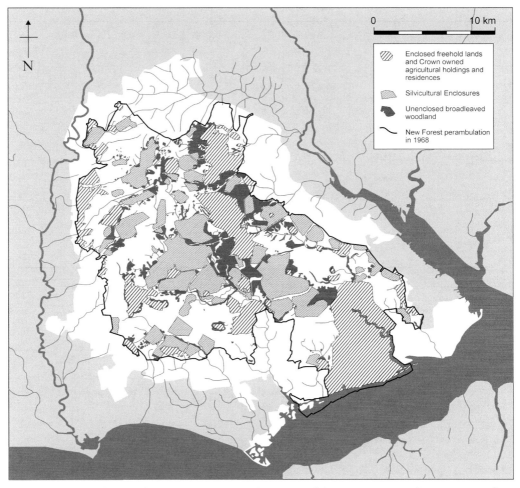

Figure 10.2. Land use in the New Forest in 1968. (Courtesy of the New Forest Ninth Century Trust).

again in 1970 when the verderers' effective control over recreational facilities, among other measures was realised.

Peter Roberts, a former chair of the New Forest Association, can state:

> Throughout the thirties and forties, the FC pushed for greater powers to enclose new lands for timber production. They finally achieved their goal with the 1949 New Forest Act, making use of the perceived national need for timber following the war despite the presentations and concerns of the New Forest Association.[51]

---

[51] Friends of the New Forest/ New Forest Association (2018) <http://newforestassociation.org/our-history/> Accessed May 2020; P. Roberts (2016) *Saving the New Forest*. New Forest Association, Lyndhurst, 69-70.

The National Trust is second only to the Crown as a landowner in the New Forest, and the FC directly manages almost half of the national park area. The Hampshire and Isle of Wight Wildlife Trust is a major landowner, with its two largest reserves, Roydon Woods and Keyhaven and Lymington Marshes, both within the boundary. The National Trust also supports 'commoning', the ancient system of farming that has shaped these commons and is directly involved in management of these areas.[52]

Figure 10.2 shows land use in the mid-20th century. Enclosure is indicated as are enclosed and unenclosed areas of woodland. Within this complicated medieval landscape resides a mosaic of land covers including open forest (including holly, beech and oak woodland), enclosed ('inclosed') woodland and conifer plantation, valley mires and bogs, wet heath, dry heath, (generally acid) grassland, grassland with bracken and gorse, and 'lawns' (short grazed grassland).[53] Within this area there are Sites of Special Scientific Interest, internationally designated sites, including SPA and Ramsar sites, local nature reserves, and other, non-statutory sites.

The New Forest National Park was established in March 2005, the first national park in the south of England, and the New Forest National Park Authority took on its full duties in April 2006. It was created:

- To conserve and enhance the natural beauty, wildlife and cultural heritage of the area.
- To promote opportunities for the understanding and enjoyment of the special qualities of the National Park by the public.
- Working in partnership with other organisations it is also the Authority's duty to seek to foster the economic and social well-being of the local communities within the National Park.[54]

10.5 The Avon Gorge and Gordano Valley

At about 2.5km long, the Bristol Avon Gorge cuts through the steeply dipping strata of the Carboniferous Limestone.[55] The folding and the thrust faults of the Mendip-Clevedon-Bristol area constitutes a 'thin-skinned foreland thrust-belt of Variscan age with dominant movement to the north'.[56] This folding took place in the late Carboniferous to early Permian and it created a folded upland area in Permian time. This upland was subsequently covered by Mesozoic sediments, subsequently it was revealed again by eastward scarp retreat of these deposits. The Gorge cuts through the southern limb of the asymmetric / recumbent Westbury Anticline revealing the sequence of Carboniferous Limestone through to the underlying Devonian Old Red Sandstone in the north, at Avonmouth.[57] It is notified SSSI.

[52] H. Cook (2018) *New Forest. The Forging of a Landscape*. Windgather Press, Oxford, 194.

[53] C.R. Tubbs (2001) *New Forest, the history, ecology and conservation*. New Forest Ninth Centenary Trust, Lyndhurst, Chapter 1.

[54] New Forest National Park (2007) 1. Key Facts and figures. <https://www.newforestnpa.gov.uk/app/uploads/2018/01/aboutus1_keyfacts.pdf> Accessed May 2021.

[55] Avon Gorge and Downs Wildlife project (n.d.) Geology. <https://avongorge.org.uk/wildlife-and-geology/geology/> Accessed May 2020.

[56] G.D. Williams and T.J. Chapman (1986). The Bristol-Mendip foreland thrust belt. *Journal of the Geological Society* 143, 63-73. https://doi.org/10.1144/gsjgs.143.1.0063

[57] R. Bradshaw (1987) The Geology and Evolution of the Avon Gorge. *Proceedings of the Bristol Naturalists Society* 47, 45-64.

The Gorge has long held a fascination for geologists, being close to Bristol and its University. The geology imparts soil type dominated by permeable soils developed on Carboniferous Limestone and more acidic soils on the underlying Devonian Old Red Sandstone. Although of limited agricultural utility and close to a large urban area, the historical ecology of the woods has attracted attention.

Located opposite the Clifton and Durdham Downs (on the east side of the Gorge), Leigh Woods are famous for the Bristol whitebeam (*Sorbus bristoliensis*). This and other rare trees and plants are to be found on both sides of the Gorge, yet Leigh Woods are an intriguing survival, partly due to their relative inaccessibility. A large area of these Woods is designated NNR and managed by Forest England (formerly the Forestry Commission) and is outside the NNR but includes a part of the Paradise Bottom SSSI.[58] On the Leigh Woods side, most of the southern end is owned and managed by NT as the NNR (the rest is a thin strip along the railway). The northern part of the SSSI is mostly FC (again, excluding the railway, but also three quarries). The FC land extends beyond the SSSI. The National Trust area also includes an Iron Age hillfort (Stokeleigh Camp) and veteran trees; efforts are being made to restore coppice rotation.[59]

A parish enclosure wall was built in 1813, and it marks the boundary between two parishes, although it was established by an enclosure award, the woods remaining unenclosed. The wall served to separate Long Ashton Woods (largely pasture or wood-pasture with some coppice) from Abbot's Leigh Woods (to the north). These are possibly ancient, although there is little clear evidence that they pre-date 1600 despite the presence of indicator species (particularly small-leaved lime *Tilia cordata* and wood anemone *Anemone nemorosa*). Pollen diagrams for the nearby Gordano valley show a pollen abundance ratio around 6:3:1:1 for oak : hazel : lime : elm, but the relative abundance of spores suggests there was a lime dominated mixed deciduous forest dominating the area, perhaps 7,000 years ago.

Records suggest coppice woodland produced faggots to be used in bread ovens; the presence of coppice suggests the exclusion of grazing animals for the most part in the 17th century. In northern Abbots Leigh Woods, both lime with a little aspen were coppiced, and there was an adjacent deer park, emphasising the need to exclude animals. To the south, the Long Ashton Woods contained some wood pasture and the succession seems to have changed in the direction of unpalatability for animals. Small leaved lime (*T. cordata*), became restricted to the plateau and increasingly confined to the crags, as it is favoured by the animals. Pendunculate oak (*Quercus robur*) survived as abundant pollards, but sessile oak (*Q. petrea*) eventually predominated. The ground layer could be grass, heath or scrub and included the ancient woodland indicator species nettled-leaved bellflower (*Campanula trachelium*).[60]

On the east side of the Gorge are the Clifton and Durdham Downs, 'The Bristol Downs'. Prior to clearance in the early Neolithic, these would have been woodland including small leaved lime, sessile oak, ash, wych elm, hazel and field maple; yew was also present. Documentary evidence points to grazing in the 9th century and use as 'common pasture' in the 12th century when

---

[58]   National Trust (n.d.) Leigh Woods <https://nt.global.ssl.fastly.net/documents/maps/1431729742984-leigh-woods. pdf> Accessed May 2020.
[59]   National Trust (n.d.) Wildlife in Leigh Woods <https://www.nationaltrust.org.uk/leigh-woods/features/wildlife-in-leigh-woods > Accessed May 2020.
[60]   C.M. Lovatt (1987) The Historical Ecology of Leigh Woods Bristol. *Proceedings of the Bristol Naturalists Society* 47, 3-19.

the area (within the Manor of Clifton) was held by St Augustine's Abbey in Bristol. Boundary stones (also called merestones or meerstones) are still present and relate to the grazing area of Durdham Downs. Rights of commoners are known from the 19th century, but systematic grazing ceased in the last century. Some rare plants are known from within the sward but most are from the Gorge itself.[61]

The Gordano valley is located near to the coast between Clevedon and Portishead. It includes an NNR comprising 126ha of mainly woodland, grassland and fen meadow and is part of an SSSI. More than 130 species of flowering plant have been recorded, including the rare brown galingale, whorled water-milfoil and fen pondweed. Birdlife includes long eared owls and woodcocks. Also observed in the recent past have been the wading birds: lapwing, snipe, and redshank. Mammals include brown hare, water shrew, harvest mouse and otter. The NNR is nationally important for its incredibly rich and diverse invertebrate fauna. 16 species of dragonfly and 23 species of butterfly are recorded.[62]

Figure 10.3. Fish, chips, and ammonites. Lyme Regis looking towards the Lias Cliffs (the author).

---

[61] S.D. Micklewhite and L.C. Frost (1987) Historical land use of the Bristol Downs as common of pasture *Proceedings of the Bristol Naturalists Society* 47, 21-26.
[62] Natural England (n.d.) Gordano Valley. <https://www.gov.uk/government/publications/avons-national-nature-reserves/avons-national-nature-reserves> Accessed May 2020.

Dr William Smith (1769 - 1839)
Author of the first geological map of England. He drained land in Purbeck circa 1800 and worked in Weymouth in 1812.

Professor William Buckland (1784 - 1856)
Professor of Geology at Oxford. A series of visits to the Site. Used valleys in East Devon and Dorset as examples of the action of Noah's flood.

Professor Adam Sedgwick (1785 - 1873)
Professor of Geology at Cambridge. Visited and bought fossils from Mary Anning.

Rev William Conybeare (1787 - 1857)
Author of the first scientific description of a landslide, at Bindon in East Devon. Vicar at Axmouth between 1836-44.

Dr Gideon Mantell (1790 - 1852)
Discoverer of *Iguanodon*. Visited in 1832 and wrote accounts of excursions on the Dorset Coast.

Sir Roderick Impey Murchison (1792 - 1871)
President of the Geological and Royal Geographical Societies. Visited the coast, and invited Mary Anning to London for her only visit there in 1829.

Sir Henry de la Beche (1796 - 1855)
Founder of the British Geological Survey. Moved to Lyme Regis in 1812, writing a series of papers and publishing *Duria Antiquior,* the first diorama of a past world.

Sir Charles Lyell (1797 - 1875)
Pioneer of the theory that modern processes operated in the geological past.

Mary Anning (1799 - 1847)
A poor woman from Dorset who became known as the 'greatest fossilist that ever lived'. Finder of the first marine reptile skeletons to come to scientific attention.

Professor Richard Owen (1804 - 1892)
Superintendent of the Natural History Museum in London. Seminal paper on the early mammals of Purbeck. Defined the name dinosaur in 1842.

Professor Louis Agassiz (1807 - 1873)
Swiss pioneer of the study of glaciation. Visited in 1834, and named two fossil fish after Mary Anning - the only fossils named after her during her lifetime.

Samuel Husbands Beckles (1814 - 1890)
Excavated fossil mammals at Swanage, under the advice of Owen.

1760 1770 1780 1790 1800 1810 1820 1830 1840 1850 1860 1870 1880 1890 1900

1859 Charles Darwin's '*On the Origin of Species*' published

Figure 10.4. Scientists associated with the Jurassic Coast (T. Badman, D. Brunsden, R. Edmonds, S. King, C. Pamplin and M. Turnbull, 2003, The Official Guide to the Jurassic Coast. Coastal Publishing, Wareham).

10.6 The Jurassic Coast

The Jurassic Coast stretches from Exmouth in East Devon to Studland Bay, Dorset; virtually all of it – as far as Poole Harbour – thus forms part of the south coast of Wessex. It includes 155km of cliffs and beaches and bears witness to around 185 million years of Mesozoic earth history dating from the early Triassic to the lower Tertiary. The UNESCO conservation designation World Heritage Site (2001) denotes places of 'outstanding universal value'.[63] UNESCO seeks to build peace through international cooperation in Education, the Sciences and Culture. Its programmes contribute to the achievement of the sustainable development goals defined in Agenda 2030, adopted by the UN General Assembly in 2015.[64]

Climate change, urbanisation, natural disasters and deliberate destruction are a persistent threat to World Heritage sites around the World. The 1954 Convention for the Protection of Cultural Property in the Event of Armed Conflict explicitly forbade the deliberate targeting of culturally significant objects during war. Under stable conditions, heritage designations often boost local economies by encouraging tourism and attracting funds.[65]

Figure 10.3 shows the Georgian town of Lyme Regis and Figure 10.4 depicts scientists associated with the Jurassic Coast in the 19th century. All must have had influence on Charles Darwin. Naturally, Mary Anning and William Smith are included (see Chapter 9); scientists can also play a long-term role in economic development as countless fossil collectors of all ages annually descend on Lyme Regis and elsewhere. Now sea level rise might be attacking the cliff-line as never before. Thanks to the likes of Anning, Smith and Darwin we have a scientific framework and a sense of geological process and time with which to work on this significant problem.

10.7 Tourism

Tourism is nothing new and the development of resorts was explored in Chapter 9. Values in landscapes, in their conservation, in their built environments and in exploration at many scales are all drivers for the respectable elements of this industry. From the late 17th century, people of the 'middling sort' started to visit inland and coastal locations, taking advantage of better roads such as the turnpikes.[66] Bathing in the 18th century would lead to the rise of a sea-side fashion centred around bathing machines by Regency times.[67]

Coastal visits contrast with inland 'resorts' such as the Spa city of Bath, notorious for its nefarious activities and decadence; this was largely curtailed during the Napoleonic era. Domestic tourism was becoming widespread and was encouraged by many. Poets and writers such as William Wordsworth and his sister Dorothy, Samuel Taylor Coleridge, Robert Southey,

[63] T. Badman, D. Brunsden, R. Edmonds, S. King, C. Pamplin and M. Turnbull (2003) *The Official Guide to the Jurassic Coast*. Coastal Publishing, Wareham, 6-7.
[64] UNESCO (2019) UNESCO in brief - Mission and Mandate <https://en.unesco.org/about-us/introducing-unesco> Accessed May 2020.
[65] Encyclopaedia Britannica (n.d.) World heritage Site <https://www.britannica.com/topic/World-Heritage-site> Accessed May 2020.
[66] R. Sweet (n.d.) Domestic tourism in Great Britain. <https://www.bl.uk/picturing-places/articles/domestic-tourism-in-great-britain> Accessed May 2020.
[67] Jane Austen Centre (2011) Sea bathing, in Georgian style <https://www.janeausten.co.uk/seabathing-georgian-style/> Accessed May 2020.

Percy Bysshe Shelley, Charles Kingsley, Henry Williamson, and R. D. Blackmore - of varying levels of respectability – were all at some points associated with Exmoor.[68] A group of writers of the Romantic period, such as William Wordsworth, Percy Bysshe Shelley and John Keats are referred to as 'nature poets' imparting a strong sense of the value placed on the natural order.

The 18th century was a time of exploration and elites were able to partake of the Grand Tour. This was a period of foreign travel, commonly undertaken by gentlemen (and sometimes ladies) to finish off their education. It was popular from the mid-17th century and available only to the rich; typically people were away for two to three and a half years. The impact on artistic style, architecture, science, and culture was considerable.

The extract from the famous poem 'The Rime of the Ancient Mariner' that began this Chapter evokes the age of exploration, depicting travel southwards towards the equator from a likely embarkation point of Watchet in Somerset. Coleridge would have been as familiar with the Quantocks as he was with the harbours and Exmoor. He, like Southey and Wordsworth, shared a sympathy with the ideals of the French Revolution and was opposed to slavery, although his youthful opposition modified over time, indeed idealism of the romantic poets overall waned over time.[69] Southey and Coleridge relocated to the Lake District where the Wordsworths were a fixture, and the outlook of the 'Lake Poets' was shaped among fellow de-radicalising wordsmiths.

From Paris, Grand Tour travellers usually proceeded to the Alps and then by boat on the Mediterranean to Italy. They would usually visit Rome and Venice, otherwise their tour might also include Spain, Portugal, Germany, Eastern Europe, the Balkans, and the Baltic. Tourism from Britain was thus greatly stimulated, but with the outbreak of the French Revolution, then the Reign of Terror prior to the rise of Napoleon Bonaparte and his associated wars, travel was severely curtailed. There was also nervousness around the continental powers. The war with France depressed demand and banks made poor speculative loans. One consequence was the economic collapse of Bath, which started with the failure of Bath Bank in 1793, putting tradesmen out of business.

Tourism saw a revival in the early 19th century after peace was restored in Europe. As travel became cheaper and easier, and in particular with the advent of the railways, visiting Europe ceased to be the preserve of the elite – the days of the Grand Tour were over.[70] It is also the arrival of railways from 1840 onwards (see Chapter 2) that really brought tourists of lesser means to Wessex towns and ports, including Weston-Super-Mare, Clevedon, Minehead and more on the Dorset Coast (Chapter 9). By the 1930s there was a mature rail network with many branch lines that survived until the 1960s (Figure 2.12). This development was linked with development of the road network into the 20th century as increasing numbers of people could afford this mode of private transport.

---

[68]   Exmoor Literary Links (2020) Literary Links <https://www.exmoor-nationalpark.gov.uk/Whats-Special/culture/literary-links> Accessed May 2020.
[69]   National Trust (n.d.) Three extraordinary Years <https://www.nationaltrust.org.uk/coleridge-cottage/features/three-extraordinary-years>; B. T. Paul-Emile (1974) Samuel Taylor Coleridge as Abolitionist *Ariel* 5(2), 59-75.
[70]   R. Knowles (2013) The Grand Tour <https://www.regencyhistory.net/2013/04/the-grand-tour.html> Accessed May 2020.

Until the advent of mass foreign holidays in the 1960s, a combination of vastly improved transport, better standards of living and more leisure time meant the west of England had long been open for tourism business. 'Farm diversification' (into non-agricultural activities) saw farmhouses converted into accommodation;[71] it is envisaged that this commenced in the late 19th century when (from 1879) the agricultural economy was hit by foreign competition (see Chapter 3).

Yet it is clear then and now, that maintaining the countryside as an attractive landscape was important in encouraging tourism.[72] The implication is clear: public policy has to be formulated to maintain countryside structure in accordance with the expectations of visitors. This may accord with notions of the English rural idyll, but it has real economic implications for both the local and regional economies.[73] In 2018, the entire south-west region of England (which would also include Devon and Cornwall) attracted 19.1 million overnight visits, that is 16% of the Great Britain total, with around £4.3 billion spent in the region as a result.[74]

The total tourism expenditure in the UK in 2013 was approximately £140 billion, and the proportional output of all industries in the UK that can be directly attributed to tourism spend was 3.7%. Wales and the South West have the highest proportions of their economic output that can be directly attributed to tourism spend (4.9% and 4.5% respectively), followed by London with 4.3%.[75]

10.8 Historic monuments and value of landscape

Picking up on the points explored in Sections 10.2 and 10.3, it should be emphasised that Wessex abounds with historic monuments. Conservation of prehistoric monuments, of more recent historic remains and the built environment runs in parallel with policies for conservation of biological and geological sites. Concern for the protection of historic monuments dates back with certainty to Elizabethan times. Early institutional arrangements are varied and complex but relate to the Office of Works (1378-1832). Its origins lay in the medieval royal household. Royal clerks were assigned responsibility for the construction and maintenance of royal castles and fortifications, royal residences and a range of other 'king's works' for successive monarchs.[76]

A statutory conservation ethos emerged over time. For example, Stonehenge was acquired by the Ministry of Works in 1918.[77] The Ministry of Works was responsible for buildings and

---

[71] G. Busby (2000) The Transition from Tourism on Farms to Farm Tourism. *Tourism Management* 21(6), 635-642. DOI: 10.1016/S0261-5177(00)00011-X

[72] About Britain (n.d.) Discover the South West of England <https://about-britain.com/regional/south-west.htm> Accessed May 2020

[73] J. Burchardt (2011) Rethinking the Rural Idyll: The English Rural Community Movement, 1913–26 *Cultural and Social History* (20)1, 73-94. https://doi.org/10.2752/147800411X12858412044438

[74] House of Commons Library (2019) Tourism: statistics and policy no. 06022 <https://commonslibrary.parliament.uk/research-briefings/sn06022/> Accessed May 2020

[75] Office for National Statistics (2016) The regional value of tourism in the UK: 2013. <https://www.ons.gov.uk/peoplepopulationandcommunity/leisureandtourism/articles/theregionalvalueoftourismintheuk/2013> Accessed May 2020

[76] The National Archives (n.d.) Records of the successive Works departments, and the Ancient Monuments Boards and Inspectorate <https://discovery.nationalarchives.gov.uk/details/r/C260> Accessed May 2020

[77] English Heritage (n.d.) History of Stonehenge <https://www.english-heritage.org.uk/visit/places/stonehenge/history-and-stories/history/> Accessed May 2020

archaeological remains, 1943-1962, then the Ministry of Public Building and Works, 1962-1970, becoming the Department of the Environment, 1970-1997.

In common with nature conservation, the 19th century saw the creation of campaigning organisations and protests to promote heritage preservation until the 1882 (first) Ancient Monuments Protection Act, which:

- Established a schedule of 50 state 'protected' monuments, all of which were prehistoric. Roman, medieval and occupied buildings were not yet included on the list.
- These were not directly protected, but if the owner wished to dispose of them, the government could acquire the monuments for caretaking by the Office of Works
- Archaeologist General Pitt-Rivers[78] was appointed as the first Inspector of Ancient Monuments.[79]

The Royal Commission on the Historical Monuments of England (RCHME) dates from 1908 when commissioners were appointed in England, in Scotland and in Wales. The Town and Country Planning Act 1932 introduced and allowed Building Preservation Orders to be served by local authorities on threatened historic buildings, with compensation to be paid if the Minister of Works failed to uphold the order. This included occupied dwelling houses for the first time.

Then, in 1937, the National Trust introduced a list of historically important buildings; eventually under the Town and Country Planning Act 1947 the Minister of Works became obliged to compile an official list to advise local authorities, the beginnings of the graded listing system, and also now required local authority to issue a Building Preservation Order if a historic building were threatened by development. The Town and Country Planning Act 1968 gave all buildings on the list statutory protection for the first time.[80]

Another landmark Act of Parliament in the immediate post-Second World War era was the Historic Buildings and Ancient Monuments Act 1953:

> 'An Act to provide for the preservation and acquisition of buildings of outstanding historic or architectural interest and their contents and related property, and to amend the law relating to ancient monuments and other objects of archaeological interest.'

The National Heritage Act 1983 established English Heritage as the government's lead advisor on the built historic environment in England and obliged the Secretary of State for the Environment to consult English Heritage on listing matters, and to refer certain applications for listed building consent for advice. The Act established grant schemes for the repair of historic buildings and ancient monuments, to be administered by the Historic Buildings Council and the Ancient Monuments Board.[81]

---

[78] Augustus Pitt-Rivers (1827-1900) was a pioneering archaeologist who excavated widely in Wessex. He pioneered modern archaeological methods, founding collections in both Oxford and Salisbury museums.
[79] Historic England (2020) Timeline of Conservation Catalysts and Legislation. < https://historicengland.org.uk/whats-new/features/conservation-listing-timeline/> Accessed May 2020.
[80] Historic England (2020) Timeline of Conservation Catalysts and Legislation. < https://historicengland.org.uk/whats-new/features/conservation-listing-timeline/> Accessed May 2020.
[81] National Archives (n.d.) Historic Buildings and Ancient Monuments Act 1953. <http://www.legislation.gov.uk/

In 1999, RCHME and English Heritage merged and in 2005 English Heritage was given direct responsibility for the administration of the listing system, which now included formal notification of owners for the first time. Then, on 1st April 2015, English Heritage was separated into Historic England and the English Heritage Trust, a new independent charity. This looks after the National Heritage Collection of monuments, while Historic England continues as an arms-length body that looks after the wider historic environment, including 'listing, planning, grants, research, advice and public information'.[82]

One key designation is that of Scheduled Ancient Monument (SAM). Scheduling began in 1913, although the earlier 1882 Ancient Monuments Protection Act had concerned largely prehistoric monuments deemed deserving of state protection (legally speaking being 'scheduled') was first compiled.[83] Hence, a scheduled monument is an historic building or site that is included in the Schedule of Monuments kept by the Secretary of State for Digital, Culture, Media and Sport. The regime is set out in the Ancient Monuments and Archaeological Areas Act 1979.[84] The protection extends not just to known structures or remains but also to the soil under or around them. This is in order to protect any archaeological interest in the site, but the extent of the protection is not dependant on there being such an interest.

This may be appropriate for such as the remains of Corfe Castle, Danebury Hillfort, Stonehenge, Avebury or Stanton Drew, but for features such as canals and water meadows that require regular infrastructural attention, other designations including SSSIs may be better suited. Designation of an SAM, as that of SSSI does not indicate ownership: they can be in private, public or charity sector ownership. Like SSSIs, SAMs require management. So does a lot more.

10.9 (Re)wilding: Dilemma or paradox?

Rewilding (sometimes termed 'non-intervention management approaches' or just 'wilding') implicitly seeks to restore ecosystems for plants, birds and animals with minimal intervention. But this is problematic from a philosophical point of view. We must ask when did humans begin to modify the environment? Much as emergence of photosynthetic cells, dinosaurs and grasses at different times affected great changes to past planetary ecosystems, so humans are changing the planet. In Wessex, humans came and went throughout the Pleistocene.

The notion of rewilding is frequently proposed as a radical approach to solving landscape-wide conservation issues. It is inevitably linked to contemporary issues of habitat conservation, species extinction and climate change. Changing land use presents opportunities for modifying the hydrological environment (including flood mitigation) and, naturally, the sequestration of carbon into re-vitalised wetlands, soils and biomass in general including woodland and forest.

We may presume it is imperative to feed the human population worldwide and that agriculture for better or worse is thus here to stay in our post-Mesolithic world. The debate may be cast

---

ukpga/Eliz2/1-2/49> Accessed May 2020.
[82] Historic England (2020) Timeline of Conservation Catalysts and Legislation. < https://historicengland.org.uk/ whats-new/features/conservation-listing-timeline/> Accessed May 2020.
[83] Historic England (n.d.) <https://historicengland.org.uk/listing/what-is-designation/scheduled-monuments/> Accessed May 2021.
[84] Historic England (2020) Scheduled Monuments <https://historicengland.org.uk/advice/hpg/has/ scheduledmonuments/> Accessed May 2020.

Figure 10.5. The view from above Ebbor Gorge NNR on Mendip above Wells, Somerset looking southwards. (The author).

as 'land-sparing *vs.* land sharing' and has considerable international attention. For example, some argue that more intensive food production in designated areas would improve species diversity elsewhere or alternatively that by abandoning large tracts of land, species habitat restoration and hence species diversification might occur.[85]

The problem with ill-defined and controversial ideas is that positions may quickly become entrenched. These can be held and maintained without really appreciating the complexity of 'wicked problems'; solving climate change would be an example.[86] Such are the current problems of an environmental nature that face the planet. It can be argued that rewilding in fact presents humanity with a prospect for *compounding* wicked problems, for widespread land abandonment is the enemy of agriculture and managed habitats ('semi-natural' ecosystems) and begs enormous questions for such necessities as food or timber production.

---

[85] F. Pearce (2018) Sparing vs Sharing: The Great Debate Over How to Protect Nature. *Yale Environment 360.* <https://e360.yale.edu/features/sparing-vs-sharing-the-great-debate-over-how-to-protect-nature> Accessed May 2020.
[86] Wicked problems are sometimes defined as problems that are difficult or impossible to solve because of incomplete, contradictory and changing requirements that are often difficult to recognise. A wicked problem refers to a complex problem for which there is no simple method or solution. Wicked problems are often socially complex, and they have to deal with changing behaviours and outcomes that are unforeseen. See E.J. Moors (2017) Water-wrestling with wicked problems. Inaugural lecture IHE Delft Institute for Water Education Delft, The Netherlands 5th Oct. 2017. <https://www.un-ihe.org/sites/default/files/inaugural_lecture_eddy_moors_5_october_2017.pdf> Accessed May 2020.

Figure 10.5 shows the view of the Somerset landscape from above Ebbor Gorge NNR. While there is no active re-wilding in the areas, the panorama gives an idea of the contrast between what a rewilded landscape (foreground) could look like in contrast with the farmed landscape, including the Somerset Levels and Moors beyond.

It is probably not misleading to say that re-wilding is a 'Big Idea' from regional-scale nature conservation interests but generates less enthusiasm from those with intimate knowledge of local semi-natural habitats and landscapes. Scrutiny of much of the available literature and writings of polemicists does not always make it clear just where or how this Big Idea is to be applied. Supporting arguments perhaps remind us of a once-predicted UK land surplus due to the overproduction of food, especially of cereals. However, what now seems to have been side-lined is an older debate, one around 'land targeting' for specific outcomes. These outcomes might be reduction of water pollution or soil erosion, habitat re-creation, flood impact and more.[87] Nor does a search of the available information really reveal much about the role of humans in the creation of such problems borne of poor environmental management, nor indeed the landscape features they have constructed.

What has been called a 'fortress' or 'hotspot' conservation approach can lead to conserved habitats existing in splendid isolation,[88] but there remains the ever-present risk of poor (and hence damaging) agricultural activities outside protection zones. The hope is that a mix of good practice, voluntary compliance and appropriate institutional regulation (including ENGO action) will minimise environmental damage.[89] To those of us who study the interaction of humans, their societies and people's immediate needs (economic or otherwise) rewilding misses many key points. It is not about 'sustainable development' for this requires much thought and investment of time and appropriate resources. Unless great care is taken, it lies somewhere between theorising and giving up, viz:

> Rewilding is the large-scale restoration of ecosystems where nature can take care of itself. It seeks to reinstate natural processes and, where appropriate, missing species – allowing them to shape the landscape and the habitats within.

> Rewilding encourages a balance between people and the rest of nature where each can thrive. It provides opportunities for communities to diversify and create nature-based economies; for living systems to provide the ecological functions on which we all depend; and for people to re-connect with wild nature.[90]

It is implied by its proponents that rewilding applies largely or wholly to upland Britain. In Wessex this might apply to the Mendips, Quantock hills and Exmoor. However, (for example) success is claimed for introduction of the beaver across a number of topographies across

---

[87] H.F. Cook and C. Norman (1996) Targeting agri-environmental policy: an analysis relating to the use of GIS. *Land Use Policy* 13(3), 217-228.

[88] H. Cook, L. Couldrick and L. Smith (2017) An assessment of intermediary roles in payments for ecosystem services schemes in the context of catchment management: An example from South West England. *J. Environmental Assessment, Policy and Management* 19(1), 1-31.

[89] H.F. Cook (2017) *The Protection and Conservation of Water Resources*. Wiley-Blackwell, Chichester, Chapter 10.

[90] Rewilding Britain (2020) Principles of Rewilding <https://www.rewildingbritain.org.uk/rewilding/> Accessed May 2020.

Britain.[91] Elsewhere in our region the landscape is generally lower overall in elevation and while areas of Dorset and the New Forest constitute extremely poor soil, much is under agriculture with some woods and forestry. For reintroduction of the beaver is claiming successes in many environments. There is no real vision for re-wilding in lowland Britain although candidate areas would, we imagine, be on agriculturally poor soils. However, many such areas include heathlands and coppiced or other woods of considerable ecological importance that require careful management.

The farmed landscape of Britain is overwhelmingly human-influenced, and today there is considerable environmental information to enable us to identify areas where targeted land cover change will benefit society and environment. We may ask where (and when) was there no human intervention? In notions of the idea of the Anthropocene, scientists propose a geological epoch whereby the planet has been drastically altered by us. These changes include the composition of the atmosphere, the chemistry of the oceans and changes to land cover and potentially stretch back for some time.

Primal societies hunting and gathering on the tundra, in forests and woodland likely affected species composition, selectively impacting on environmental genetics. Early farmers cleared trees, created soil erosion, and impacted rivers. It is exceedingly difficult to determine whether a piece of 'natural' looking wood is primal or re-grown. For example, beneath the 'virgin' rainforests of Amazonia is clear evidence of exceptionally large settlements indeed.

Processing LiDAR images of Savernake Forest in Wiltshire has revealed earthworks beneath the modern canopy that range in time over thousands of years (Chapter 1, Figure 1.13). In historic times the vegetation cover would have been largely grazed wood pasture and not unaltered wildwood. There are many other similar examples. The notion of 'wilderness' existing historically aside from human agency is therefore difficult to substantiate.

Neither, however, should we see those supporters of dark green policies as Jeremiahs lamenting the state of our planet, for we face enormous challenges from such as climate change, loss of species or flooding. But in Wessex we may be ahead of the game. The countryside evolved through accumulated knowledge and experience, through the interactions of humans with a rich and varied natural environment.

Rivers and floodplains produced floral diversity, water meadows, floodwater detention features, milling and fish stocks. Woods and forests have been cleared, re-grown, and managed for hunting and timber production as well as being valued for recreation, biodiversity, and habitats. As we face an uncertain policy future for rural areas, we do well to reflect on the experience in countryside management handed down to us from history and find the courage (and money) to place our expertise alongside any notions of wilderness. It is true that wetland restoration may be important in biodiversity, water management (both quality and flood defence) and carbon sequestration.[92]

---

[91] T. Begum (2001) Natural History Museum News <https://www.nhm.ac.uk/discover/news/2021/april/record-numbers-of-beavers-are-being-introduced-to-the-uk.html> accessed Aug 2021.
[92] H.F. Cook (2010). Floodplain agricultural systems: functionality, heritage and conservation *Journal of Flood Risk Management* 3, 192-200 DOI:10.1111/j.1753-318X.2010.01069.x

Restoring water meadows gives not only heritage benefits, but also improvements to sediment trapping, phosphorus reduction and oxygen status in the associate river alongside sustainable grass production.[93] Broadleaved woodlands, to be biodiverse, need management that produces both biodiversity of plants, animals and invertebrates as well as providing amenity for communities and livelihoods for woodland workers.[94] Any rewilding scheme has to be targeted where it too will deliver multiple benefits. One example might be the restoration of broadleaved woodland across a tributary valley liable to flooding, thereby protecting human infrastructure as well as providing habitat and landscape amenity value.

Within the landscape is a co-evolved system that delivers sustainability (especially of food production and hence rural livelihoods), heritage and biodiversity in equal measures. An important part of sustainability is to maintain a farmed landscape that permits not only the supply of quality food, but also maximises the benefits of any soil type and is enjoyable in that it attracts tourists. Careful planning should be able to address the issue of 'island habitats' by providing for linking wildlife corridors.

10.10 Climate change

With two coasts, both with substantially variable tidal conditions and strong currents; extensive coastal lowlands (see Chapter 4) are known to have flooding problems; varied geologies supporting river systems that mostly require flood defence or mitigation measures somewhere; an important agricultural sector; very important semi-natural habitats; and last but not least large centres of population, climate change has to be of concern in Wessex.

Concern there may be, but Wessex, with its possibilities of carbon sequestration in wetlands, woodlands and some soils as well as modifying agriculture to reduce greenhouse gas emission, presents the classic land use options towards solving global problems. Other challenges may come not so much from heavy industry but from transport in the motorways and other major roads of Wessex and from a thriving tourist industry – all present points of carbon -and other greenhouse gas- production.

The global increase in temperature of 0.85°C since 1880 is mirrored in the UK climate. Average annual UK temperatures over land and the surrounding seas have increased in line with global observations, with a trend towards milder winters and hotter summers. Sea levels globally and around the UK have risen by 150-200 mm since 1900. Whilst natural variability in the climate will continue to have a large influence on individual weather events, the recent episodes of severe and sustained rainfall are consistent with projections of climate change.[95] And globally speaking, the summer of 2023 appeared to be the hottest on record to date.[96]

---

[93] H.F. Cook, R.L. Cutting. and E. Valsami-Jones (2015). Impacts of meadow irrigation on temperature, oxygen, phosphorus and sediment relations for a river. *Journal of Flood Risk Management* 10 (4), 463–473. DOI: 10.1111/jfr3.12142

[94] H. Cook and K. Stearne (2021) Rural crafts: a study in south Wiltshire *Crafts Research* 12(1), 105-125. https://doi.org/10.1386/crre_00042_1

[95] Committee on Climate change (2016) UK Climate Change Risk Assessment 2017 <https://www.theccc.org.uk/wp-content/uploads/2016/07/UK-CCRA-2017-Synthesis-Report-Committee-on-Climate-Change.pdf> Accessed May 2020.

[96] https://climate.copernicus.eu/summer-2023-hottest-record

The semi-natural habitats of Wessex will also be vulnerable to climate change. While there may be a replacement of coastal grazing marsh by renewed saltmarsh creation at the coast, inland certain valued habitats, especially lowland heathlands and associate wetlands may be especially vulnerable (see Chapter 7). This is something that environmental managers will have to address; impacts may include the loss of some species altogether and the introduction of others that may prove problematic in their invasive nature. Nutrient cycling will be altered through changes in temperature and site hydrology. One immediate response will be in the hydrological cycle, with changes in flood behaviour and management of water resources.

Compared with the climatic reference baseline period of 1961-1990, Wessex can expect the average summer to be drier and warmer, the average winter to be milder and wetter, and for extreme events to happen with greater frequency. Overall, changes may be serious for the water industry in terms of bulk supply. Water resource quality is more likely to be compromised with warmer summers triggering biological activity, while heavy rainfall in prolonged episodes or short, sharp spells can result in contaminants being washed into reservoirs or groundwater sources. Regarding sewerage, sewage treatment and sludge, the highest risks relate to inundation of sewers during intense or prolonged rainfall, with adverse impacts on both consumers and receiving watercourses. Other expected problems include odour during warm weather; reduced dilution in receiving waters during drought; and sedimentation in sewers, also due to drought.[97] Unfortunately, events of recent years seem to have borne much of these predictions out.

10.11 Sustainable Wessex?

Achieving sustainable development in Wessex presents the same problems and issues as elsewhere. On the one hand, diversity of landscape presents many opportunities, and overall levels of air pollution are lower than elsewhere outside urban areas.[98] However, water pollution results from both agricultural activities and poor sewage management.[99] There are a plethora of conservation designations in the region overseen by statutory bodies (mainly Natural England and the Environment Agency, see sections 10.2 and 10.3), Wildlife Trusts, and several NGO or charitable organisations aimed at environmental management and improvement.[100]

It has been demonstrated that environmental campaigning is at another high point. The focus has moved away from concern about unplanned development and conservation of built and archaeological heritage, and protection of particular landmarks or historically significant landscapes towards local manifestations of specifically *global* issues.

Global climate change concern (or the Climate Emergency) has provided a framework upon which to hang most, if not all environmental concerns. During the decades before global

---

[97] Wessex Water (2015) Wessex Water's second report to DEFRA under the climate change adaptation reporting <https://assets.publishing.service.gov.uk/government/uploads/system/uploads/attachment_data/file/474352/climate-adrep-wessex-water.pdf> Accessed May 2020.

[98] BBC (2018) UK's most polluted towns and cities revealed <https://www.bbc.co.uk/news/health-43964341> Accessed May 2020.

[99] Wessex Water (2020) Reporting Pollutions <https://www.wessexwater.co.uk/help-and-advice/emergencies/reporting-pollutions> Accessed May 2020.

[100] Wessex Rivers Trust (2020) <https://www.wessexrt.org.uk/>; Harnham Water Meadows Trust (n.d.) Welcome <https://www.salisburywatermeadows.org.uk/> Accessed May 2020.

warming became in turn a new (or revived) theory, a debate, then scientific and political orthodoxy, transnational pollution problems gave us international agreements. Fluids do not respect national or regional boundaries, therefore measures to tackle marine pollution, agreements around rivers crossing borders and then the successes in curbing acid rain in Europe and in North America paved the way in terms of international co-operation.

Global environmental consciousness is reaching maturity because humans released geological carbon to the atmosphere alongside propellants, refrigerants, aeroplane and motor vehicle exhaust gasses and emissions from agriculture. 'Think global, act local' is the hackneyed cliché and so Wessex has its range of ways and means to halt climate change, reduce water pollution – and of course improve biodiversity. We can start with Agenda 21, made ready for the new century. With its origins in the UN Conference on Environment and Development held in Rio de Janeiro in June 1992, it is a comprehensive plan of action to be taken globally, nationally and locally by relevant organisations in every area in which humanity impacts on the environment. Agenda 21, the Rio Declaration on Environment and Development, and the Statement of Principles for the Sustainable Management of Forests were adopted by more than 178 Governments in 1992.

The Programme for Further Implementation of Agenda 21 and the Commitments to the Rio principles, were strongly reaffirmed at the World Summit on Sustainable Development in September 2002.[101] Interestingly Agenda 21 saw combatting poverty as important in achieving sustainable development. Local groups with local authority support were spawned by this new global initiative in Gloucestershire (Gloucester), Dorset (Poole and Bournemouth) and Wiltshire (Salisbury).[102]

The new campaigning era has given us the Transition Network (2006). The Transition movement recognises an urgent need to reduce carbon dioxide emissions, greatly reduce our reliance on fossil fuels and make wise use of precious resources. By empowering communities through action and education it aims to increase the chances of all groups in society to live well, healthily and with sustainable livelihoods.[103] Transition is organised in towns and cities, so for Salisbury:

'A dynamic, non-political, local organisation, Salisbury Transition City seeks to encourage alternative energy sources, reduce our reliance on fossil fuels such as oil, revitalise local business and maximise the use of neighbourhood resources. And this in turn will help sustain our city and make it an even nicer place to live in.'[104]

Air pollution is never far from the public eye (or nose). Emissions from diesel vehicles are problematic in the extreme. Diesel powered cars and lorries, aeroplanes and shipping all contribute to atmospheric particulate loading and greenhouse gas emissions in Wessex. The

---

[101] UNCED (1992) Agenda 21 <https://sustainabledevelopment.un.org/outcomedocuments/agenda21> Accessed May 2020.
[102] South West regional Assembly (2002) A sustainable Future for the South West <http://www.southwest-ra.gov.uk/media/SWRA/Sustainable%20Development/Moving_in_the_Right_Direction.pdf> Accessed May 2020.
[103] Transition Network (2020) A movement of communities coming together to reimagine and rebuild our world. <https://transitionnetwork.org/> Accessed May 2020.
[104] Transition Network (2020) Salisbury Transition City <https://transitioninitiative.org/initiatives/salisbury-transition-city/ > Accessed May 2020.

problems are at their worst in confined spaces, urban areas and close to major roads. However, aside from greenhouse gasses (such as carbon dioxide and nitrous oxide) and carbon monoxide, particulates are demonstrably bad for human health[105]. Bristol is one of several cities with illegal levels of air pollution from diesel traffic. The UK government has been ordered by the courts to bring air pollution levels down to legal limits in the shortest possible time in order to meet its obligations on air pollution under UK and EU law. A total ban is to be placed on all privately-owned diesel vehicles during the day in the city centre. The scheme, which requires government approval, is due to start in 2021.[106]

Akin to the Transition Movement in its emphasis on inclusion and persuasion, a slightly older idea, similarly beginning in the NGO sector, has grown from grassroots concern around the state of rivers and for fish populations in particular. Rivers Trusts belong to the last three decades.[107] Locally to Wessex, the Wessex Chalk Streams (subsequently the 'Wessex Rivers Trust') and Rivers Trust can state:

> Throughout England, Catchment Partnerships are coming together to collectively agree on the priorities for the local water environment. Many are developing, in partnership, realistic 'catchment action plans' which will guide the work of the partnerships and also help to inform larger-scale strategic River Basin Management plans that the Environment Agency is putting in place for each of the 11 River Basins Districts in England & Wales. These River Basin Management Plans will facilitate UK reporting to Europe on its progress in delivering WFD improvements.

> The Catchment Based Approach must be about much more than just complying with the Water Framework Directive. It allows local communities, businesses, organisations and other stakeholders to come together to undertake actions or develop projects which incorporate local priorities such as flood risk management, fisheries and biodiversity.[108]

It must furthermore be remembered that heavy regulation (for example by the Environment Agency) and legal action is costly, time consuming and generates ill-will. A long-standing principle for non-governmental 'third sector' bodies is the trust they potentially build. Those with membership systems, such as the National Trust are considered legitimated by public enrolment rather than merely the threat of legal action.[109]

Farmers' markets are an enduring private sector initiative that has a great potential to move towards sustainable development and are commonplace throughout Wessex. While the motives for these are many and include a will to link producer with consumers, and particularly consumers of food products in urban areas, the notion of localism is strong. Many

---

[105] IOSH (n.d.) Diesel Engine Exhaust Fumes: The facts. IOSH Wigston Leicestershire <http://www.iosh.co.uk/~/media/NTTL%20files/POL2531%20-%20Diesel%20Fact%20Sheet%20WEB.ashx> Accessed May 2020.

[106] Bristol City Council (2019) Clean Air for Bristol<https://democracy.bristol.gov.uk/documents/s42633/BCC%20CAZ%20OBC%202%20-%20Executive%20Summary.pdf> Accessed May 2020.

[107] Westcountry Rivers Trust (n.d.) Early Beginnings <https://wrt.org.uk/the-westcountry-rivers-trust-story/ > Accessed May 2020.

[108] Wessex Chalk Streams and Rivers Trust (n.d) Partnerships <https://www.wcsrt.org.uk/catchment-partnerships> Accessed January 2020.

[109] H. Cook and A. Inman (2012) The voluntary sector and conservation for England: achievements, expanding roles and uncertain future. *Journal of Environmental Management* 112, 170-177.

boast locally sourced food and the spread of towns and cities involved is impressive. CPRE report that Stroud and Winchester rank among the 'Top Ten Farmers' markets in England.[110]

The rise of farmers Markets has involved the formation of an association (FARMA), certification and a record of core principles to which individual enterprises should aspire and to which compliance is required for certification.

Essentially:

> The primary aim of Farmers' Markets is to support local farmers and producers, who sell produce they have grown or made to their local community.
>
> The Farmers' Market movement started in the UK in the 1990s as a result of poor farm prices, leading to many family farms needing a viable local route to market their produce. Today there are markets across the UK providing farmers with an important 'lifeline.[111]

Newer, radical, and more spontaneous groups are emerging. While Extinction Rebellion ('XR') compromised the operation of central London at times in 2019, in Wessex local XR protestors groups disrupted traffic in Bristol, demonstrated in Bath, Portsmouth and Southampton and are organised in Swindon and Salisbury.

XR supported the strikes by UK Schools Climate Network and attracted wide support.[112] Dominated by school students, strikes relating to the Climate Emergency have taken place across the Wessex region, here as elsewhere inspired by the young Swedish activist Greta Thunberg. Thousands of students have walked out of classrooms calling on the Government to make tackling climate change a priority.[113] Since 2020, there might have been more political protests aimed towards environmental activism were it not for the shutdown associated with covid-19.

10.12 Conclusions

The rise of a conservation ethic, be it for nature conservation, archaeological heritage or building and urban conservation has been strong in Wessex and well-organised since the 19th century. As new problems and perceptions arose, the very English notion of campaigning that was evident well over a century ago evolved to influence all economic sectors. Interestingly, the statutory / public sector response took some time to emerge beyond basic but essential nineteenth and early twentieth legal frameworks. The driver lies in voluntary and in non-governmental action.

---

[110] M. Brown (n.d.) Ten Top farmers markets CPRE <https://www.dev2.cpre.org.uk/magazine/item/3011-ten-top-farmers-markets> Accessed May 2020.

[111] Save-Money-Guide (n.d.) UK Farmers Markets <http://www.save-money-guide.com/UK-farmers-markets.html> Accessed May 2020.

[112] E Berkeley and U. Irfan (2019) Why the climate protests that disrupted London were different Vox. <https://www.vox.com/energy-and-environment/2019/4/24/18511491/climate-change-protests-london-extinction-rebellion> Accessed May 2020.

[113] Global Justice Now (n.d.) Global Climate Strike <https://www.globaljustice.org.uk/events/global-climate-strike?> Accessed May 2020.

Resurgent campaigning in the post-Second World War era belonged either side of the 1960s: CND began in 1958 and continues to the present day; Friends of the Earth and the more edgy Greenpeace became active in England in the early 1970s. While few would deny the role of social media in recent years, a re-politicisation of UK society, with strong echoes of the 1960s and 1970s radicalism of their parents and grandparents, has largely been the response of young people to the perceived theft of their future due to environmental degradation. Other issues may feature, but climate change provides the modern framework.

As if to react against the growing statism of the post-war era, localism, democratisation, and the entire language of inclusion resound in the phases of protest manifest in post-Second World War Britain. As the above account seeks to demonstrate, city councils have a role to play in seeking sustainable development in Wessex, supporting local and spontaneous 'grass roots' action within communities. Farmers' markets are a private sector response; the rise of autonomy among communities intuitively moves towards smaller organisations and the independent producer and trader.

Back in the natural world (which provides inspiration for all the above), it is to be celebrated that Wessex possesses such a range of heritage and biodiversity. Only time will tell how effective private, public and NGO sector action will prove. ENGOs have the capability to bring together different organisations, generally with differing ultimate objectives. For example, the Rivers Trusts work not only with private sector farmers but with statutory organisations such as the Environment Agency and Natural England and with any voluntary sector organisation that shares conservation objectives. The key word here is indeed 'share' for there remains the possibility of differing approaches (they are after all value-driven) even when ultimate objectives seem compatible.

A well-resourced, appropriately staffed, and outward looking ENGO should have the capacity to anticipate, bridge and if necessary resolve differences of approach. There is, after all, a long and noble tradition of collective action influencing the statutory processes. Important decisions must be made within this patchwork of governance arrangements. One example would be coastal management. The managed retreat and coastal habitat change at Porlock is a kind of policy experiment with, economically, little to lose but much to gain in terms of institutional learning and habitat change and gain.

Elsewhere the cost of maintaining sea defences along the coast of the Somerset Levels may be brought into question, but without them the threat to property and livelihood would be on a massive scale. One expects the cost of sea defences will be met for a long time yet. Opportunities for energy development in the Severn estuary remain contentious on both economic and environmental grounds. Inland the excesses of 'production agriculture' will need to be restrained, yet the growing importance of local markets, food and environmental quality may yet bring sustainable food production to the fore. Famers are self-evidently central to landscape management that considers the environmental goods of water quality, flood defence/ mitigation, carbon sequestration and habitat and biodiversity management. We are all in the frame for the reduction of harmful atmospheric emissions.

Figure 10.6. Wessex's prehistory re-imagined. Part of the stone circle at Ham Hill Country Park, near Yeovil, Somerset with the author. He understands this was completed in AD 2000 ± 0 years (the author).

Index